CAMBRIDGE LIBRARY COLLECTION

Books of enduring scholarly value

Maritime Exploration

This series includes accounts, by eye-witnesses and contemporaries, of voyages by Europeans to the Americas, Asia, Australasia and the Pacific during the colonial period. Driven by the military and commercial interests of powers including Britain, France and the Netherlands, particularly the East India Companies, these expeditions brought back a wealth of information on climate, natural resources, topography, and distant civilisations. Their detailed observations provide fascinating historical data for climatologists, ecologists and anthropologists, and the accounts of the mariners' experiences on their long and dangerous voyages are full of human interest.

A Voyage Round the World, in the Years MDCCXL, I, II, III, IV.

Compiled by the naval chaplain Richard Walter (1717–85), though the extent of his editorial contribution is not certain, this 1748 publication documents the extraordinary circumnavigation accomplished by the British naval officer George Anson (1697–1762) between 1740 and 1744. During the Anglo-Spanish conflict which Thomas Carlyle later described as the War of Jenkins' Ear, Anson was chosen to command a squadron to raid and plunder the Pacific coast of South America. After a delayed departure, the expedition struggled with terrible weather, rough seas and outbreaks of scurvy as it rounded Cape Horn. Hundreds of men were lost and eventually only the warship *Centurion* remained, badly battered and undermanned. Despite the disaster, the expedition became famous for its capture in 1743 of a Spanish treasure galleon laden with silver. Anson won much acclaim for this feat, and he entered into politics. This account, meanwhile, became a bestseller.

Cambridge University Press has long been a pioneer in the reissuing of out-of-print titles from its own backlist, producing digital reprints of books that are still sought after by scholars and students but could not be reprinted economically using traditional technology. The Cambridge Library Collection extends this activity to a wider range of books which are still of importance to researchers and professionals, either for the source material they contain, or as landmarks in the history of their academic discipline.

Drawing from the world-renowned collections in the Cambridge University Library and other partner libraries, and guided by the advice of experts in each subject area, Cambridge University Press is using state-of-the-art scanning machines in its own Printing House to capture the content of each book selected for inclusion. The files are processed to give a consistently clear, crisp image, and the books finished to the high quality standard for which the Press is recognised around the world. The latest print-on-demand technology ensures that the books will remain available indefinitely, and that orders for single or multiple copies can quickly be supplied.

The Cambridge Library Collection brings back to life books of enduring scholarly value (including out-of-copyright works originally issued by other publishers) across a wide range of disciplines in the humanities and social sciences and in science and technology.

A Voyage Round the World,

in the Years MDCCXL, I, II, III, IV.

*Compiled from Papers and Other Materials
of the Right Honourable George Lord Anson,
and Published under his Direction,
by Richard Walter, Chaplain
to his Majesty's Ship the Centurion*

GEORGE ANSON
EDITED BY RICHARD WALTER

CAMBRIDGE
UNIVERSITY PRESS

CAMBRIDGE
UNIVERSITY PRESS

University Printing House, Cambridge, CB2 8BS, United Kingdom

Cambridge University Press is part of the University of Cambridge.
It furthers the University's mission by disseminating knowledge in the pursuit of
education, learning and research at the highest international levels of excellence.

www.cambridge.org
Information on this title: www.cambridge.org/9781108074995

© in this compilation Cambridge University Press 2014

This edition first published 1748
This digitally printed version 2014

ISBN 978-1-108-07499-5 Paperback

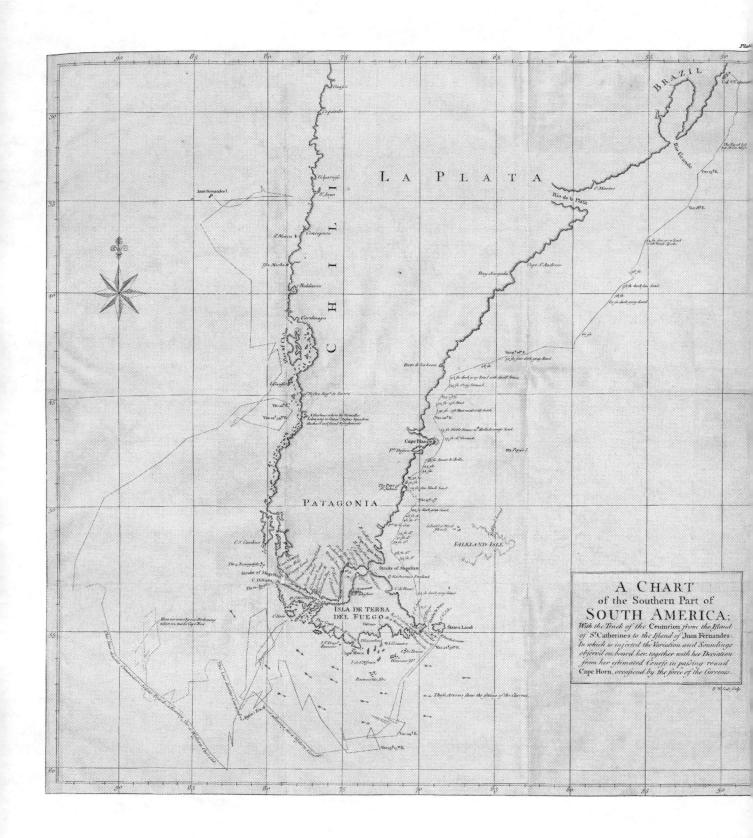

The material originally positioned here is too large for reproduction in this reissue. A PDF can be downloaded from the web address given on page iv of this book, by clicking on 'Resources Available'.

A

VOYAGE

ROUND THE

WORLD,

In the Years MDCCXL, I, II, III, IV.

BY

GEORGE ANSON, Esq;

Commander in Chief of a Squadron of His Majesty's
Ships, sent upon an Expedition to the *South-Seas*.

COMPILED

From Papers and other Materials of the Right Honourable
GEORGE Lord *ANSON*, and published under his Direction,

By RICHARD WALTER, M. A.

Chaplain of his MAJESTY's Ship the *Centurion*, in that Expedition.

Illustrated with Forty-Two COPPER-PLATES.

LONDON:
Printed for the AUTHOR;
By JOHN and PAUL KNAPTON, in Ludgate-Street. MDCCXLVIII

TO

HIS GRACE

JOHN,

DUKE of *BEDFORD,*

MARQUIS of *TAVISTOCK,*

EARL of *BEDFORD,*

BARON *RUSSEL,*

BARON *RUSSEL* of *Thornhaugh,*

AND

BARON *HOWLAND* of *Streatham;*

One of His Majefty's Principal Secretaries of State; and Lord-Lieutenant, and Cuftos Rotulorum of the County of BEDFORD.

My LORD,

THE following Narrative of a very fingular naval atchievement is addreffed to Your GRACE, both on account of the infinite obligations which the Commander in Chief at all times profeffes

to

DEDICATION.

to have received from your Friendſhip; and alſo, as the
Subject itſelf naturally claims the patronage of One, under
whoſe direction, the *Britiſh* Navy has reſumed its ancient
Spirit and Luſtre, and has in one ſummer ennobled itſelf by two victories, the moſt deciſive, and (if the
ſtrength and number of the captures be conſidered) the
moſt important, that are to be met with in our Annals.
Indeed, an uninterrupted ſeries of ſucceſs, and a manifeſt ſuperiority gained univerſally over the enemy,
both in commerce and glory, ſeem to be the neceſſary
effects of a revival of ſtrict diſcipline, and of an un-
biaſſed regard to merit and ſervice. Theſe are marks
that muſt diſtinguiſh the happy period of time in
which Your GRACE preſided, and afford a fitter ſubject for hiſtory, than for an addreſs of this nature.
Very ſignal advantages of rank and diſtinction, obtained and ſecured to the naval profeſſion by Your
GRACE's auſpicious influence, will remain a laſting
monument of Your unwearied zeal and attachment to
it, and be for ever remembred with the higheſt
gratitude, by all who ſhall be employed in it. As
theſe were the generous rewards of paſt exploits,
they will be likewiſe the nobleſt incentives, and ſureſt
pledges of the future. That Your GRACE's eminent talents, magnanimity, and diſintereſted zeal,
whence

DEDICATION.

whence the Public has already reaped such signal benefits, may in all times prove equally successful in advancing the prosperity of *Great-Britain*, is the ardent wish of,

My LORD,

Your GRACE'S

Most obedient,

Most devoted,

AND

Most humble Servant,

RICHARD WALTER.

A

LIST

OF THE

SUBSCRIBERS.

A.

* HIS Grace the D. of Argyll.
* Rt. Hon. the Earl of Al-
bemarle.
* Rt. Hon. the Countefs of Albemarle.
* Rt. Hon. the Earl of Afhburnham.
* Rt. Hon. the Earl of Ancram.
* Rt. Hon. the Countefs of Ancram.
* Rt. Hon. the Lord Andover.
* Rt. Rev. and Hon. the Lord Bp. of
St. Afaph.
* Rt Hon. the Lord Anfon.
* Hon. Mr. Juftice Abney.
* Hon. Richard Arundel, Efq;
Sir Thomas Dyke Acland.
* Sir William Abdy, Bart.
* Sir Jacob Aftley, Bart.
Sir Robert Abdy, Bart.
* Sir Richard Atkyns, Bart.
* Thomas Anfon, Efq;
Jofeph Andrews, Efq;
George Armftrong, Efq;
* George Atkyns, Efq;
Rev. Richard Allin, B. D.
Richard Afhby, Efq;

* John Alderfon, Efq;
Rev. George Afhby, M. A.
Charles Allanfon, Efq;
Rev. ———— Andrews, D. D.
Rev. ———— Annefly, D. D.
Mr. John Adington.
Adam Afkew, M. D.
Mr. Jofeph Airey.
Mr. Jeremiah Andrews.
Mr. Samuel Allin.
Rev. Mr. Addenbrooke.
* William Pritchard Afhurft, Efq;
* Mr. William Ayleworth.
Francis Andrews, Efq;
John Abery, Efq;
Richard Allen, Efq;
Rowland Aynfworth, Efq;
* Abraham Atkins, Efq;
Rev. Mr. Anguifh.
Rev. Dr. Ayerft.
Richard Aftell, Efq;
Mr. Charles Allen.
Rev. Mr. Edmund Arnold.
John Auften, Efq;
John Anftis, Efq;
Rev. Mr. Apthorp.

Samuel Atkins, Efq;
* Ralph Allen, Efq;
* William Adair, Efq;
Mr. Richard Auften.
Mr. Robert Arthur.
Mr. James Ayfcough.
Samuel Atkinfon, Efq;
James Abercrombie, Efq;
Captain Arnot.
Mr. John Ackers.
Mr. Henry Alleyne.
Richard Adams, Efq;
* William Adams, Efq;
William Windham Afhe, Efq;
George Atkins, Efq;
Rev. Mr. Allot.
* Mr. George Aufrere.
Lillie Smith Aynfcombe, Efq;
Rev. Dr. Atwell.
Mr. Henry Alcroft.
Rev. Mr. Allen.
John Afhwood, Efq;
Mr. Stephen Auften.
Mr. Robert Akenhead.
Mr. John Addy.

HIS

B

* HIS Grace the Duke of Beaufort.
* His Grace the D. of Bedford.
* Rt. Hon. the Earl of Burlington.
Rt. Hon. the Earl of Bute.
* Rt. Hon. the Earl Brooke.
* Rt. Hon. the Lord Burghley.
* Rt. Hon. the Lord Bury.
* Rt. Hon. the Lord Boyne.
* Rt. Hon. the Lady Bateman.
Rt. Rev. the Lord Bishop of Bangor.
Rt. Hon. the Lord Berkeley of Strattone.
* Rt. Hon. the Lord Baltimore.
* Rt. Hon. Count Charles Bentink.
* Rt. Hon. the Lord Chief Baron Bowes.
Hon. and Rev. Dr. Booth Dean of Windsor.
* Hon. Admiral John Byng.
Sir Edmund Bacon, Bart.
* Sir Walter Bagot, Bart.
Sir Griffith Boynton, Bart.
Sir John Bernard, Bart.
Sir Robert Baylis, Kt.
Sir John Bosworth, Kt.
William Blackbourn, Esq;
John Bloss, Esq;
Mr. Thomas Brant.
Mr. Claude Bosanquet.
Mr. John Burges.
Robert Bennet, Esq;
Samuel Bosanquet, Esq;
John Bamber, M. B.
Mr. William Belchier.
Thomas Browne, M. D.
Rev. Mr. Philip Birt.
Mr. Thomas Bigg.
Mr. William Biffett.
Mr. William Branson.
Mr. Thomas Bigge.
Colonel John Brown.
Thomas Bradshaw, Esq;
Mr. John Badcock.
Mr. Daniel Booth.
Hugh Barker Bell, Esq;
Mrs. Burton.
Rev. Mr. Bostock.
Captain Patrick Baird.
Rev. Dr. Bridges.
Rev. Mr. John Branthwaite.
Mr. Stephen Barbut.
John Boyd, Esq;
Mr. John Blackstone.
Rev. Mr. Edward B. Blackett.

Thomas Ryley Blankley, Esq;
Mr. William Branson.
Mr. John Baxter.
Rev. Mr. Barnes.
Robert Banks, M. D.
Claudius Bennet, Esq;
Christian Henrich Brande, Esq;
Mr. James Backhouse.
William Beecher, Esq;
Mr. John Bailey.
Mr. Barnaby Backwell.
* Mr. William Backwell.
Mr. Luke Bennett.
John Becher, Esq;
Mr. John Boyfield.
Mr. Josiah Boyfield.
Richard Buckley, Esq;
Fitz William Barrington, Esq;
Mr. Jonas Butler.
Mr. John Butler.
Captain Hans Bredeson.
Captain Jochum Bugge.
Mrs. Butler.
John Buissiere, Esq;
Mr. William Belchier.
Rev. Mr. Samuel Baker.
—— Bird, Esq;
Nicholas Bonfoy, Esq;
Richard Bull, Esq;
Rev. Richard Buckby, B. A.
Mr. Francis Beilby.
Mr. James Bucknall.
Rev. William Bedford, M. A.
Mr. John Bateman.
Daniel Burgess, Esq;
* Richard Badcock, Esq.
Mr. Baynam.
Mr. Barnardiston.
Thomas Blofeld, Esq;
Mr. James Bourdieu.
Mr. Alexander Black.
Benjamin Bond, junr. Esq;
Mr. Noah Blisson.
Mr. Brooke Bridges.
Mrs. Bridges.
—— Bridgman, Esq;
Mr. William Birch.
Rev. Edward Bell, M. A.
Mr. Francis Best.
Charles Brown, Esq;
James Best, Esq;
Buck Nutcombe Bluett, Esq;
Rev. Martin Barnes, B. D.
Rev. John Buxton, M. A.
Rev. Robert Buxton, M. A.
William Barbor, M. B.
Rev. —— Berridge, M. A.

Chambers Bate, Esq;
Rev. George Brough, M. A.
The Book-Society at Stowmarket.
Rev. —— Bringhurst, M. A.
George Bond, Esq;
Rev. Mr. Breese.
Fotherby Baker, Esq;
* Edmund Brown, Esq;
Rev. Thomas Buttonshaw, M. A.
Samuel Barker, Esq;
Mr. Brent.
John Buxton, Esq;
Mr. John Button.
Lieutenant George Blankley.
Mr. Richard Blackham.
Rev. —— Balguy, M. A.
Rev. —— Barnard, M. A.
* Timothy Brett, Esq;
Mr. William Boawre, A. B.
Mr. —— Bloodworth.
* Thomas Bonfoy, Esq;
* Brooke Boothby, Esq;
* Captain Peter Bruce.
* George Blondall, Esq;
* John Bigg, Esq;
* William Bodvell, Esq;
* John Bonnel, Esq;
Mr. Benjamin Blyth.
Roger Blount, Esq;
Major Balladine.
Mr. Thomas Burdett.
Elliot Bishop, Esq;
Micheal Bidulph, Esq;
James Brockman, Esq;
Rev. Mr. William Beels.
Rev. Mr. Bagshaw,
Mr. Bell.
Mrs. Barlowe.
Mrs. Ann Bridges.
John Bacon, Esq;
Mr. Henry Burgess.
Christopher Batt, Esq;
Mr. William Bowden.
William Brinck, Esq;
Thomas Barlow, Esq;
Benjamin Barlow, Esq;
Rev. Mr. John Bourn.
Mr. Booth.
Robert Ball, Esq;
* Thomas Beckford, Esq;
Captain Matthew Bookey.
Captain William Earl Benson.
* Captain Blomfield Barradale.
Thomas Brand, Esq;
Mr. William Bulmer.
Ralph Bell, Esq;
Edward Blacket, Esq;

Mr.

A LIST of the SUBSCRIBERS.

* Harry Burrard, Efq;
Mr. Peter Barbar.
Mr. Briggs.
Bryan Bentham, Efq;
Thomas Bate, Efq;
Mr. Henry Bullock.
Mr. Barton.
Mr. Bailey.
* George Breton, Efq;
Mr. Nathaniel Baffnet.
* Peter Burrel, Efq;
Rev. Mr. Edward Barnard.
Mr. George Bailey.
Mr. James Bogle.
* Rev. Ofmin Beauvoir, A. M.
Mr. William Bowes.
Richard Beranger, Efq;
* Pucey Brooke, Efq;
Mr. George Baker.
Mr. George Brown.
Mr. Stephen Barbier.
Mr. George Bifhop.
Mr. Thomas Browne.
Mr. Bedford.
* Mr. Michael Barber.
Mr. John Berry.
* Mr. Henton Brown.
* Norborne Berkeley, Efq;
* Mr. Herman Berens.
John Browning, Efq;
* Thomas Brand, Efq;
* John Briftow, Efq;
Mr. Richard Bailye.
Mr. John Brotherton.
Mr. Samuel Birt.
Mr. Buckland.
Mr. James Brackftone.
Mr. William Bonner.
Mr. Samuel Baker.
Mr. John Beecroft.

C

HIS Grace the Lord Archbp. of Canterbury.
Rt. Hon. the Earl of Cardigan.
Rt. Hon. the Earl Cowper.
Rt. Hon. Lord Charles Cavendifh.
Rt. Hon. Lord George Cavendifh.
* Rt. Hon. the Lord Craven.
* Rt. Hon. the Lord Chetwynd.
* Hon. Robert Coke, Efq;
* Hon. John Chichefter, Efq;
* Hon. Mr. Baron Clarke.
* Sir John Chichefter, Bart.
* Sir Walter Calverley, Bart.
Sir John Chapman, Bart.
Hon. Sir Henry Calthrope, Kt. of the Bath.

Hon. William Conolly, Efq;
* Lady Conyers.
Lady Cullen.
Lady Cairnes.
Sir William Calvert, Kt.
* Sir James Creed, Kt.
* Sir Richard Chafe, Kt.
Samuel Craghead, Efq;
William Cuft, Efq;
Rev. Mr. Corbert.
Nathaniel Cooper, Efq;
Mr. Henry Coulthurft.
Mr. Nicholas Crifpe.
Rev. Mr. Nicholas Cholwel.
Nathaniel Cole, Efq;
—— Calthrope, Efq;
Sabine Chandler, Efq;
* Richard Crop, Efq;
* Phillip Crefpigny, Efq;
Mr. Richard Cowlam.
Mr. Chaloner Cooper.
Mr. Clarke.
Mr. Thomas Creed.
George Lewis Coke, Efq;
John Chafe, Efq;
Mr. John Camp.
Colonel Cracherode.
Mifs Elizabeth Collier.
Mr. William Coulthurft.
* William Cowper, Efq;
Mr. Charles Child.
* Thomas Corbett, Efq;
* William Corbett, Efq;
Rev. Mr. Caftlefranc.
Matthew Concannen, Efq;
Rev. Mr. Thomas Colefax.
Mr. John Crellius.
Mr. Martin Capron.
James Cocks, junr. Efq;
James Chetham, Efq;
Godfrey Clarke, Efq;
* William Clayton, Efq;
* John Cay, Efq;
Mr. John Mort Cowbert.
* John Cleveland, Efq;
John Conyers, Efq;
Rev. Mr. Carr.
Mr. John Cotton.
Mr. Cook.
Dr. Cockburn.
Rev. Anthony Chefter, M. A.
Thomas Colclough, Efq;
Mr. William Henry Coppard.
John Cole, Efq;
Mr. Cartwright.
Mr. Edward Carter.
* Nathaniel Clements, Efq;
* Mr. James Cowfat.

John Copeland, Efq;
Mr. Copeland.
Dr. Conyers.
Mr. Robert Crew.
Mr. George Coltman.
Mr. Jofeph Commins.
Bryan Cooke, Efq;
Rev. Edward Chapman, A. M.
Jofhua Churchill, Efq;
William Caftle, Efq;
William Crowe, Efq;
Rev. Mr. Caley.
Thomas Chafe, Efq;
Daniel Cox, M. D.
Charles Chauncy, M. D.
Mr. Richard Charlton, junr.
* Walter Carey, Efq;
Henry Carrington, A. B.
* Capt. Alexander Crowden.
* Charles Carleton, M. D.
Clare-Hall Library, Cambridge.
Rev. —— Courtail, M. A.
Rev. Mr. Cumberland.
Society of Clergy at Bofton.
John Creed, Efq;
Mr. John Chambers.
* Caius College Library, Cambridge.
Mr. Peter Chefter.
Catherine-Hall Library, Cambridge.
Rev. Henry Clofe, M. A.
Rev. John Cuel, M. A.
Corpus Chrifti College Library, Camb.
Samuel Clarke, Efq;
Mrs. Chamberlayne.
Rev. Benjamin Crowe, M. A.
Rev. ——Chapman, M. A.
Rev. Mr. Cole.
Mr. Cotton.
Pryfe Campbell, Efq;
Rev. Mr. Charles Cowper.
Mr. Carrington.
Captain Richard Clarke.
William Cole, Efq;
Mr. Paul Corbett.
Rev. Mr. Clayton.
Charles Cholmondley, Efq;
* Captain David Cheap.
Jeremiah Cray, Efq;
Mr. Ralph Carr.
Mr. John Cookfon.
Rev. ——Cooper, M. A.
Mr. William Colfon.
Mr. Norrifon Coverdale.
John Carter, Efq;
Thomas Crowe, M. D.
Mr. Robert Carr.
Mr. William Cooke.
Samuel Collett, Efq;

(a)

Samuel

Samuel Child, Efq;
Mr. Francis Child.
Mr. John Chandler.
Mr. Peter Campart.
Mr. Henry Cavendifh.
Rev. Mr. Cooke.
Rev. Mr. Chapman.
Mrs. Mary Child.
Mr. John Cathcart.
Mr. James Cooke.
Mr. Thomas Cawne.
Rev. Mr. Thomas Chamberlayne.
Mr. John Coltman.
Mr. Edward Collis.
* John Spencer Colepeper, Efq;
Rev. Mr. Martin Clare, F. R. S.
James Compton, Efq;
Richard Cambridge, Efq;
Mr. Jofiah Colebroke.
Jofeph Crefwick, Efq;
* Mr. Anthony Cooper.
* Henry Crewys, Efq;
Mr. John Collett.
Mr. Thomas Cotfell.
Mr. Richard Clark.
* James Carr, Efq;
* John Chefhyre, Efq;
Captain Fiefeild Coe.
Mr. John Crofts.
Mr. William Craighton.
Mr. John Clay.
Mr. Richard Clements.

D

* HIS Grace the Duke of De-
vonfhire.
* Rt. Hon. the Earl of Dalkeith.
* Rt. Hon. the Lord De Lawarr.
* Rt. Hon. the Lord Ducie Moreton
* Rt. Hon. the Lord Duncannon.
Sir Francis Henry Drake, Bart.
* Sir James Dafhwood, Bart.
Sir George Dalfton, Bart.
Lady Dafhwood.
John Difney, Efq;
John Dias, Efq;
Mr. Thomas Dummer.
Robert Darrell, Efq;
* Roger Drake, Efq;
Mr. Thomas Dimfdale.
Mr. Ely Dyfon.
Rev. Mr. Edward Darell.
Mr. John Denyer.
John Darell, Efq;
* Peter Delme, Efq;
* John Delme, Efq;
Robert Dinwiddie, Efq;
Captain Thomas Defaguliers.

Mr. Theophilus Dorrington.
Colonel Durand.
John Damer, Efq;
William Domvile, Efq;
Rev. Lewis Devifme, A. M.
Mr. Stephen Devifme.
Edward Dixon, Efq;
Mr. Thomas Delafield.
Rev. Mr. Charles Dalton.
James Ducarel, Efq;
* Arthur Dobbs, Efq;
* Jofeph Damer, Efq;
* Mr. Charles Denis.
* Rev. Mr. Dampier.
* Mrs. De Gols.
William Dillingham, Efq;
* Mofes Mendez Da Cofta, Efq;
Mr. Thomas Day.
Mr. John Dick.
William Dixon, Efq;
Mr. Dixon.
Mrs. De Grey.
Rev. —— Dixon, B. D.
Rev. —— Daddo, M. A.
Rev. Mr. Stephen Duck.
Mr. Davies.
Richard Du Horty, Efq;
Rev. John Dawney, M. A.
Mr. Thomas Dawfon.
Rev. Mr. Dovey.
Mr. Daniel Debat, A. B.
* Mr. William Darwent.
* Alexander Davie, Efq;
Rev. Mr. John Downes.
Mr. David Delavau.
James Dalbiac, Efq;
Mr. John Dell.
Rev. Mr. Michael Dorfet.
Captain Duff.
* Captain Peter Denis.
Rev. Mr. Thomas Dampier.
Mr. David Dalrymple.
Peter Ducane, Efq;
Mr. William Duncombe.
* Francis Douce, M. D.
Mr. Robert Duncan.
Mr. Benjamin Denham.
Matthew Duane, Efq;
* Captain Philip Durell.
* Arthur Dobbins, Efq;
Mr. James Dornford.
Rev. Mr. Dunftan.
* Mr. Peter Dobree.
* Mr. Ifaac Dobree.
Captain Lars Anderfon Dahl.
* James Debeauvoir, Efq;
Mr. Robert Dodfley.
Mr. Lockyer Davis.

E

* RIGHT Hon. the Lord Edge-
cumbe.
Sir John Evelyn, Bart.
Edward Everard, Efq;
* Humphry Edwin, Efq;
Walter Edwards, Efq;
John Eld, Efq;
Anthony Ewer, Efq;
Mr. John Ellicott.
Mr. Thomas Eames.
Auguftine Earle, Efq;
Emanuel College Library, Cambridge.
* Welbore Ellis, Efq;
Francis Eld, Efq;
Mr. Martin Eelking.
Thomas Edwards, Efq;
Mr. William Eames.
Mr. John Ellis.
Robert Eddowes, Efq;
Thomas Elyott, Efq;
Mr. Edmund Elyott, B. A.
Rev. Mr. Evans.
Rev. Mr. Englifh
* Richard Edwards, Efq;
Mr. Charles Eafton.
John Edwards, junr. Efq;
* Mifs Ecclefton.
Rev. Mr. Entick.
Rev. Dr. Eden.
Mr. William Eaton.
Mr. Edward Eafton.
Meffrs. George and Alex. Ewing.

F

* RIGHT Hon. the Earl Fitz-
walter.
Rt. Hon. the Lord Falkland.
* Hon. Henry Finch, Efq;
* Rt. Hon. William Fortefcue, Efq;
Mafter of the Rolls.
Hon. Mr. Juftice Fofter.
Hon. John Forbes, Efq;
* Hon. Richard Fitz-Patrick, Efq;
* Hon. Mrs. Fortefcue.
* Hon. Mrs. Charlotte Fane.
Sir Thomas Farnaby, Bart.
* Sir Cordel Firebrace, Bart.
Sir Matthew Featherftone, Bart.
* Sir John Frederick, Bart.
William Freeman, Efq;
Rev. Mr. William Friend.
Robert Freeman, Efq;
* Simon Fanfhaw, Efq;
* William Fauquier, Efq;

2

Rowland

A LIST of the SUBSCRIBERS.

Rowland Frye, Efq;
Mr. George Farrell.
Rev. Mr. William Friend.
Henry Flitcroft, Efq;
Richard Frank, Efq;
Sloane Fowler, Efq;
Mr. Fitzhugh.
Mr. George Fleming.
Mr. Alexander Forbes.
Mr. John Fox.
* Frederic Frankland, Efq;
Mr. Franks.
* Francis Fane, Efq;
Robert Fleetwood, Efq;
Rev. Peter Freeman, B. D.
Mr. Jeffery Fifher.
Rev. Mr. Fuller.
John Frederick, Efq;
Dennis Farrer, Efq;
Rev. —— Foley, M. A.
Rev. Mr. Fearon.
Rev. Mr. George Ferne.
Rev. Mr. Fetherfton.
Rev. William Foord, M. A.
Mr. John Fifh.
John Farfide, Efq;
William Farfide, Efq;
John Fuller, Efq;
Mr. Stephen Fuller.
Rev. John Fitzherbert, M. A.
* William Fitzherbert, Efq;
* Martin Folkes, Efq;
Rev. Mr. Edward Francklin.
George Fitzgerald, Efq;
Mr. Fuzard
Mr. Baruch Fox.
* Captain Thomas Frankland.
Mr. Edward Frewen.
Mrs. Mary Frankland.
Mr. Richard Fetherfton.
Mr. John Fifher.
Mr. Luke Franklin.
Mr. Richardfon Ferrand.
Mr. William Frears.
Mr. Richard Fofter.
Lieutenant Jofeph Frame.
Mr. George Faulkner.
Mr. William Frederick.

G

* HIS Grace the Duke of Grafton.
* Moft Hon. the Marquis of Granby.
* Rt. Hon. the Earl of Gainfborough.
* Rt. Hon. the Lord Galway.
* Rt. Hon. the Lord Gowran.
Sir William Gage, Bart.

Sir Henry Grey, Bart.
Sir Reginald Graham, Bart.
* Rear Admiral John Gafcoyne.
* Francis Gafhry, Efq;
Philip Garbrand, Efq;
William Guildas, Efq;
Mr. Thomas Gamull.
Mrs. Griffith.
Richard Graham, Efq;
Mr. Nathaniel Green.
Captain Martin Groundman.
Major William Gardner.
Thomas Gee, Efq;
Mr. Goodacre.
Mr. Godwin.
Mr. Robert Griffin.
Mr. Peter Garrick.
David Garrick, Efq;
Mr. John Gill.
Mr. William Gerrard Galpine.
* Jofeph Gulfton, Efq;
Dr. Garnier.
Richard Gem, M. D.
* Mr. Sampfon Gideon.
Mr. Gawler.
* Mr. Edward Grofe.
* John Green, Efq;
Captain Charles Gough.
* James Gynander, Efq;
* George Greenville, Efq;
Mr. John Gurdon.
Ifaac Gale, Efq;
Rev. Mr. Gay.
Nathanael Gurdon, Efq;
* William Gery, Efq;
Major George Gray.
——Green, Efq; M. A.
Rev. John Garnett, B. D.
Rev. John Green, B. D.
George Grey, Efq;
Richard Green, Efq;
Charles Gray, Efq;
Rev. Peter Goddard, M. A.
Rev. Mr. Goodall.
Mr. James Gautier.
* Captain William Gordon.
Mr. John Gregory,
Rev. William Greenwood, D. D.
Rev.——Griffith, D. D.
Mr. Henry Gregory.
Captain James Gambier.
Mr. Jofeph Gayland.
Mr. Richard George.
Mr. James Gaylard.
Mr. Thomas Gent.
John Geekie, Efq;
John Gore, Efq;
Mr. Edward Grace.

Mr. James Godin.
* Phillip Gello, Efq;
Mr. Richard Gleed.

H

* RIGHT Honourable Philip Lord Hardwicke, Lord High-Chancellor.
* Moft Hon. the Marquis of Hartington.
* Rt. Hon. the Earl of Holdernefle.
Rt. Hon. the Countefs of Home.
Rt. Hon. and Rt. Rev. the Lord Bp. of Hereford.
Hon. Thomas Hervey, Efq;
* Hon. Capt. Hamilton.
Hon. and Rev. Mr. How, M. A.
Sir William Halford, Bart.
Sir Thomas Head, Bart.
Sir Robert Hildyard, Bart.
Sir William Heathcote, Bart.
Sir John Heathcote, Bart.
Sir Edward Hulfe, Bart. M. D.
* Hon. Sir Philip Honywood, Knight of the Bath.
* Hon. Sir William Harbord, Knight of the Bath.
* Hon. Sir Edward Hawke, Knight of the Bath.
* Hon. Lieutenant-General Handafyd.
* Hon. Lieutenant-General Hufke.
* Hon. Lieutenant-General Howard.
Baron Hardenberg.
Sir Richard Hoare, Kt.
Sir Jofeph Hankey, Kt.
John Hayes, Efq;
Mr. Hancock.
Benjamin Hall, Efq;
Colonel Hodges.
James Henckell, Efq;
* Richard Heath, Efq;
Mr. Robert Harrifon.
Mr. Monins Hollingbury.
Mr. Thomas Hort.
Mr. Lacy Hawes.
Mr. Chriftopher Hedges.
Rev. Dr. Hughes.
Edward Hopegood, Efq;
Hugh Hamerfley, Efq;
* Mrs Rebecca Houblon.
Mrs. Elizabeth Houblon.
William Hanbury, Efq;
Jacob Harvey, Efq;
Mr. Harper.
Samuel Hellier, Efq;
Mr. John Hammond.
Rev. Richard Hurd, M. A.

Mr.

A LIST of the SUBSCRIBERS.

Mr. William Hodſhon.
Jeffery Hetherington, Eſq;
Henry Hutchins, Eſq;
Mr. Marmaduke Hilton.
Thomas Hart, Eſq;
Mr. Iſaac Heaton.
Mr. John Holmes.
Mr. Hedgſon.
* Alexander Hume, Eſq;
* Abraham Hume, Eſq;
* Thomas Orby Hunter, Eſq;
Lieutenant-Colonel Philip Honywood.
Richard Honywood, Eſq;
John Honywood, Eſq;
Mr. Walter Holt.
Mr. Thomas Hurdman.
Mr. Thomas Hirſt.
Rev. Mr. Harling, A. M.
Mr. John Horſley.
* Mr. James Horne.
Robert Holford, Eſq;
William Hay, Eſq;
* John Hooke, Eſq;
* Edward Hooper, Eſq;
* John Lewis Harſcher, junr. Eſq;
Edward Hopſon, Eſq;
* Mr. Thomas Hitt.
Edmund Hoſkins, Eſq;
* George Haldane, Eſq;
Miſs Holden.
Richard Hopwood, M. D.
Mr. Hollier.
Fowke Huſſy, Eſq;
* Mr. Samuel Herring
——Heberdeen, M. D.
Rev. Samuel Henry Healy, B. A.
Mr. Haſwell.
Rev. Mr. Hetherington.
Mrs. Harding.
Rev. Henry Herring, A. M.
Robert Hotchkin, Eſq;
Mr. Halſhide.
John Hudſon, Eſq;
Mann Horsfield, Eſq;
Mr. Samuel Haſsle.
Rev.——Harriſon, M. A.
Rev. Mr. Hyly.
Rev. William Herring, M. A.
Thomas Hill, Eſq;
Mr. Edward Holden.
Cheſter Moor Hall, Eſq;
Edmond Hornby, Eſq;
Rev. Richard Hayes, M. A.
William Higginſon, Eſq;
George Hartly, Eſq;
Rev. Mr. Henry Haſcard.
Richard Houlditch, Eſq;
Captain Samuel Hobſon.

Samuel Horſey, Eſq;
Edward Hulſe, Eſq;
Captain Thomas Harriſon.
Lieutenant James Hume.
Mr. John Harvey.
Mr. John Heaton.
Mr. Hall, M. A.
Mrs. Sarah Hill.
* Rev. Mr. Harriſon.
Abraham Hall, M. D.
Mrs. Hayes.
Mr. Thomas Hopkins.
Mr. Henry Holcombe.
Mr. Francis Harling.
Thomas Haſſel, Eſq;
Mr. Henley.
Mr. Richard Holland.
Charles Hoſkins, Eſq;
Mr. Edward Harding.
John Halls, Eſq;
Rev. John Hume, D. D.
Edward Hooker, Eſq;
* Mr. Iſaac Hunter.
Mr. Adam Hayes.
Tilman Henckel, Eſq;
Rev. Mr. Mark Hildeſley.
Philip Harcourt, Eſq;
Mr. William Hatch.
* Charles Hubert Herriot, Eſq;
Samuel Heathcote, junr. Eſq;
Harman Hoburg, Eſq;
Mr. John Jacob Heldt.
Mr. James Hunter.
Mr. Nathaniel Hillier.
Mr. William Hawkins.
Mr. William Holloway.
Mr. Mileſon Hingeſton.
Mr. Samuel Heathcote.
Mr. Samuel Hartly.
Henry Hoare, Eſq;
Rev. Mr. James Hotchkis.
Mr. Thomas Heſkins.
Mr. Nathaniel Harris.
Peter Hinde, Eſq;
Mr. Peter Hemet.
Mr. Thomas Hopwood.
Mrs. Elizabeth Holden.
Mr. John Harriſon.
William Hammond, Eſq;
Captain Peter Johanſen Holts.
Mr. Jaſper Horn.
Mr. John Hammerſley.
Mr. Charles Hitch.
Mr. John Hinton.
Mr. John Hildyard.
Mr. James Hodges.
Mr. Hayhow.

I

* RIGHT Hon. the Earl of Jerſey.
* Rt. Hon. the Lord Ilcheſter.
* Hon. and Rev. Dr. Ingram.
Baron Ilton.
* Sir Hildebrand Jacob, Bart.
* Sir William Irby, Bart.
Major John Johnſton.
Swynfen Jervis, Eſq;
J. Jenkinſon, Eſq;
Rev. Mr. Iſaac Johnſon.
Theodore Jacobſon, Eſq;
Rev. Mr. John Jeffreys.
* Richard Jackſon, Eſq;
* William Jones, Eſq;
* Rev. Richard Jackſon, M. A.
Robert Johnſon, Eſq;
* John Jefferies, Eſq;
John Jones, Eſq;
Thomas Johnſon, Eſq;
Mr. Charles Jewſon.
Mr. Joſeph Jones.
Alexander Johnſon, Eſq;
Mr. Joſeph Jackſon.
Laſcelles Ironmonger, Eſq;
Jeſus College Library, Cambridge.
Rev. George Inman, A. M.
Rev. Mr. Jones.
Rev. Mr. Jones, junr.
Rev. Mr. Jackſon.
Rev.——Jackſon, B. D.
Mr. James Jackſon.
James Jurin, M. D.
Rev. John Johnſon, D. D.
Mr. Simon Julins.
Mr. Simon Jackſon.
John Innys, Eſq;
Pelham Johnſton, M. D.
Mrs. Lucy Jacombe.
Mr. Thomas Jervis.
Cheret Jones, Eſq;
John Jeſs, Eſq;
Mr. James Jones.
William Jenkins, Eſq;
Erneſt Auguſt Jager, Eſq;
Mr. John Jeſſer.
* Edward Jackſon, Eſq;
Mr. Benjamin Johnſon.
Mr. Daniel Jones.
Samuel Jones, Eſq;

K

* HER Grace the Ducheſs of Kent.

A LIST of the SUBSCRIBERS.

* Rt. Rev. the Lord Bp. of Kildare.
* Hon. Auguftus Keppel, Efq;
* Hon. Frederick Keppel, Efq;
Ralph Knox, Efq;
Mrs. Mary Kingman.
Mr. Samuel Kiliken.
Colonel Kerr.
Mrs. Kien.
John Kerrick, M. D.
Rev. Mr. Keen.
Mr. Robert Keymour.
Colonel King.
Mr. William Kinlefide.
Rev. Richard Kitchingman, M. A.
Rev.——Keller, A. M.
Mr. Robert Kitchingman, L L B.
Clement Kent, Efq;
Mrs. Kent.
Rev. Samuel Knight, M. A.
Haylock Kingfley, Efq;
Jafper Kingfman, Efq;
Robert Key, M. D.
Mr. William Key.
Rev. Mr. Knipe.
Mr. Chriftopher Keating.
Edward Knipe, Efq;
Mr. William Knipe.
Randolph, Knipe, Efq;
Mr. John Kent.
Mr. William Ifaac Kops.
Mr. William Cooper Keating.

L.

RIGHT Hon. the Earl of Lich-field.
* Rt. Hon. the Earl of Leicefter.
Rt. Hon. the Lord Lonfdale.
* Rt. Rev. the Lord Bp. of London.
Rt. Rev. the Lord Bifhop of Lichfield.
Rt. Rev. the Lord Bifhop of Landaff.
* Rt. Hon. the Lord Langdale.
* Hon. Henry Legge, Efq;
* Hon. Mr. Baron Legge.
Sir William Lowther, Bart.
Sir Thomas Littleton, Bart.
* Sir Edward Littleton, Bart.
Sir Atwill Lake, Bart.
Sir Robert Ladbroke, Knight, Lord-Mayor of the City of London.
* Sir Richard Lloyd, Kt.
* Mr. Charles Lowth.
Robert Legard, Efq;
Thomas Lifter, Efq;
* Charles Long, Efq;
* Beefton Long, Efq;
Ellis Lloyd, Efq;;
Nicholas Linwood, Efq;
* George Lyttelton, Efq;

Richard Lockwood, junr. Efq;
Mr. William Lake.
Nathaniel Lloyd, Efq;
Matthew Lee, M. D.
Richard Lindfey, Efq;
Mr. John Liddefdale.
James Lever, Efq;
William Locke, Efq;
Mr. Jofhua Locke.
Henry Lowther, Efq;
Mifs Elizabeth Lewis.
Mifs Sufannah Lewis.
* Rev. Dr. Lynch.
Mr. Lancafter.
Mr. Peter Lathbury.
Mr. George Lillington.
Mr. John Larpent.
Daniel Peter Layard, M D. F. R. S.
Mr. Thomas Leech.
Mr. Henry Lewis.
Temple Laws, Efq;
Jofeph Letherland, M. D.
*Edwin Lafcelles, Efq;
Henry Lafcelles, Efq;
Mr. Langley.
* Thomas Lowfeild, Efq;
* Charles Lowndes, Efq;
Mr. John Lowther.
* Benjamin Lethieullier, Efq;
* Chriftopher Lethieullier, Efq;
Mr. John Leech.
Thomas Lifter, Efq;
Rev. John Linton, B. A.
* Rev. Mr. Langbaine.
Francis Long, Efq;
Mr. Thomas Light.
Thomas Luck, Efq;
Rev. Mr. Law.
Mr. Lloyd.
Richard Long, Efq;
Robert Long, Efq;
Richard Langley, Efq;
Mr. Lonfdale.
Rev. Mr. Lamplugh.
Rev. Mr. Lally.
* Thomas Liell, Efq;
Mr. Thomas Ludlam.
Rev. William Lamb, M. A.
Mr. Charles Lidgould.
* John Lloyd, Efq;
Mr. Edward Lowry.
Manning Lethieullier, Efq;
* Captain Limburner.
* Theophilus Lane, Efq;
Mr. Samuel Lankford.
Rev. Mr. Lydiatt.
Rev. Mr. Laurence.
Mr. Thomas Langley.

Mr. Griffith Loyd.
Mr. James Lardant.
Mrs. Judith Lyford.
Mr. Lewis.
Mr. John Luke Landon.
Mr. Daniel Le Voutier.
Mr. John Lee.
Mr. Edward Lee.
* Richard Lateward, Efq;
Mr. Roujat Le Hook.
Thomas Lempriere, Efq;
Rev. Mr. Lyne.
Mr. John Lucas.
Mr. Charles Le Grys.
* Captain Edward Lecras.
Mr. James Le Marchant.
* Mr. Robert Lee.
Mr. William Loxham.
Mr. Edward Lowry.
Mr. Peter Le Keux.
* Lieutenant William Langdon
Mr. Edward Langton.
Rev. Mr. James Leaver.
Mr. Thomas Longman.
Mr. James Leake.

M

HIS Grace the Duke of Marl-borough.
* His Grace the Duke of Manchefter.
* Rt. Hon. the Earl of Morton.
* Rt. Hon. the Earl of Macclesfield.
* Rt. Hon. the Earl of Malton.
* Rt. Hon. the Lord James Manners.
* Rt. Hon. the Lord Maynard.
* Rt. Hon. the Lord Monfon.
* Rt. Hon. the Lord Montfort.
Rt. Hon. the Lord Middleton.
Rt. Hon. the Lady Middleton.
Hon. and Rev. Dr. Henry More.
Hon. and Rev. Gideon Murray, M.A.
Hon. Colonel Thomas Murray.
Sir William Maynard, Bart.
Sir Charles Molloy, Bart.
Sir Ralph Milbank, Bart.
Sir Edward Manfel, Bart.
* Chriftopher Mole, Efq; for the Hon. the Directors of the Eaft-India Company, 31 Books.
Valentine Morris, Efq;
William Milford, Efq;
Mr. Richard Martyn.
Mr. Thomas Mofely.
* James Mytton, Efq;
Dr. Mitchell.
Benjamin Martyn, Efq;

David

A LIST of the SUBSCRIBERS.

David Mitchell, Efq;
Mr. Edmund Monk.
Robert Moore, Efq;
Dr. Munckley.
Mr. William May.
Mr. Ifrael Mauduit.
* Andrew Mitchell, Efq;
* Nicholas Mann, Efq;
Cutts Maydwell, Efq;
Mr. Peter Motteaux, junr.
Mr. Samuel More.
Jofiah Martyn, Efq;
Mr. William Markes.
Sydenham Malthus, Efq;
Marm. Middleton, Efq;
Mr. Jofeph Maffee.
* John Macarell, Efq;
Mr. Maffey.
Rev. John Murgahodd, M. A.
Mr. Jofeph Martin.
Mr. Marfden.
Mr. John Mort.
Rev. Edward Mufgrave, A. B.
Rev. Mr. Morgan.
Charles Moore, Efq;
* George Montgomery, Efq;
Mr. Edward Moore.
George Maxwell, Efq;
* Lieutenant Thomas Moore.
Rev. Roger Moftyn, M. A.
Dr. M'Atty.
Edward Mann, Efq;
William Metcalfe, Efq;
Rev. Conyers Middleton, D. D.
Henry Morley, Efq;
* Charles Maneftee, Efq;
Rev. John May, M. A.
Mr. Robert Mackay.
Rev. John Manning, M. A.
Mr. Moody.
Rev. Roger Mortlock, M. A.
Mr. Thomas Manley, M. A.
George Macartney, Efq;
Captain James Mercer.
Captain Montolieu.
Dr. Mabb.
* Rev. Dr. Henry Miles, F. R. S.
Lieutenant Jofeph Myers.
Mr. George Murray.
Mr. Henry Milburne.
Andrew Mitchel, Efq;
* Mr. William Mills.
* Mr. Thomas Mills.
* Mr. Laurence Millechamp.
Hon. Major Maccarty.
* Richard Mead, M. D.
Richard Mead, Efq;

James Mead, Efq;
Rev. Mr. Robert Morgan.
Rev. Mr. William Murray.
Captain William Morris.
Hugh Marriott, Efq;
Captain Benjamin Mafon.
Mr. Harman Myer.
Peter Darnell Muilman, Efq;
Mr. Henry Martel.
Mr. Henry Mufgrave.
Mr. John Michel.
Norman Macleod, Efq;
John Martin, M. D.
Mr. John Milnes.
* Mr. Robert Mann, junr.
Mr. Mills.
Mr. Leonard Martin.
Mr. James Maze, junr.
* Mr. Thomas Mofely, junr.
Mr. Roger Matthews.
Robert Marfh, Efq;
Mr. Samuel Rogers Mansfield.
Mr. James Maze.
Mr. John Manfhip, junr.
Mr. Daniel Meffman.
Humphrey Monoux, Efq;
Rev. Mr. Lewis Monoux.
Captain James Millefon.
* Captain John Montagu.
Mr. Thomas Metcalfe.
Mr. John Mayfon.
Harward Martin, Efq;
Captain Andrew Mow.
Captain Hans Mow.
Mr. Jofeph Mofs.
Mr. John Mace.
Mr. John Millan.
Mr. Andrew Millar.

N

* HIS Grace the Duke of Nor-folk.
* Her Grace the Duchefs of New-caftle.
* Rt. Hon. the Earl of Northefk.
Rt. Hon. the Lord North and Guilford.
Mr. Henry Norris.
Mr. Matthew Nafh.
* Albert Nefbit, Efq;
William Northey, Efq;
Robert North, Efq;
* Robert Nefbit, M. D.
Colonel Robert Napier.
Rev. Dr. Richard Newcombe.
Mr. William Naylor.
Mr. William Neale.

Mr. Nelfon.
Rev. Mr. George North.
Rev. Mr. William Nunns.
* Captain Juftinian Nutt.
* James Naifh, Efq;
Rev. John Nesfield, M. A.
Rev. Mezifon Newton, M. A.
Thomas Nutting, Efq;
Mr. Thomas Nevile, A. B.
Rev. —— Newcome, D. D.
Mr. Nathaniel Newberry.
Rev. —— Neale, M. A.
* Francis Nailour, Efq;
Mr. John Nuttall.
Mr. Robert New.
Mr. Henry Napton.
Mr. Abraham Newhoufe.
Mr. Sandford Nevile.
Mrs. Mary Newdigate.
Mr. John Needham.
* Thomas Afhbourn Newton. Efq;
Captain Sofren Nielfon.
Mr. John Noone.
Mrs. Needham.

O

* RIGHT Honourable the Lord Onflow.
Hon. Percy Windham Obrien, Efq;
Sir John Oglander, Bart.
Sir Danvers Osborn, Bart.
Peter Ofborn, Efq;
James Orlebar, Efq;
Captain Lucius Obryen.
Lieutenant John Obryen.
Old Society of Ringers at York.
Robert Osborn, Efq;
William Osbaldifton, Efq;
George Osbaldifton, Efq;
Robert Ord, Efq;
Rev. —— Obyns, D. D.
William Ockenden, Efq;
Timothy Ottbie, Efq;
* Leek Okeover, Efq;
* Rev. —— Owen, D. D.
Henry Ord, Efq;
Mr. Francis Ogier.
Henry Osborn, Efq;
Rev. Thomas Osborn, LL. D.
Mr. Abraham Ogier.
Mr. John Osbourn.
Mr. John Orme.
Rev. Mr. Robert Oakeley.
* Mr. John Overy.
Mr. John Oxley.

MOST

A LIST of the SUBSCRIBERS.

P

MOST Rev. John Potter, D.D. late Lord Archbishop of Canterbury.
* His Grace the Duke of Portland.
Rt. Hon. the Earl of Plymouth.
* Rt. Hon. the Countess of Portland.
* Rt. Hon. Henry Pelham, Esq; Chancellor of the Exchequer.
Rt. Hon. Stephen Pointz, Esq;
Sir Edward Pickering, Bart.
Sir Thomas Parkins, Bart.
* Hon. Colonel Pelham.
Thomas Potter, Esq;
* David Polhill, Esq;
Eliakim Palmer, Esq;
Mr. George Purvis.
Reginald Pole, Esq;
Mr. Thomas Pennant.
Mr. Samuel Parmenter.
Rev. Mr. John Pennington.
Mr. George Paynter.
Robert Purse, Esq;
Rev. Mr. Pickering.
John Periam, Esq;
Mr. Poirier.
Mrs. Pulteney.
Mr. John Poyner.
* Dr. Charles Pinfold.
Mr. Resta Patching.
Mr. Edward Clarke Parish.
Mr. Pardoe.
John Plumptree, junr. Esq;
Rev. Mr. James Parker.
Rev. Mr. Thomas Prowse.
Rev. Mr. Pudsey.
William Purcas, Esq;
Mr. William Palmer.
Captain Dean Pointz.
Thomas Pulleyn, Esq;
Harrison Pilkington, Esq;
Pembroke-Hall Library, Cambridge.
Rev. Mr. Plumtree.
Mr. John Partridge.
Rev. Mr. Francis Pyle.
Mr. Purt.
Mr. John Pine.
Philip Parsons, Esq;
Jocelyne Pickard, Esq;
Thomas Percival, Esq;
* Mr. John Porter.
* Captain Jervis Henry Porter.
* Thomas Penn, Esq;
——Pringle, M. D.
Rev. Robert Piper, B. D.
* Mr. Arthur Pond.

Mr. Thomas Parker.
——Peel, Esq; M. A.
Captain John Pritchard.
Rev. Henry Prescot, B. D.
——Plumtree, M. D.
Thomas Sawyer Parris, Esq;
* John Proby, Esq;
* Mr. William Porter.
* Robert Pulleyn, Esq;
John Plumtree, Esq;
Lieutenant Colonel Pattison.
Mr. Robert Plumtree, M. A.
Captain Thomas Proby.
Charles Poole, Esq;
* Mr. Baptist Proby.
* George Morton Pit, Esq;
* Rev. Francis Sawyer Parris, D. D.
Rev.——Powel, M. A.
* John Philipson, Esq;
* Mr. Page.
* Mrs. Page.
John Palmer, Esq;
Mr. John Palmer.
Colonel Francis Peirson.
Mr. Robert Palmer.
Mr. Thomas Pinkard.
Mr. Daniel Pocock.
William Parker, Esq;
Mr. William Pawson.
Mr. John Price.
Henry Pennyman, Esq;
* Thomas Powys, Esq;
Mr. Thomas Plumer.
Azariah Pinney, Esq;
Mr. William Powell.
Thomas Parr, Esq;
Samuel Pye, M. D.
Captain Richard Peirson.
Mr. Edward Payne.
Mr. Francis Prime.
Mr. Percival Pott.
Mr. John Purling.
Mr. Thomas Palmer.
Rev. Mr. Richard Palmer.
Captain Jarvis Porter.
Captain Thomas Parker.
Mr. John Powdich.
Captain George Petterson.
* Isaac Preston, Esq;
Peter Peirson, Esq;
Mr. William Percy.
Mr. Herman Pohlman.
* John Payne, Esq;
* Mr. John Parr.
Mr. Richard Percy.
Mr. George Petzold.
Thomas Pomfret, Esq;

Mrs. Pitt.
Mr. Nicholas Pentony.
Mr. Samuel Parrish.
Mr. Joseph Pote.

Q

QUEEN's College Library, Cambridge.
* John Quick, Esq;
Mr. Nutcombe Quick.

R

* HIS Grace the Duke of Rutland.
* Most Hon. the Marquis of Rockingham.
* Rt. Hon. the Lord Romney.
* Rt. Hon. the Lord Ravensworth.
* Rt. Hon. the Lady Ravensworth.
Hon. John Robartes, Esq;
* Sir John Robinson, Bart.
Sir John Rous, Bart.
Sir Robert Rich, Bart.
Sir Tancred Robinson, Bart.
Samuel Reynardson, Esq;
Mr. John Ryan.
Mr. Jer. Roe.
Mr. Henry Ryall.
Rev. Mr. Arthur Robinson.
Mr. James Royston.
Robert Roane, Esq;
Robert Robinson, M. D.
* Mr. Rogers.
* Jones Raymond, Esq;
Rev. Mr. Herbert Randolph.
Thomas Robinson, Esq;
* Richard Roderick, Esq;
George Ruck, Esq;
Mr. Charles Reynoldson.
Mr. Charles Radcliff.
Mr. Morton Rockcliff.
Samuel Rush, Esq;
Bisse Richards, Esq;
Nicholas Roberts, Esq;
John Ranby, Esq;
Mr. Benjamin Robinson.
Mr. Samuel Rhodes.
* John Rush, Esq;
Rev. Mr. Rutter.
Mr. John Rowe.
* Mr. John Rule.
Lancelot Rolleston, Esq;
Matthew Robinson, Esq;
Rev. Dr. Richardson.

Mr.

A LIST of the SUBSCRIBERS.

Mr. John Rooke.
Rev. Henry Rooke, D. D.
Richard Ray, Efq; M. A.
Rev. Mr. Ray.
Brigadier Edward Richbelle.
Henry Reade, Efq;
Rev. Thomas Rutherforth, D. D.
Mr. Peter Ruffel.
Mr. Henry Richmond.
Jofeph Rea, Efq;
Mr. Richardfon, M. A.
* Mr. T. Rowney.
Rev. Mr. Rayner.
* William Rivet, Efq;
Thomas Reave, M. D.
Henry Rowe, Efq;
Mr. Thomas Rodbard.
Mr. John Rayner.
Mr. William Reddall.
Captain John Redman.
John Ruffell, Efq;
Matthew Rollifton, Efq;
Chriftopher Rawlinfon, Efq;
Rev. Mr. Richard Reddall.
Nathaniel Ryder, Efq;
Mr. Edward Rufhworth.
Thomas Rudd, Efq;
Mr. Richard Romman.
Rifley Rifley, Efq;
Mr. Thomas Rodber.
Walter Robertfon, Efq;
Mrs. Mary Roffey.
Mrs. Anne Roffey.
Mr. Andrew Ram.
Ruffell Revell, Efq;
Mr. Roberts.
* Thomas Roycroft, Efq;
* Thomas Ryvas, Efq;
* Captain Edward Rycaut.
Mr. Andrew Rogers.
Mr. Daniel Radford.
Mr. James Reade.
* Mr. John Reepe.
Mr. Jofeph Reynardfon.
Mr. John Rigg.
Captain Daniel Ruffell.
Mr. Nathaniel Rothery.
Mr. William Roberts.
John Reeve, Efq;
Mr. Jacob Robinfon.
Meff. John and James Rivington.
Mr. Caleb Ratten.

S

HER Grace the Duchefs of So-
merfet.

* Rt. Hon. the Earl of Sandwich.
* Rt. Hon. the Countefs of Shaftfbury.
* Rt. Hon. the Earl of Strafford.
Rt. Hon. the Lord Vifcount St. John.
* Rt. Rev. the Lord Bp. of Sodor and Man.
Hon. Edwin Sandys, Efq;
Sir William St. Quintin, Bart.
Sir George Savile, Bart.
Sir Thomas Style, Bart.
Sir John Strange, Kt.
Humphry Senhoufe, Efq;
James Stonehoufe, M. D.
Nicholas Styleman, Efq;
* Frederick Ulrick Schicke. Efq;
Mr. Richard Stevens.
Lovel Stanhope, Efq;
Mr. Daniel Scott.
Edmond Sawyer, Efq;
Thomas Sadler, Efq;
Mr. Symons.
Charles Smyth, Efq;
Richard Skrine, Efq;
Mr. John Smith.
Mr. Richard William Seale.
Mr. John Simmons.
* Charles Stanhope, Efq;
Mr. Thomas Smallwood,
Roger Sedgwick, M. B.
Mr. Henry Siffon.
R. Simms, Efq;
Harvey Sparkes, Efq;
Henry Sandys, Efq;
Mr. Savill.
Henry Richard Scudamore, Efq;
Rev. Mr. George Shakerley.
Philip Stephens, Efq;
* Mr. James Spragg.
* Captain Thomas Stanhope.
* Henry Stuart Stevens, Efq;
Mr. Thomas Speed.
Mr. Sharp.
Mr. Richard Sheldon.
* Admiral Smith.
John Sutton, Efq;
Mr. Edmund Stevens.
Mr. Thomas Smith.
Mr. Jonathan Scott.
Henry Shelley, Efq;
William Spicer, Eq;
* Captain Stevenfon.
Francis Say, Efq;
Mr. George Scott.
Alexander Stuart, Efq;
* Charles Smith, Efq;
* Rev. Dr. Arthur Smyth.
* Captain Arthur Scot.

* ——Sloan, Efq;
* Jacob Salvador, Efq;
Rev. —— Skottowe, B. D.
Rev. Mr. George Sykes.
Major Sawyer.
Mr. George Scott.
Richard Symons, Efq;
Robert Salusbury, Efq;
Rev. John Scott, A. M.
Mr. Thomas Smyth.
——Sparkes, Efq;
Rev. Mr. Smith.
Rev. ——Sedgwick, B. D.
Rev. ——Shuter, M. A.
Hervey Spragg, B. A.
Mr. John Spirker.
Rev. Henry Stebbing, D. D.
Rev. Henry Stebbing, M. A.
Robert Sutton, Efq;
Rev. Charles Soan, LL. B.
Henry Kynafton Southoufe, Efq
* ——Simpfon, Efq;
Mifs Sally Sewell.
Henry Snooke, Efq;
Henry Spencer, Efq;
Mr. John Sherwood.
Charles Scrivener, Efq;
Rev. Thomas Smythies, M. A.
Mr. Benjamin Sabbarton.
Mr. Walter Scott.
Mr. George Stanyford.
* Captain Charles Saunders.
Mr. Slingelandt.
Mr. St. Quintin.
Rev. Ralph Sneyd, D. D.
Major Sneyd.
Rev. ——Saunderfon, M. A.
* William Stanley, Efq;
Thomas Stack, M. D.
* Edward Spragg, Efq;
* Mifs Sabbarton.
* Sidney College Library, Cambridge.
Rev. Mr. Charles Squire.
* Mr. Doyley Stevens.
Mr. Samuel Bennett Smith.
Mr. John Sabatier.
Mr. Jenner Swaine.
William Selwin, Efq;
Mr. Edmund Stevens.
William Skinner, Efq;
* Mr. Laurence Singleton.
Mr. James Scot.
Sontley South, Efq;
Mr. Charles Salkeld.
* John Sharpe, Efq;
Mr. Stewart.
Rev. Mr. Snow.

* Captain

A LIST of the SUBSCRIBERS.

* Captain Philip Saumarez.
Rev. ——Sleech, D. D.
Mr. Henry Simeon.
Lieutenant Colonel John Stewart.
Mr. Stephen Simpson.
Joseph Simpson, Esq;
Mr. Charles Simpson.
Richard Symes, Esq;
* George Scott, Esq;
Mrs. Martha Steuart.
Rev. Mr. Edward Saul.
Rev. Mr. Philip Sone.
Rev. Mr. Smart.
Captain Mollineux Shaldham.
Mr. John Strettell.
Mr. Joseph Smith.
Rev. ——Sumner, D. D.
Rev. Mr. Swinden.
Lieutenant James Smith.
* Mr. Robert Smith.
Mr. James Stent.
Richard Stevens, Esq;
Mrs. Sloper.
* John Saumarez, Esq;
Mr. Henry Sleach.
* Captain Samuel Scot.
* Henry Swan, Esq;
Mr. Richard Smith.
Mr. William Sone.
Mr. Richard Stanford.
* Captain Nathaniel Stevens.
* Captain John Storr.
Robert Salkeld, Esq;
* Captain Skeffington.
* Mr. Thomas Smith.
Mess. Stabler and Barstow.
Mrs. Elizabeth Smithurst.
Mr. Edward Smith.
Mr. John Shuckburgh.

T

* RIGHT Hon. the Earl of Traquair.
* Rt. Hon. the Lord Torrington.
Rt. Hon. the Lord Tyrconnel.
* Rt. Hon. the Lord Trevor.
* Rt. Hon. the Lord Talbot.
* Hon. Captain George Townshend.
Sir Charles Keymeys Tynte, Bart.
Sir John Thompson, Kt.
Sir Peter Thompson. Kt.
Mr. Samuel Trymmer.
Thomas Thornbury, Esq;
Mr. William Thomas.
Captain Samuel Thornton.
Trinity-House in Hull.
* Rev. Mr. Tough.

* Robert Thompson, Esq;
Rev. Mr. Herbert Taylor.
* Robert Taylor, M. D.
Richard Tylden, Esq;
Samuel Theyer, Esq;
Coel Thornhill, Esq;
John Tucker, Esq;
* Rev. Mr. Nicholas Tindal.
John Twisleton, Esq;
Rev. Mr. Tyson.
Marmaduke Tunstall, Esq;
——Turner, Esq;
Blayney Townley, Esq;
Mr. Francis Tregagle.
Andrew Taylor, Esq;
George Trenchard, Esq;
Mr. Jacob Thibou.
John Tilson, Esq;
Mr. James Turner.
* Rev. Mr. Talham.
* William Trumbull, Esq;
* Arthur Trevor, Esq;
* Abraham Taylor, Esq;
Rev. Mr. Terry.
Captain George Tindal.
John Turner, Esq;
William Thomson, Esq;
John Tuckfield, Esq;
Rev. —— Taylor, B. D.
Trinity-College Library, Cambridge.
Joseph Tudor, Esq;
Rev. George Tilson, M. A.
Edmund Tyrrel, Esq;
* Thomas Tickell, Esq;
Mr. James Thornton.
Mr.——Tristram.
——Tatham, Esq;
John Thurston, M. D.
Rev. Gustavus Thompson, M. A.
Rev. Robert Thomlinson, D. D.
Mr. Peregrine Tyzack.
Rev. James Tunstall, D. D.
Rev. Mr. Traherne.
Richard Tyson, M. D.
Thomas Tower, Esq;
* Mr. Robert Tunstal.
Joseph Tily, Esq;
Mr. Barnard Townsend.
Rev. Mr. John Taylor.
Rev. Mr. George Tymms.
Christopher Toncer, Esq;
George Tash, Esq;
Ralph Towne, Esq;
Mr. William Thomas.
Abraham Tucker, Esq;
Clement Tudway, Esq;
Mr. Hugh Tomlins.
Mr. John Townsend.

Captain William Tayler.
Oliver Tilsen, Esq;
Mr. Joshua Toft.
Mrs. Anne Toft.
Mr. Thomas Tyndall.
Captain Christen Tideman.
Captain Hans Tysch.
Rev. Mr. Trant.
Mr. William Thurlbourn.
Mr. Robert Taylor.
Mr. Barnabas Thorne.

V

* HON. Thomas Villiers, Esq;
Mr. George Virgoe.
Mr. Robert Vincent.
* S. Vilett, Esq;
* William Vigor, Esq;
* Captain Philip Vincent.
——Vandeval, Esq;
* Thomas Voughon, Esq;
Mr. John Van Rixtel.
Thomas Uthwat, Esq;
Mr. James Unwin.
Mr. Isaac Vanaffendelft.
Mr. John Vokes.

W

* RIGHT Hon. the Earl of Warwick.
* Rt. Hon. the Earl of Warrington.
* Rt. Hon. the Earl Waldegrave.
* Rt. Rev. the Lord Bp. of Worcester.
Rt. Hon. the Lord Willoughby of Parham.
* Rt. Hon. the Lord Ward.
Lady Williams.
* Hon. Colonel Waldegrave.
Hon. Francis Willoughby, Esq;
Rev. and Hon. Mr. Wandesford.
Hon. Mr. Justice Wright.
* Sir Watkin Williams Wynn, Bart.
Sir Edward Worsley, Bart.
Sir Anthony Westcombe, Bart.
Sir Randal Ward, Bart.
Mr. Edward Woodcock.
Richard Wilkes, M. D.
Mr. Richard Whilock.
John Wilkes, Esq;
Ralph Willet, Esq;
Philip Corbet Webb, Esq;
Gilbert Walmsley, Esq;
Daniel Wray, Esq;
John Wowen, Esq;
Francis Wollaston, Esq;
* Mr. William Watson.

Mr.

A LIST of the SUBSCRIBERS.

Mr. John Ward.
Thomas Wilbraham, LL. D.
Mr. Samuel Wyatt.
Nathaniel Wettenhall, Efq;
Anthony Walburgh, Efq;
Thomas White, Efq;
Thomas Whittington, Efq;
Mr. Harbord Wright.
John Williams, Efq;
Mr. Thomas Whifker.
Mr. Thomas Welch.
Mr. Whifker.
Thomas Wilfon, Efq;
Mr. Samuel Wilfon.
Mr. William Webb.
Thomas Weftern, Efq;
Captain Temple Weft.
Mr. Charles Wildbore.
James Wallace, Efq;
Edward Wright, Efq;
Mr. Warkman.
Rev. Mr. Witton.
Francis Wace, Efq;
Rev. Mr. Jeffery Walmfley.
William Wilkinfon, Efq;
Jofhua Winder, Efq,
Rev. Dr. Wright.
——Watkins, Efq;
Ed. Wilmot, M. D.
Mr. William Ware.
* Rev. Mr. William Warburton.
* Mr. Wildey.
William Woolball, Efq;
* ——Wollafton, Efq;
* William Saltren Willet, Efq;
* Francis Woodhoufe, Efq;
* William Watts, Efq;
Mr. Watts.
Mr. Whiftler.
Rev. Mr. Robert Wilfon.

Mr. Wilfon.
Captain Weller.
John Wynn, Efq;
——White, Efq;
Rev. John Withers, M. A.
Mr. Waterland.
Rev. Dr. Henry Waterland.
Mr. William Welfitt.
Rev. Mr. Battie Worfop.
* John Walton, Efq;
Rev.——Warcopp, LL. B.
Henry Lee Warner, Efq;
Rev. Mr. Wheeler.
* Philip Ward, Efq;
Rev. Mr.——Wrangham.
Mr.——Wollafcot.
Rev. Stephen Whiffon, M. A.
Mr. Robert Willis.
Mr. Robert Waiftfield.
Rev. Mr. Wright.
William Whitehead, Efq;
Mr. John Wibberfley.
Rev.——Whalley, D. D.
Rev.——Wilfon, B. D.
Thomas Whately, Efq;
Mr. Edward Williams.
Mr. Henry Wigley.
Walter Walker Ward, D. D.
William Wollafton, Efq;
John Wedgewood, Efq;
* Mr. Arthur Walter.
* Chriftopher Walter, Efq;
* Mr. Arthur Walter, junr.
Mr. David Wharàm.
Mr. John Warrall.
Mr. Thomas Whyte.
Captain John Williams.
Matthew Wildbore, Efq;
Mr. Thomas Wright.

2

Rev. Mr. Edmund Williamfon.
Rev. Mr. Henry Watkins.
Matthew Wife, Efq;
Mr. Jofeph Walton.
Colonel Wardour.
Ralph Whiftler, Efq;
Samuel Wegg. Efq;
* William Welby, Efq;
Mr. Robert Watfon.
* Mr. Ifrael Wilkes.
Mr. John Willett.
Mrs. White.
Mr. Job Wilkes.
Mr. Henry Woodcock.
Benjamin Woodward, Efq;
Edward Wheeler, Efq;
* Mr. Peter Waldo.
Thomas Waters, Efq;
George Weller, Efq;
Mr. John Watfon.
Ifaac Whittington, Efq;
Mr. Samuel Whitmore.
Rev. Mr. William Wilfon.
Captain John Williams.
Lieutenant Edward Wheeler.
Mr. William Ward.
Mr. Richard Ware.
Mr. Henry Whitridge.
Mr. John Whifton.

Y

* HON. Henry Yelverton, Efq;
   Hon. Philip Yorke, Efq;
William Young, Efq;
Rev. Philip Yonge, M. A.
Mr. Talbot Young.
Mr. John Young.

CON-

CONTENTS.

(c)   CHAP.

CONTENTS.

CHAP.

CONTENTS.

BOOK III.

(c 2)   CHAP.

CONTENTS.

ERRATA.

*P*AGE 40. *line* 6. *for* about, *read* about to. P. 88. *l.* 34. *f.* heighth, *r.* height. P. 97. *l.* 20. *f.* rout, *r.* route ; *l.* 21, 22, 27. *f.* tract, *r.* track. P. 114. *l.* 32. *f* A *r. a* ; *l.* 33. *f.* B *r. b* ; *f.* C *r. c* ; *l.* 34. *f.* D *r. d.* P. 115. l. 1. *f.* E *r. e.* P. 200. l. 34. *f.* Cur, *r.* Our. P. 216. *l.* 9. *f.* Eaft-end Ifland, *r.* Eaft end of the Ifland. P. 254. *l.* 33. *f.* D, *r.* C. P. 255. *l.* 12. *f.* I I, *r.* H H. P. 267. *l.* 33. *f.* no, *r. a.* P. 282. *l.* 3. *f.* longitude, *r.* latitude. P. 305. *l.* 17. *f.* fower, *r.* four. P. 355. *l.* 6. *f.* metaorphofis, *r.* metamorphofis.—In the plan of *Chequetan, f.* Bath, *r* Path. In fome impreffions of the *Chinefe* veffels, the fore-fheet in the veffel A is placed on the wrong fide of the maft.

The Reader is defired to excufe the feveral falfe fpellings in the Plates, as they are none of them of moment ; and the erafing and correcting them would have coft much time and trouble.

INTRODUCTION.

NOTWITHSTANDING the great improvement of navigation within the laſt two Centuries, a Voyage round the World is ſtill conſidered as an enterprize of a very ſingular nature; and the Public have never failed to be extremely inquiſitive about the various accidents and turns of fortune, with which this uncommon attempt is generally attended: And though the amuſement expected in a narration of this kind, is doubtleſs one great ſource of this curioſity, and a ſtrong incitement with the bulk of readers, yet the more intelligent part of mankind have always agreed, that from theſe relations, if faithfully executed, the more important purpoſes of navigation, commerce, and national intereſt may be greatly promoted: For every authentic account of foreign coaſts and countries will contribute to one or more of theſe great ends, in proportion to the wealth, wants, or commodities of thoſe countries, and our ignorance of thoſe coaſts; and therefore a Voyage round the World promiſes a ſpecies of information, of all others the moſt deſirable and intereſting; ſince great part of it is performed in ſeas, and on coaſts, with which we are as yet but very imperfectly acquainted, and in the neighbourhood of a country renowned for the abundance of its wealth, though it is at the ſame time ſtigmatiſed for its poverty, in the neceſſaries and conveniencies of a civilized life.

Theſe conſiderations have occaſioned the publication of the enſuing work; which, in gratifying the inquiſitive turn of mankind, and contributing to the ſafety and ſucceſs of future navigators, and to the extenſion of our commerce and power, may doubtleſs vie with any narration of this kind hitherto made public: Since the circumſtances of this undertaking already known to the world, may

be

INTRODUCTION.

be fuppofed to have ftrongly excited the general curiofity; for whether we confider the force of the fquadron fent on this fervice, or the diverfified diftreffes that each fingle fhip was feparately involved in, or the uncommon inftances of varying fortune, which attended the whole enterprize, each part, I conceive, muft, from its rude well known outlines, appear worthy of a compleater and more finifhed delineation : And if this be allowed with refpect to the narrative part of the work, there can be no doubt about the more ufeful and inftructive parts, which are almoft every where interwoven with it; for I can venture to affirm, without fear of being contradicted on a comparifon, that no voyage I have yet feen, furnifhes fuch a number of views of land, foundings, draughts of roads and ports, charts, and other materials, for the improvement of geography and navigation, as are contained in the enfuing volume ; which are of the more importance too, as the greateft part of them relate to fuch Iflands or Coafts, as have been hitherto not at all or erroneoufly defcribed, and where the want of fufficient and authentic information might occafion future enterprizes to prove abortive, perhaps with the deftruction of the men and veffels employed therein.

And befides the number and choice of thefe marine drawings and defcriptions, there is another very effential circumftance belonging to them, which much enhances their value ; and that is, the great accuracy they were drawn with. I fhall exprefs my opinion of them in this particular very imperfectly, when I fay, that they are not exceeded, and perhaps not equalled by any thing of this nature hitherto made public : For they were not copied from the works of others, or compofed at home from imperfect accounts, given by incurious and unfkilful obfervers, as hath been frequently the cafe in thefe matters ; but the greateft part of them were drawn on the fpot with the utmoft exactnefs, by the direction, and under the eye of Mr. *Anfon* himfelf ; and where (as is the cafe in three or four of them) they have been done by lefs fkilful hands, or were found in poffeffion of the enemy, and confequently their juft-
                                                                                      nefs

ness could be less relied on, I have always taken care to apprize the reader of it, and to put him on his guard against giving entire credit to them; although I doubt not, but these less authentic draughts, thus cautiously inserted, are to the full as correct as those, which are usually published on these occasions. For as actual surveys of roads and harbours, and nice and critical delineations of views of land, take up much time and attention, and require a good degree of skill both in planning and drawing, those who are defective in industry and ability, supply these wants by bold conjectures, and fictitious descriptions; and as they can be no otherwise confuted than by going on the spot, and running the risque of suffering by their misinformation, they have no apprehensions of being detected; and therefore, when they intrude their suppositious productions on the Public, they make no conscience of boasting at the same time, with how much skill and care they are performed. And let not those who are unacquainted with naval affairs imagine, that impositions of this kind are of an innocent nature; for as exact views of land are the surest guide to a seaman, on a coast where he has never been before, all fictions in so interesting a matter must be attended with numerous dangers, and sometimes with the destruction of those who are thus unhappily deceived.

Besides these draughts of such places as Mr. *Anson* or the ships under his command have touched at in the course of this expedition, and the descriptions and directions relating thereto, there is inserted, in the ensuing work, an ample description, with a chart annexed to it, of a particular navigation, of which hitherto little more than the name has been known, except to those immediately employed in it: I mean the track described by the *Manila* ship, in her passage to *Acapulco*, through the northern part of the *Pacific* Ocean. This material part is collected from the draughts and journals met with on board the *Manila* galeon, founded on the experience of more than a hundred and fifty years practice, and corroborated in its principal circumstances by the concurrent evidence of all the *Spanish* prisoners taken in that vessel. And as many of their journals,

which

INTRODUCTION.

which I have examined, appear to have been not ill kept, I prefume, the chart of that northern Ocean, and the particulars of their
route through it, may be very fafely relied on by future Navigators.
The advantages, which may be drawn from an exact knowledge
of this navigation, and the beneficial projects that may be formed
thereon, both in war and peace, are by no means proper to be difcuffed in this place: But they will eafily offer themfelves to the
fkilful in maritime affairs. However, as the *Manila* fhips are the
only ones which have ever traverfed this vaft ocean, except a *French*
ftraggler or two, which have been afterwards feized on the coaft of
Mexico, and as during near two ages, in which this trade has been
carried on, the *Spaniards* have, with the greateft care, fecreted all
accounts of their voyages from the reft of the world; thefe reafons
alone would authorize the infertion of thofe papers, and would recommend them to the inquifitive, as a very great improvement in
geography, and worthy of attention from the fingularity of many
circumftances recited therein. I muft add too, (what in my opinion is far from being the leaft recommendation of thefe materials)
that the obfervations of the variation of the compafs in that Ocean,
which are inferted in the chart from thefe *Spanifh* journals, tend
greatly to compleat the general fyftem of the magnetic variation, of
infinite import to the commercial and feafaring part of mankind.
Thefe obfervations were, though in vain, often publickly called for
by our learned countryman the late Dr. *Halley*, and to his immortal reputation they confirm, as far as they extend, the wonderful
hypothefis he had entertained on this head, and very nearly corref-
pond in their quantity, to the predictions he publifhed above fifty
years fince, long before he was acquainted with any one obfervation
made in thofe feas. The afcertaining the variation in that part of
the world is juft now too of more than ordinary confequence, as the
Editors of a new variation-chart lately publifhed, have, for want of
obfervations in thofe parts, been mifled by an erroneous analogy,
and have miftaken the very fpecies of variation in thofe northern
feas; for they make it wefterly where it is eafterly, and have laid it
down 12° or 13° fhort of its real quantity.   Thus

4

INTRODUCTION.

Thus much it has been thought neceffary to premife with regard to the hydrographical and geographical part of the enfuing work; which it is hoped the reader will, on perufal, find much ampler and more important than this flight fketch can well indicate. But as there are hereafter occafionally interfperfed fome accounts of *Spanifh* tranfactions, and many obfervations on the difpofition of the *American Spaniards*, and on the condition of the countries border-ing on the *South-Seas*, and as herein I may appear to differ greatly from the opinions generally eftablifhed, I think it incumbent on me particularly to recite the authorities I have been guided by on this occafion, that I may not be cenfured, as having given way either to a thoughtlefs credulity on one hand, or, what would be a much more criminal imputation, to a wilful and deliberate mifreprefentation on the other.

Mr. *Anfon*, before he fet fail upon this expedition, befides the printed journals to thofe parts, took care to furnifh himfelf with the beft manufcript accounts he could procure of all the *Spanifh* fet-tlements upon the coafts of *Chili*, *Peru* and *Mexico*: Thefe he carefully compared with the examinations of his prifoners, and the informations of feveral intelligent perfons, who fell into his hands in the *South-Seas*. He had likewife the good fortune, in fome of his captures, to poffefs himfelf of a great number of letters and pa-pers of a public nature, many of them written by the Viceroy of *Peru* to the Viceroy of *Santa Fee*, to the Prefidents of *Panama* and *Chili*, to Don *Blafs de Lezo*, Admiral of the galeons, and to di-vers other perfons in public employments; and in thefe letters there was ufually inferted a recital of thofe they were intended to anfwer; fo that they contained a confiderable Part of the correfpon-dence between thefe officers for fome time previous to our arrival on that coaft: We took befides many letters fent from perfons em-ployed by the Government to their friends and correfpondents, which were frequently filled with narrations of public bufinefs, and fometimes contained undifguifed animadverfions on the views and conduct of their fuperiors. From thefe materials thofe accounts

(d)   of

INTRODUCTION.

of the *Spanish* affairs are taken, which may at firſt ſight appear the moſt exceptionable. In particular, the hiſtory of the various caſualties which befel *Pizarro*'s ſquadron, is for the moſt part compoſed from intercepted letters: Though indeed the relation of the inſurrection of *Orellana* and his followers, is founded on rather a leſs diſputable authority: For it was taken from the mouth of an *Engliſh* Gentleman then on board *Pizarro*, who often converſed with *Orellana*; and it was, on enquiry, confirmed in its principal circumſtances by others who were in the ſhip at the ſame time: So that the fact, however extraordinary, is, I conceive, not to be conteſted.

And on this occaſion I cannot but mention, that though I have endeavoured, with my utmoſt care, to adhere ſtrictly to truth in every article of the enſuing narration; yet I am apprehenſive, that in ſo complicated a work, ſome overſights muſt have been committed, by the inattention to which at times all mankind are liable. However, I know of none but literal miſtakes, ſome of which are corrected in the table of Errata: And if there are other errors which have eſcaped me, I flatter myſelf they are not of moment enough to affect any material tranſaction, and therefore I hope they may juſtly claim the readers indulgence.

After this general account of the contents of the enſuing work, it might be expected, perhaps, that I ſhould proceed to the work itſelf, but I cannot finiſh this Introduction, without adding a few reflexions on a matter very nearly connected with the preſent ſubject; and, as I conceive, neither deſtitute of utility, nor unworthy the attention of the Public; I mean, the animating my countrymen both in their public and private ſtations, to the encouragement and purſuit of all kinds of geographical and nautical obſervations, and of every ſpecies of mechanical and commercial information. It is by a ſettled attachment to theſe ſeemingly minute particulars, that our ambitious neighbours have eſtabliſhed ſome part of that power, with which we are now ſtruggling: And as we have the means in our hands of purſuing theſe ſubjects more effectually, than they can, it would be a diſhonour to us longer to neglect ſo eaſy and

beneficial

INTRODUCTION.

beneficial a practice: For, as we have a Navy much more numerous than theirs, great part of which is always employed in very distant stations, either in the protection of our colonies and commerce, or in assisting our allies against the common enemy, this gives us frequent opportunities of furnishing ourselves with such kind of materials, as are here recommended, and such as might turn greatly to our advantage either in war or peace: For, not to mention what might be expected from the officers of the Navy, if their application to these subjects was properly encouraged, it would create no new expence to the Goverment to establish a particular regulation for this purpose; since all that would be requisite, would be constantly to embark on board some of our men of war, which are sent on these distant cruises, a person, who with the character of an engineer, and the skill and talents necessary to that profession, should be employed in drawing such coasts, and planning such harbours, as the ship should touch at, and in making such other observations of all kinds, as might either prove of advantage to future Navigators, or might any ways tend to promote the Public service. Besides, persons habituated to this employment (which could not fail at the same time of improving them in their proper business) would be extremely useful in many other lights, and might serve to secure our Fleets from those disgraces, with which their attempts against places on shore have been often attended: And, in a Nation like ours, where all sciences are more eagerly and universally pursued, and better understood than in any other part of the world, proper subjects for such employments could not long be wanting, if due incouragement were given to them. This method here recommended is known to have been frequently practised by the *French*, particularly in the instance of Monsieur *Frezier*, an Engineer, who has published a celebrated voyage to the *South-Seas*: For this person in the year 1711, was purposely sent by the *French* King into that country on board a merchantman, that he might examine and describe the coast, and take plans of all the fortified places, the better to enable the *French* to prosecute their illicit trade, or, in case of a rupture with the

court

INTRODUCTION.

court of *Spain*, to form their enterprizes in thofe feas with more readinefs and certainty. Should we purfue this method, we might hope, that the emulation amongft thofe who were thus employed, and the experience, which even in time of peace, they would hereby acquire, might at length procure us a proper number of able Engineers, and might efface the national fcandal, which our deficiency in that fpecies of men has fome times expofed us to : And furely, every ftep to encourage and improve this profeffion is of great moment to the Public ; as no perfons, when they are properly inftructed, make better returns in war, for the encouragement and emoluments beftowed on them in time of peace. Of which the advantages the *French* have reaped from their dexterity (too numerous and recent to be foon forgot) are an ample confirmation.

And having mentioned Engineers, or fuch as are fkilled in drawing, and the other ufual practices of that profeffion, as the propereft perfons to be employed in thefe foreign enquiries, I cannot (as it offers itfelf fo naturally to the fubject in hand) but lament, how very imperfect many of our accounts of diftant countries are rendered by the relators being unfkilled in drawing, and in the general principles of furveying ; even where other abilities have not been wanting. Had more of our travellers been initiated in thefe acquirements, and had there been added thereto fome little fkill in the common aftronomical obfervations, (all which a perfon of ordinary talents might attain, with a very moderate fhare of application) we fhould by this time have feen the geography of the globe much correcter, than we we now find it ; the dangers of navigation would have been confiderably leffened, and the manners, arts and produce of foreign countries would have been much better known to us, than they are. Indeed, when I confider, the ftrong incitements that all travellers have to acquire fome part at leaft of thefe qualifications, efpecially drawing ; when I confider how much it would facilitate their obfervations, affift and ftrengthen their memories, and of how tedious, and often unintelligible, a load of defcription it would rid them, I cannot but wonder that any perfon, that intends to vifit diftant countries, with a

view

INTRODUCTION.

view of informing either himself or others, should be unfurnished with so useful a piece of skill. And to inforce this argument still further, I must add, that besides the uses of drawing, which are already mentioned, there is one, which, though not so obvious, is yet perhaps of more consequence than all that has been hitherto urged; and that is, that those who are accustomed to draw objects, observe them with more distinctness, than others who are not habituated to this practice. For we may easily find, by a little experience, that in viewing any object however simple, our attention or memory is scarcely at any time so strong, as to enable us, when we have turned our eyes away from it, to recollect exactly every part it consisted of, and to recal all the circumstances of its appearance; since, on examination, it will be discovered, that in some we were mistaken, and others we had totally overlooked: But he that is employed in drawing what he sees, is at the same time employed in rectifying this inattention; for by confronting his ideas copied on the paper, with the object he intends to represent, he finds in what manner he has been deceived in its appearance, and hence he in time acquires the habit of observing much more at one view, and retains what he sees with more correctness than he could ever have done, without his practice and proficiency in drawing.

If what has been said merits the attention of Travellers of all sorts, it is, I think, more particularly applicable to the Gentlemen of the Navy; since, without drawing and planning, neither charts nor views of land can be taken; and without these it is sufficiently evident, that navigation is at a full stand. It is doubtless from a persuasion of the utility of these qualifications, that his Majesty has established a drawing Master at *Portsmouth*, for the instruction of those, who are presumed to be hereafter intrusted with the command of his Royal Navy: And though some have been so far misled, as to suppose that the perfection of Sea-officers consisted in a turn of mind and temper resembling the boisterous element they had to deal with, and have condemned all literature and science as effeminate, and derogatory to that ferocity, which, they would falsely persuade us,

was

INTRODUCTION.

was the moſt unerring characteriſtic of courage: Yet it is to be
hoped, that ſuch abſurdities as theſe have at no time been authoriſed
by the Public opinion, and that the belief of them daily diminiſhes.
If thoſe who adhere to theſe miſchievous poſitions were capable of
being influenced by reaſon, or ſwayed by example, I ſhould think
it ſufficient for their conviction to obſerve, that the moſt valuable
drawings inſerted in the following work, though done with ſuch
a degree of ſkill, that even profeſſed artiſts can with difficulty imitate them, were taken by Mr. *Peircy Brett*, one of Mr. *Anſon*'s
Leiutenants, and ſince Captain of the *Lion* man of war; who, in
his memorable engagement with the *Elizabeth* (for the importance
of the ſervice, or the reſolution with which it was conducted, inferiour to none this age has ſeen) has given ample proof, that a proficiency in the arts I have been here recommending is extremely
conſiſtent with the moſt exemplary bravery, and the moſt diſtinguiſhed ſkill in every function belonging to the duty of a Sea-officer.
Indeed, when the many branches of ſcience are conſidered, of
which even the common practice of navigation is compoſed, and the
many improvements, which men of ſkill have added to this practice
within theſe few years, it would induce one to believe, that the advantages of reflection and ſpeculative knowledge were in no profeſſion more eminent than in that of a ſea-officer: For, not to mention
ſome expertneſs in geography, geometry and aſtronomy, which it
would be diſhonourable for him to be without, (as his journal and his
eſtimate of the daily poſition of the ſhip are no more than the practice of particular branches of theſe arts) it may be well ſuppoſed,
that the management and working of a ſhip, the diſcovery of her
moſt eligible poſition in the water, (uſually ſtiled her Trim) and the
diſpoſition of her ſails in the moſt advantageous manner, are articles,
wherein the knowledge of mechanics cannot but be greatly aſſiſtant :
And perhaps the application of this kind of knowledge to naval ſubjects may produce as great improvements in ſailing and working a
ſhip, as it has already done in many other matters conducive to the
eaſe and convenience of human life : For when the fabric of a ſhip,

and

INTRODUCTION.

and the variety of her fails are confidered, together with the artificial
contrivances of adapting them to her different motions, as it cannot
be doubted, but thefe things have been brought about by more than
ordinary fagacity and invention, fo neither can it be doubted but that
a fpeculative and fcientific turn of mind may find out the means of
directing and difpofing this complicated mechanifm much more ad-
vantageoufly than can be done by mere habit, or by a fervile copying of what others may perhaps have erroneoufly practifed in the
like emergency : But it is time to finifh this digreffion and to leave the
reader to the perufal of the enfuing work ; which, with how little
art foever it may be executed, will yet, from the importance of the
fubject, and the utility and excellence of the materials, merit fome
fhare of the Public attention.

A VOYAGE

A

VOYAGE

ROUND THE

WORLD,

BY

GEORGE ANSON, Efq;

Commander in Chief of a Squadron of his
MAJESTY's Ships.

BOOK I.

CHAP. I.

Of the equipment of the fquadron: The incidents relating thereto, from its firft appointment to its fetting fail from St. *Helens*.

THE fquadron under the Command of Mr. *Anfon* (of
which I here propofe to recite the moft material proceedings) having undergone many changes in its deftination,
its force, and its equipment, in the ten months between
its firft appointment and its final failing from St. *Helens* ; I conceive
the hiftory of thefe alterations is a detail neceffary to be made public, both for the honour of thofe who firft planned and promoted
this enterprize, and for the juftification of thofe who have been en-

B   trufted

trufted with its execution. Since it will from hence appear, that the accidents the expedition was afterwards expofed to, and which prevented it from producing all the national advantages the ftrength of the fquadron, and the expectation of the public, feemed to prefage, were principally owing to a feries of interruptions, which delayed the Commander in the courfe of his preparations, and which it exceeded his utmoft induftry either to avoid or to get removed.

When in the latter end of the fummer of the year 1739, it was forefeen that a war with *Spain* was inevitable, it was the opinion of feveral confiderable perfons then trufted with the Adminiftration of affairs, that the moft prudent ftep the Nation could take, on the breaking out of the war, was attacking that Crown in her diftant fettlements; for by this means (as at that time there was the greateft probability of fuccefs) it was fuppofed that we fhould cut off the principal refources of the enemy, and reduce them to the neceffity of fincerely defiring a peace, as they would hereby be deprived of the returns of that treafure by which alone they could be enabled to carry on a war.

In purfuance of thefe fentiments, feveral projects were examined, and feveral refolutions taken in Council. And in all thefe deliberations it was from the firft determined, that *George Anfon*, Efq; then Captain of the *Centurion*, fhould be employed as Commander in Chief of an expedition of this kind : And he then being abfent on a cruize, a veffel was difpatched to his ftation fo early as the beginning of *September*, to order him to return with his fhip to *Portfmouth*. And foon after he came there, that is, on the 10th of *November* following, he received a letter from Sir *Charles Wager*, ordering him to repair to *London*, and to attend the board of Admiralty : Where, when he arrived, he was informed by Sir *Charles*, that two Squadrons would be immediately fitted out for two fecret expeditions, which however would have fome connexion with each other : That he, Mr. *Anfon*, was intended to command one of them, and Mr. *Cornwall* (who hath fince loft his life glorioufly in the defence of his Country's honour) the other :

That

That the fquadron under Mr. *Anfon* was to take on board three Independent Companies of a hundred men each, and *Bland's* regiment of Foot: That Colonel *Bland* was likewife to imbark with his regiment, and to command the land-forces: And that, as foon as this fquadron could be fitted for the fea, they were to fet fail, with exprefs orders to touch at no place till they came to *Java Head* in the *Eaft-Indies*: That there they were only to ftop to take in water, and thence to proceed directly to the city of *Manila*, fituated on *Luconia*, one of the *Philippine* Iflands: That the other fquadron was to be of equal force with this commanded by Mr. *Anfon*, and was intended to pafs round Cape *Horn* into the *South-Seas*, and there to range along that coaft; and after cruizing upon the enemy in thofe parts, and attempting their fettlements, this fquadron in its return was to rendezvous at *Manila*, and there to join the fquadron under Mr. *Anfon*, where they were to refrefh their men, and refit their fhips, and perhaps receive further orders.

This fcheme was doubtlefs extremely well projected, and could not but greatly advance the Public Service, and at the fame time the reputation and fortune of thofe concerned in its execution; for had Mr. *Anfon* proceeded for *Manila* at the time and in the manner propofed by Sir *Charles Wager*, he would, in all probability, have arrived there before they had received any advice of the war between us and *Spain*, and confequently before they had been in the leaft prepared for the reception of an enemy, or had any apprehenfions of their danger. The city of *Manila* might be well fuppofed to have been at that time in the fame defencelefs condition with all the other *Spanifh* fettlements, juft at the breaking out of the war: That is to fay, their fortifications neglected, and in many places decayed; their cannon difmounted, or ufelefs by the mouldring of their carriages; their magazines, whether of military ftores or provifion, all empty; their garrifons unpaid, and confequently thin, ill-affected, and difpirited; and the royal chefts in *Peru*, whence alone all thefe diforders could receive their redrefs, drained to the very bottom: This, from the intercepted letters of their Viceroys

and

and Governors, is well known to have been the defencelefs ftate of
Panama, and the other *Spanifh* places on the coaft of the *South-
Sea*, for near a twelvemonth after our declaration of war. And it
cannot be fuppofed that the city of *Manila*, removed ftill farther by
almoft half the circumference of the globe, fhould have experienced
from the *Spanifh* Government, a greater fhare of attention and concern for its fecurity, than *Panama*, and the other important ports in
Peru and *Chili*, on which their poffeffion of that immenfe Empire depends. Indeed, it is well known, that *Manila* was at that
time incapable of making any confiderable defence, and in all probability would have furrendered only on the appearance of our
fquadron before it. The confequence of this city, and the ifland it
ftands on, may be in fome meafure eftimated, from the healthinefs of
its air, the excellency of its port and bay, the number and wealth
of its inhabitants, and the very extenfive and beneficial commerce
which it carries on to the principal Ports in the *Eaft-Indies*, and
China, and its exclufive trade to *Acapulco*, the returns for which,
being made in filver, are, upon the loweft valuation, not lefs than
three millions of Dollars.*per annum.*

And on this Scheme Sir *Charles Wager* was fo intent, that in a
few days after this firft conference, that is, on *November* 18,
Mr. *Anfon* received an order to take under his command the *Argyle*, *Severn*, *Pearl*, *Wager*, and *Tryal Sloop*; and other orders were
iffued to him in the fame month, and in the *December* following,
relating to the victualling of this fquadron. But Mr. *Anfon* attending the Admiralty the beginning of *January*, he was informed by
Sir *Charles Wager*, that for reafons with which he, Sir *Charles*, was
not acquainted, the expedition to *Manila* was laid afide. It may be
conceived, that Mr. *Anfon* was extremely chagrined at the lofing the
command of fo infallible, fo honourable, and in every refpect, fo defirable an enterprize, efpecially too as he had already, at a very
great expence, made the neceffary provifion for his own accommodation in this voyage, which he had reafon to expect would prove a
very long one. However, Sir *Charles*, to render this difappointment

in

in some degree more tolerable, informed him that the expedition to the *South-Seas* was still intended, and that he, Mr. *Anson*, and his squadron, as their first destination was now countermanded, should be employed in that service. And on the 10th of *January* he received his commission, appointing him Commander in Chief of the forementioned squadron, which (the *Argyle* being in the course of their preparation changed for the *Gloucester*) was the same he sailed with above eight months after from St. *Helens*. On this change of destination, the equipment of the squadron was still prosecuted with as much vigour as ever, and the victualling, and whatever depended on the Commodore, was so far advanced, that he conceived the ships might be capable of putting to sea the instant he should receive his final orders, of which he was in daily expectation. And at last, on the 28th of *June* 1740, the Duke of *Newcastle*, Principal Secretary of State, delivered to him his Majesty's instructions, dated *January* 31, 1739, with an additional instruction from the Lords Justices, dated *June* 19, 1740. On the receipt of these, Mr. *Anson* immediately repaired to *Spithead*, with a resolution to sail with the first fair wind, flattering himself that all his delays were now at an end. For though he knew by the musters that his squadron wanted three hundred seamen of their complement, (a deficiency which, with all his assiduity, he had not been able to get supplied) yet, as Sir *Charles Wager* informed him, that an order from the board of Admiralty was dispatched to Sir *John Norris* to spare him the numbers which he wanted, he doubted not of his complying therewith. But on his arrival at *Portsmouth*, he found himself greatly mistaken, and disappointed in this persuasion: for on his application, Sir *John Norris* told him, he could spare him none, for he wanted men for his own fleet. This occasioned an inevitable and a very considerable delay; for it was the end of *July* before this deficiency was by any means supplied, and all that was then done was extremely short of his necessities and expectation. For Admiral *Balchen*, who succeeded to the command at *Spithead*, after Sir *John Norris* had sailed to the west-

ward

ward, inftead of three hundred able failors, which Mr. *Anfon* wanted of his complement, ordered on board the fquadron a hundred and feventy men only ; of which thirty-two were from the hofpital and fick quarters, thirty-feven from the *Salifbury*, with three officers of Colonel *Lowther*'s regiment, and ninety-eight marines, and thefe were all that were ever granted to make up the forementioned deficiency.

But the Commodore's mortification did not end here. It has been already obferved, that it was at firft intended that Colonel *Bland*'s regiment, and three independent companies of a hundred men each, fhould embark as land-forces on board the fquadron. But this difpofition was now changed, and all the land-forces that were to be allowed, were five hundred invalids to be collected from the out-penfioners of *Chelfea* college. As thefe out-penfioners confift of foldiers, who from their age, wounds, or other infirmities, are incapable of fervice in marching regiments, Mr. *Anfon* was greatly chagrined at having fuch a decrepid detachment allotted him ; for he was fully perfuaded that the greateft part of them would perifh long before they arrived at the fcene of action, fince the delays he had already encountered, neceffarily confined his paffage round Cape *Horn* to the moft rigorous feafon of the year. Sir *Charles Wager* too joined in opinion with the Commodore, that invalids were no ways proper for this fervice, and follicited ftrenuoufly to have them exchanged; but he was told that perfons, who were fuppofed to be better judges of foldiers than he or Mr. *Anfon*, thought them the propereft men that could be employed on this occafion. And upon this determination they were ordered on board the fquadron on the 5th of *Auguft* : But inftead of five hundred, there came on board no more than two hundred and fifty-nine; for all thofe who had limbs and ftrength to walk out of *Portfmouth* deferted, leaving behind them only fuch as were literally invalids, moft of them being fixty years of age, and fome of them upwards of feventy. Indeed it is difficult to conceive a more moving fcene than the imbarkation of thefe unhappy veterans : They were them-

felves

felves extremely averfe to the fervice they were engaged in, and fully apprized of all the difafters they were afterwards expofed to; the apprehenfions of which were ftrongly mark'd by the concern that appeared in their countenances, which was mixed with no fmall degree of indignation, to be thus hurried from their repofe into a fatiguing employ, to which neither the ftrength of their bodies, nor the vigour of their minds, were any ways proportioned, and where, without feeing the face of an enemy, or in the leaft promoting the fuccefs of the enterprize they were engaged in, they would in all probability ufeleffly perifh by lingring and painful difeafes; and this too, after they had fpent the activity and ftrength of their youth in their Country's fervice.

And I cannot but obferve, on this melancholy incident, how extremely unfortunate it was, both to this aged and difeafed detachment, and to the expedition they were employed in; that amongft all the outpenfioners of *Chelfea* Hofpital, which were fuppofed to amount to two thoufand men, the moft crazy and infirm only fhould be culled out for fo fatiguing and perilous an undertaking. For it was well known, that however unfit, invalids in general might be for this fervice, yet by a prudent choice, there might have been found amongft them five hundred men who had fome remains of vigour left: And Mr. *Anfon* fully expected, that the beft of them would have been allotted him; whereas the whole detachment that was fent to him, feemed to be made up of the moft decrepid and miferable objects, that could be collected out of the whole body; and by the defertion abovementioned, thefe were a fecond time cleared of that little health and ftrength which were to be found amongft them, and he was to take up with fuch as were much fitter for an infirmary, than for any military duty.

And here it is neceffary to mention another material particular in the equipment of this fquadron. It was propofed to Mr. *Anfon*, after it was refolved that he fhould be fent to the *South-Seas*, to take with him two perfons under the denomination of Agent Victuallers. Thofe who were mentioned for this employment had formerly been

in

in the *Spanish West-Indies*, in the *South-Sea* Company's service, and it was supposed that by their knowledge and intelligence on that coast, they might often procure provisions for him by compact with the inhabitants, when it was not to be got by force of arms: These Agent Victuallers were, for this purpose, to be allowed to carry to the value of 15,000 *l.* in merchandize on board the squadron; for they had represented, that it would be much easier for them to procure provisions with goods, than with the value of the same goods in money. Whatever colours were given to this scheme, it was difficult to persuade the generality of mankind, that it was not principally intended for the enrichment of the Agents, by the beneficial commerce they proposed to carry on upon that coast. Mr. *Anson*, from the beginning, objected both to the appointment of Agent Victuallers, and the allowing them to carry a cargo on board the squadron: For he conceived, that in those few amicable ports where the squadron might touch, he needed not their assistance to contract for any provisions the place afforded; and on the enemy's coast, he did not imagine that they could ever procure him the necessaries he should want, unless (which he was resolved not to comply with) the military operations of his squadron were to be regulated by the ridiculous views of their trading projects. All that he thought the Government ought to have done on this occasion, was to put on board to the value of 2 or 3000 *l.* only of such goods, as the *Indians*, or the *Spanish* Planters in the less cultivated part of the coast, might be tempted with; since it was in such places only that he imagined it would be worth while to truck with the enemy for provisions: And in these places it was sufficiently evident, a very small cargo would suffice.

But though the Commodore objected both to the appointment of these officers, and to their project; yet, as they had insinuated that their scheme, besides victualling the squadron, might contribute to settling a trade upon that coast, which might be afterwards carried on without difficulty, and might thereby prove a very considerable national advantage, they were much listened to by some considerable

derable perfons: And of the 15,000 *l.* which was to be the amount of their cargo, the Government agreed to advance them 10,000 upon impreft, and the remaining 5000 they raifed on bottomry bonds; and the goods purchafed with this fum, were all that were taken to fea by the fquadron, how much foever the amount of them might be afterwards magnified by common report.

This cargo was at firft fhipped on board the *Wager* Store Ship, and one of the Victuallers; no part of it being admitted on board the men of war. But when the Commodore was at St. *Catharine's,* he confidered, that in cafe the fquadron fhould be feparated, it might be pretended that fome of the fhips were difappointed of provifions for want of a cargo to truck with, and therefore he diftributed fome of the leaft bulky commodities on board the men of war, leaving the remainder principally on board the *Wager,* where it was loft: And more of the goods perifhing by various accidents to be recited hereafter, and no part of them being difpofed of upon the coaft, the few that came home to *England,* did not produce, when fold, above a fourth part of the original price. So true was the Commodore's prediction about the event of this project, which had been by many confidered as infallibly productive of immenfe gains. But to return to the tranfactions at *Portfmouth.*

To fupply the place of the two hundred and forty invalids which had deferted, as is mentioned above, there were ordered on board two hundred and ten marines detached from different regiments: Thefe were raw and undifciplined men, for they were juft raifed, and had fcarcely any thing more of the foldier than their regimentals, none of them having been fo far trained, as to be permitted to fire. The laft detachment of thefe marines came on board the 8th of *Auguft,* and on the 10th the fquadron failed from *Spithead* to St. *Helens,* there to wait for a wind to proceed on the expedition.

But the delays we had already fuffered had not yet fpent all their influence, for we were now advanced into a feafon of the year, when the wefterly winds are ufually very conftant, and very violent; and it was thought proper that we fhould put to fea in com-

C

pany

pany with the fleet commanded by Admiral *Balchen*, and the expedition under Lord *Cathcart*. And as we made up in all twenty-one men of war, and a hundred and twenty-four fail of merchant-men and tranfports, we had no hopes of getting out of the Channel with fo large a number of fhips, without the continuance of a fair wind, for fome confiderable time. This was what we had every day lefs and lefs reafon to expect, as the time of the equinox drew near; fo that our golden dreams, and our ideal poffeffion of the *Peruvian* treafures, grew each day more faint, and the difficulties and dangers of the paffage round Cape *Horn* in the winter feafon filled our imaginations in their room. For it was forty days from our arrival at St. *Helens*, to our final departure from thence: And even then (having orders to proceed without Lord *Cathcart*) we tided it down the Channel with a contrary wind. But this interval of forty days was not free from the difpleafing fatigue of often fetting fail, and being as often obliged to return; nor exempt from dangers, greater than have been fometimes experienced in furrounding the globe. For the wind coming fair for the firft time, on the 23d of *Auguft*, we got under fail, and Mr. *Balchen* fhewed himfelf truly folicitous to have proceeded to fea, but the wind foon returning to its old quarter, obliged us to put back to St. *Helens*, not without confiderable hazard, and fome damage received by two of the tranfports, who, in tacking, ran foul of each other: Befides this, we made two or three more attempts to fail, but without any better fuccefs. And, on the 6th of *September*, being returned to an anchor at St. *Helens*, after one of thefe fruitlefs efforts, the wind blew fo frefh, that the whole fleet ftruck their yards and topmafts to prevent their driving: And, notwithftanding this precaution, the *Centurion* drove the next evening, and brought both cables a-head, and we were in no fmall danger of driving foul of the Prince *Frederick*, a feventy-gun fhip, moored at a fmall diftance under our ftern; which we happily efcaped, by her driving at the fame time, and fo preferving her diftance: Nor did we think ourfelves fecure, till we at laft let go the fheet anchor, which fortunately brought us up.

However,

However, on the 9th of *September*, we were in some degree relieved from this lingring vexatious situation, by an Order which Mr. *Anson* received from the Lords Justices, to put to sea the first opportunity with his own squadron only, if Lord *Cathcart* should not be ready. Being thus freed from the troublesome company of so large a fleet, our Commodore resolved to weigh and tide it down Channel, assoon as the weather should become sufficiently moderate; and this might easily have been done with our own squadron alone full two months sooner, had the orders of the Admiralty, for supplying us with seamen, been punctually complied with, and had we met with none of those other delays mentioned in this narration. It is true, our hopes of a speedy departure were even now somewhat damped, by a subsequent order which Mr. *Anson* received on the 12th of *September*; for by that he was required to take under his convoy the St. *Albans* with the *Turkey* fleet, and to join the *Dragon*, and the *Winchester*, with the *Streights* and the *American* trade at *Torbay* or *Plymouth*, and to proceed with them to sea as far as their way and ours lay together: This incumbrance of a convoy gave us some uneasiness, as we feared it might prove the means of lengthening our passage to the *Maderas*. However, Mr. *Anson*, now having the command himself, resolved to adhere to his former determination, and to tide it down the Channel with the first moderate weather; and that the junction of his Convoy might occasion as little a loss of time as possible, he immediately sent directions to *Torbay*, that the fleets he was there to take under his care, might be in a readiness to join him instantly on his approach. And at last, on the 18th of *September*, he weighed from St. *Helens*; and though the wind was at first contrary, had the good fortune to get clear of the Channel in four days, as will be more particularly related in the ensuing chapter.

Having thus gone through the respective steps taken in the equipment of this squadron, it is sufficiently obvious how different an aspect this expedition bore at its first appointment in the beginning of

January,

January, from what it had in the latter end of *September*, when it left the Channel; and how much its numbers, its ftrength, and the probability of its fuccefs were diminifhed, by the various incidents which took place in that interval. For inftead of having all our old and ordinary feamen exchanged for fuch as were young and able, (which the Commodore was at firft promifed) and having our numbers compleated to their full complement, we were obliged to retain our firft crews, which were very indifferent; and a deficiency of three hundred men in our numbers was no otherwife made up to us, than by fending us on board a hundred and feventy men, the greateft part compofed of fuch as were difcharged from hofpitals, or new-raifed marines who had never been at fea before. And in the land-forces allotted us, the change was ftill more difadvantageous; for there, inftead of three independent companies of a hundred men each, and *Bland*'s regiment of foot, which was an old one, we had only four hundred and feventy invalids and marines, one part of them incapable for action by age and infirmities, and the other part ufelefs by their ignorance of their duty. But the diminifhing the ftrength of the fquadron was not the greateft inconveniency which attended thefe alterations; for the contefts, reprefentations, and difficulties which they continually produced, (as we have above feen, that in thefe cafes the authority of the Admiralty was not always fubmitted to) occafioned a delay and wafte of time, which in its confequences was the fource of all the difafters to which this enterprize was afterwards expofed: for by this means we were obliged to make our paffage round Cape *Horn* in the moft tempeftuous feafon of the year; whence proceeded the feparation of our fquadron, the lofs of numbers of our men, and the imminent hazard of our total deftruction: And by this delay too, the enemy had been fo well informed of our defigns, that a perfon who had been employed in the *South-Sea* Company's fervice, and arrived from *Panama* three or four days before we left *Portfmouth*, was able to relate to Mr. *Anfon* moft of the particulars

ticulars of the deſtination and ſtrength of our ſquadron, from what he had learnt amongſt the *Spaniards* before he left them. And this was afterwards confirmed by a more extraordinary cir-cumſtance : For we ſhall find, that when the *Spaniards* (fully ſa-tisfied that our expedition was intended for the *South-Seas*) had fit-ted out a ſquadron to oppoſe us, which had ſo far got the ſtart of us, as to arrive before us off the iſland of *Madera,* the Com-mander of this ſquadron was ſo well inſtructed in the form and make of Mr. *Anſon*'s broad pennant, and had imitated it ſo ex-actly, that he thereby decoyed the *Pearl,* one of our ſquadron, within gun-ſhot of him, before the Captain of the *Pearl* was able to diſcover his miſtake.

C H A P.

CHAP. II.

The paſſage from St. *Helens* to the Iſland of *Madera*; with a ſhort account of that Iſland, and of our ſtay there.

ON the 18th of *September*, 1740, the ſquadron, as we have obſerved in the preceding chapter, weighed from St. *Helens* with a contrary wind, the Commodore propoſing to tide it down the Channel, as he dreaded leſs the inconveniencies he ſhould thereby have to ſtruggle with, than the riſk he ſhould run of ruining the enterprize, by an uncertain, and, in all probability, a tedious attendance for a fair wind.

The ſquadron allotted to this ſervice conſiſted of five men of war, a ſloop of war, and two victualling ſhips. They were the *Centurion* of ſixty guns, four hundred men, *George Anſon*, Eſq; Commander; the *Glouceſter* of fifty guns, three hundred men, *Richard Norris* Commander; the *Severn* of fifty guns, three hundred men, the Honourable *Edward Legg* Commander; the *Pearl* of forty guns, two hundred and fifty men, *Matthew Mitchel* Commander; the *Wager* of twenty-eight guns, one hundred and ſixty men, *Dandy Kidd* Commander; and the *Tryal* Sloop of eight guns, one hundred men, the Honourable *John Murray* Commander; the two Victuallers were Pinks, the largeſt of about four hundred, and the other of about two hundred tons burthen, theſe were to attend us, till the proviſions we had taken on board were ſo far conſumed, as to make room for the additional quantity they carried with them, which, when we had taken into our ſhips, they were to be diſcharged. Beſides the complement of men born by the abovementioned ſhips as their crews, there were embarked on board the ſquadron about four hundred and ſeventy invalids and marines,

under

under the denomination of land-forces, as has been particularly mentioned in the preceding chapter, which were commanded by Lieutenant Colonel *Cracherode*. With this squadron, together with the St. *Albans* and the *Lark*, and the trade under their convoy, Mr. *Anson*, after weighing from St. *Helens*, tided it down the Channel for the first forty-eight hours; and, on the 20th, in the morning, we discovered off the *Ram-Head* the *Dragon*, *Winchester*, *South-Sea Castle*, and *Rye*, with a number of merchantmen under their Convoy: These we joined about noon the same day, our Commodore having orders to see them (together with the St. *Albans* and *Lark*) as far into the sea as their course and ours lay together. When we came in sight of this last mentioned fleet, Mr. *Anson* first hoisted his broad pennant, and was saluted by all the men of war in company.

When we had joined this last Convoy, we made up eleven men of war, and about one hundred and fifty sail of merchantmen, consisting of the *Turky*, the *Streights*, and the *American* trade. Mr. *Anson*, the same day, made a signal for all the Captains of the men of war to come on board him, where he delivered them their fighting and sailing instructions, and then, with a fair wind, we all stood towards the South-West; and the next day at noon, being the 21st, we had run forty leagues from the *Ram-Head*; and being now clear of the land, our Commodore, to render our view more extensive, ordered Captain *Mitchel*, in the *Pearl*, to make sail two leagues a-head of the fleet every morning, and to repair to his station every evening. Thus we proceeded till the 25th, when the *Winchester* and the *American* Convoy made the concerted signal for leave to separate, which being answered by the Commodore, they left us: As the St. *Albans* and the *Dragon*, with the *Turky* and *Streights* Convoy, did on the 29th. After which separation, there remained in company only our own squadron and our two victuallers, with which we kept on our course for the Island of *Madera*. But the winds were so contrary, that we had the mortification to be forty days in our passage thither from St. *Helens*, though it is known to be often

done

done in ten or twelve. This delay was a moſt unpleaſing circumſtance, productive of much diſcontent and ill-humour amongſt our
people, of which thoſe only can have a tolerable idea, who have
had the experience of a like ſituation. And beſides the peeviſhneſs
and deſpondency which foul and contrary winds, and a lingring
voyage never fail to create on all occaſions, we, in particular, had
very ſubſtantial reaſons to be greatly alarmed at this unexpected impediment. For as we had departed from *England* much later than
we ought to have done, we had placed almoſt all our hopes of ſucceſs in the chance of retrieving in ſome meaſure at ſea, the time
we had ſo unhappily waſted at *Spithead* and St. *Helens.* However,
at laſt, on *Monday, October* the 25th, at five in the morning, we,
to our great joy, made the land, and in the afternoon came to an
anchor in *Madera Road,* in forty fathom water; the *Brazen-head*
bearing from us E by S, the *Loo* N N W, and the great Church
N N E. We had hardly let go our anchor, when an *Engliſh* privateer ſloop ran under our ſtern, and ſaluted the Commodore with
nine guns, which we returned with five. And, the next day, the
Conſul of the Iſland coming to viſit the Commodore, we ſaluted
him with nine guns on his coming on board.

This Iſland of *Madera,* where we are now arrived, is famous
through all our *American* ſettlements for its excellent wines, which
ſeem to be deſigned by Providence for the refreſhment of the inhabitants of the Torrid Zone. It is ſituated in a fine climate, in the
latitude of 32 : 27 North; and in the longitude from *London* of,
by our different reckonings, from 18° ½ to 19° ½ Weſt, though laid
down in the charts in 17°. It is compoſed of one continued hill,
of a conſiderable height, extending itſelf from Eaſt to Weſt: The
declivity of which, on the South-ſide, is cultivated and interſperſed
with vineyards; and in the midſt of this ſlope the Merchants have
fixed their country ſeats, which help to form an agreeable proſpect.
There is but one conſiderable town in the whole Iſland, it is named
Fonchiale, and is ſeated on the South part of the Iſland, at the bottom of a large bay. This is the only place of trade, and indeed the

only

only one where it is poffible for a boat to land. *Fonchiale*, towards the fea, is defended by a high wall, with a battery of cannon, befides a caftle on the *Loo*, which is a rock ftanding in the water at a fmall diftance from the fhore. Even here the beach is covered with large ftones, and a violent furf continually beats upon it ; fo that the Commodore did not care to venture the fhips long boats to fetch the water off, as there was fo much danger of their being loft ; and therefore ordered the Captains of the fquadron to employ *Portuguefe* boats on that fervice.

We continued about a week at this Ifland, watering our fhips, and providing the fquadron with wine and other refrefhments. And, on the 3d of *November*, Captain *Richard Norris* having fignified by a letter to the Commodore, his defire to quit his command on board the *Gloucefter*, in order to return to *England* for the recovery of his health, the Commodore complied with his requeft ; and thereupon was pleafed to appoint Captain *Matthew Mitchel* to command the *Gloucefter* in his room, and to remove Captain *Kidd* from the *Wager* to the *Pearl*, and Captain *Murray* from the *Tryal* Sloop to the *Wager*, giving the command of the *Tryal* to Lieutenant *Cheap*. Thefe promotions being fettled, with other changes in the Lieutenancies, the Commodore, on the following day, gave to the Captains their orders, appointing St. *Jago*, one of the *Cape de Verd* Iflands, to be the firft place of rendezvous in cafe of feparation ; and directing them, if they did not meet the *Centurion* there, to make the beft of their way to the Ifland of St. *Catherine's*, on the coaft of *Brazil*. The water for the fquadron being the fame day compleated, and each fhip fupplied with as much wine and other refrefhments as they could take in, we weighed anchor in the afternoon, and took our leave of the Ifland of *Madera*. But before I go on with the narration of our own tranfactions, I think it neceffary to give fome account of the proceedings of the enemy, and of the meafures they had taken to render all our defigns abortive.

D

When

When Mr. *Anson* visited the Governor of *Madera*, he received information from him, that for three or four days, in the latter end of *October*, there had appeared, to the westward of that Island, seven or eight ships of the line, and a Patache, which last was sent every day close in to make the land. The Governor assured the Commodore, upon his honour, that none upon the Island had either given them intelligence, or had in any sort communicated with them, but that he believed them to be either *French* or *Spanish*, but was rather inclined to think them *Spanish*. On this intelligence, Mr. *Anson* sent an Officer in a clean sloop, eight leagues to the westward, to reconnoitre them, and, if possible, to discover what they were: But the Officer returned without being able to get a sight of them, so that we still remained in uncertainty. However, we could not but conjecture, that this fleet was intended to put a stop to our expedition, which, had they cruised to the eastward of the Island instead of the westward, they could not but have executed with great facility. For as, in that case, they must have certainly fallen in with us, we should have been obliged to throw overboard vast quantities of provision to clear our ships for an engagement, and this alone, without any regard to the event of the action, would have effectually prevented our progress. This was so obvious a measure, that we could not help imagining reasons which might have prevented them from pursuing it. And we therefore supposed, that this *French* or *Spanish* squadron was sent out, upon advice of our sailing in company with Admiral *Balchen* and Lord *Catchcart*'s expedition: And thence, from an apprehension of being over-matched, they might not think it adviseable to meet with us, till we had parted company, which they might judge would not happen, before our arrival at this Island. These were our speculations at that time; and from hence we had reason to suppose, that we might still fall in with them, in our way to the *Cape de Verd* Islands. And afterwards, in the course of our expedition, we were many of us persuaded, that this was the

Spanish

Spanish fquadron commanded by *Don Jofeph Pizarro*, which was fent out purpofely to traverfe the views and enterprizes of our fquadron, to which, in ftrength, they were greatly fuperior. As this *Spanish* armament then was fo nearly connected with our expedition, and as the cataftrophe it underwent, though not effected by our force, was yet a confiderable advantage to this Nation, produced in confequence of our equipment, I have, in the following chapter, given a fummary account of their proceedings, from their firft fetting out from *Spain* in the year 1740, till the *Afia*, the only fhip which returned to *Europe* of the whole fquadron, arrived at the *Groyne* in the beginning of the year 1746.

D 2   CHAP.

C H A P. III.

The hiſtory of the ſquadron commanded by Don Jo-ſeph Pizarro.

THE ſquadron fitted out by the Court of *Spain* to attend our motions, and traverſe our projects, we ſuppoſed to have been the ſhips ſeen off *Madera*, as mentioned in the pre-ceding chapter. And as this force was ſent out particularly againſt our expedition, I cannot but imagine, that the following hiſtory of the caſualties it met with, as far as by intercepted letters and other information the ſame has come to my knowledge, is a very eſſential part of the preſent work : For by this it will appear we were the oc-caſion, that a conſiderable part of the naval power of *Spain* was diverted from the proſecution of the ambitious Views of that Court in *Europe* ; and the men and ſhips, loſt by the enemy in this under-taking, were loſt in conſequence of the precautions they took to ſecure themſelves againſt our enterprizes. This ſquadron (beſides two ſhips intended for the *Weſt-Indies*, which did not part company till after they had left the *Maderas*) was compoſed of the follow-ing men of war, commanded by Don *Joſeph Pizarro* :

The *Aſia* of ſixty-ſix guns, and ſeven hundred men ; this was the Admiral's ſhip.
The *Guipuſcoa* of ſeventy-four guns, and ſeven hundred men.
The *Hermiona* of fifty-four guns, and five hundred men.
The *Eſperanza* of fifty guns, and four hundred and fifty men.
The St. *Eſtevan* of forty guns, and three hundred and fifty men.
And a Patache of twenty guns.

Theſe ſhips, over and above their complement of ſailors and ma-rines, had on board an old *Spaniſh* regiment of foot, intended to

reinforce

reinforce the garrifons on the coaft of the *South-Seas*. When this fleet had cruifed for fome days to the leeward of the *Maderas*, as is mentioned in the preceding chapter, they left that ftation in the beginning of *November*, and fteered for the river of *Plate*, where they arrived the 5th of *January*, O. S. and coming to an anchor in the bay of *Maldonado*, at the mouth of that river, their Admiral *Pizarro* fent immediately to *Buenos Ayres* for a fupply of provifions; for they had departed from *Spain* with only four months provifions on board. While they lay here expecting this fupply, they received intelligence, by the Treachery of the *Portuguefe* Governor of St. *Catherine's*, of Mr. *Anfon's* having arrived at that Ifland on the 21ft of *December* preceding, and of his preparing to put to fea again with the utmoft expedition. *Pizarro*, notwithftanding his fuperior force, had his reafons (and as fome fay his orders likewife) for avoiding our fquadron any where fhort of the *South-Seas*. He was befides extremely defirous of getting round Cape *Horn* before us, as he imagined that ftep alone would effectually baffle all our defigns; and therefore, on hearing that we were in his neighbourhood, and that we fhould foon be ready to proceed for Cape *Horn*, he weighed anchor with the five large fhips, (the Patache being difabled and condemned, and the men taken out of her) after a ftay of feventeen days only, and got under fail without his provifions, which arrived at *Maldonado* within a day or two after his departure. But notwithftanding the precipitation, with which he departed, we put to fea from St. *Catherine's* four days before him, and in fome part of our paffage to Cape *Horn*, the two fquadrons were fo near together, that the *Pearl*, one of our fhips, being feparated from the reft, fell in with the *Spanifh* Fleet, and miftaking the *Afia* for the *Centurion*, had got within gun-fhot of *Pizarro*, before fhe difcovered her error, and narrowly efcaped being taken.

It being the 22d of *January* when the *Spaniards* weighed from *Maldonado*, (as has been already mentioned) they could not expect to get into the latitude of Cape *Horn* before the equinox; and as they had reafon to apprehend very tempeftuous weather in doubling

it

it at that feafon, and as the *Spanifh* failors, being for the moft par accuftomed to a fair weather country, might be expected to be very averfe to fo dangerous and fatiguing a navigation, the better to encourage them, fome part of their pay was advanced to them in *European* goods, which they were to be permitted to difpofe of in the *South-Seas,* that fo the hopes of the great profit, each man was to make on his fmall venture, might animate him in his duty, and render him lefs difpofed to repine at the labour, the hardfhips and the perils he would in all probability meet with before his arrival on the coaft of *Peru.*

Pizarro with his fquadron having, towards the latter end of of *February,* run the length of Cape *Horn,* he then ftood to the weftward in order to double it; but in the night, of the laft day of *February, O. S.* while with this view they were turning to windward, the *Guipufcoa,* the *Hermiona,* and the *Efperanza,* were feparated from the Admiral; and, on the 6th of *March* following, the *Guipufcoa* was feparated from the other two; and, on the 7th (being the day after we had paffed *Streights le Maire*) there came on a moft furious ftorm at N W, which, in defpight of all their efforts, drove the whole fquadron to the eaftward, and obliged them, after feveral fruitlefs attempts, to bear away for the river of *Plate,* where *Pizarro* in the *Afia* arrived about the middle of *May,* and a few days after him the *Efperanza* and the *Eftevan.* The *Hermiona* was fuppofed to founder at fea, for fhe was never heard of more; and the *Guipufcoa* was run a-fhore, and funk on the coaft of *Brazil.* The calamities of all kinds, which this fquadron underwent in this unfuccefsful navigation, can only be paralleled by what we ourfelves experienced in the fame climate, when buffeted by the fame ftorms. There was indeed fome diverfity in our diftreffes, which rendered it difficult to decide, whofe fituation was moft worthy of commiferation. For to all the misfortunes we had in common with each other, as fhattered rigging, leaky fhips, and the fatigues and defpondency, which necefsarily attend thefe difafters, there was fuperadded on board our fquadron

dron the ravage of a moſt deſtructive and incurable diſeaſe, and on board the *Spaniſh* ſquadron the devaſtation of famine.

For this ſquadron, either from the hurry of their outſet, their preſumption of a ſupply at *Buenos Ayres,* or from other leſs obvious motives, departed from *Spain,* as has been already obſerved, with no more than four months proviſion, and even that, as it is ſaid, at ſhort allowance only; ſo that, when by the ſtorms they met with off Cape *Horn,* their continuance at ſea was prolonged a month or more beyond their expectation, they were thereby reduced to ſuch infinite diſtreſs, that rats, when they could be caught, were ſold for four dollars a-piece; and a ſailor, who died on board, had his death concealed for ſome days by his brother, who, during that time, lay in the ſame hammock with the corpſe, only to receive the dead man's allowance of proviſions. In this dreadful ſituation they were alarmed (if their horrors were capable of augmentation) by the diſcovery of a conſpiracy among the marines, on board the *Aſia,* the Admiral's ſhip. This had taken its riſe chiefly from the miſeries they endured: For though no leſs was propoſed by the conſpirators than the maſſacring the officers and the whole crew, yet their motive for this bloody reſolution ſeemed to be no more than their deſire of relieving their hunger, by appropriating the whole ſhips proviſions to themſelves. But their deſigns were prevented, when juſt upon the point of execution, by means of one of their confeſſors, and three of their ringleaders were immediately put to death. However, though the conſpiracy was ſuppreſſed, their other calamities admitted of no alleviation, but grew each day more and more deſtructive. So that by the complicated diſtreſs of fatigue, ſickneſs and hunger, the three ſhips which eſcaped loſt the greateſt part of their men: The *Aſia,* their Admiral's ſhip, arrived at *Monte Vedio* in the river of *Plate,* with half her crew only; the St. *Eſtevan* had loſt in like manner half her hands, when ſhe anchored in the bay of *Barragan;* the *Eſperanza,* a fifty gun ſhip, was ſtill more unfortunate, for of four hundred and fifty hands which ſhe brought from *Spain,* only fifty-eight remained alive, and the whole regiment of foot periſhed

except

except fixty men. But to give the reader a more diſtinct and particular idea of what they underwent upon this occaſion, I ſhall lay before him a ſhort account of the fate of the *Guipuſcoa*, from a letter written by Don *Joſeph Mendinuetta* her Captain, to a perſon of diſtinction at *Lima*; a copy of which fell into our hands afterwards in the *South-Seas.*

He mentions, that he ſeparated from the *Hermiona* and the *Eſperanza* in a fog, on the 6th of *March*, being then, as I ſuppoſe, to the S. E. of *Staten-Land*, and plying to the weſtward; that in the night after, it blew a furious ſtorm at N. W, which, at half an hour after ten, ſplit his mainſail, and obliged him to bear away with his foreſail; that the ſhip went ten knots an hour with a prodigious ſea, and often ran her gangway under water; that he likewiſe ſprung his main-maſt; and the ſhip made ſo much water, that with four pumps and bailing he could not free her. That on the 19th it was calm, but the ſea continued ſo high, that the ſhip in rolling opened all her upper works and ſeams, and ſtarted the butt ends of her planking and the greateſt part of her top timbers, the bolts being drawn by the violence of her roll: That in this condition, with other additional diſaſters to the hull and rigging, they continued beating to the weſtward till the 12th : That they were then in ſixty degrees of ſouth latitude, in great want of proviſions, numbers every day periſhing by the fatigue of pumping, and thoſe who ſurvived, being quite diſpirited by labour, hunger, and the ſeverity of the weather, they having two ſpans of ſnow upon the decks : That then finding the wind fixed in the weſtern quarter, and blowing ſtrong, and conſequently their paſſage to the weſtward impoſſible, they reſolved to bear away for the river of *Plate* : That on the 22d, they were obliged to throw overboard all the upper-deck guns, and an anchor, and to take ſix turns of the cable round the ſhip to prevent her opening : That on the 4th of *April*, it being calm but a very high ſea, the ſhip rolled ſo much, that the mainmaſt came by the board, and in a few hours after ſhe loſt, in like manner, her fore-maſt and her mizen-maſt; and that, to accumu-

late

late their misfortunes, they were foon obliged to cut away their bowfprit, to diminifh, if poffible, the leakage at her head: That by this time he had loft two hundred and fifty men by hunger and fatigues; for thofe who were capable of working at the pumps, (at which every Officer without exception took his turn) were allowed only an ounce and half of bifcuit *per diem*; and thofe who were fo fick or fo weak, that they could not affift in this neceffary labour, had no more than an ounce of wheat; fo that it was common for the men to fall down dead at the pumps: That, including the Officers, they could only mufter from eighty to a hundred perfons capable of duty: That the South Weft winds blew fo frefh, after they had loft their mafts, that they could not immediately fet up jury mafts, but were obliged to drive like a wreck, between the latitudes of 32 and 28, till the 24th of *April,* when they made the coaft of *Brazil* at *Rio de Patas,* ten leagues to the fouthward of the Ifland of St. *Catherine's;* that here they came to an anchor, and that the Captain was very defirous of proceeding to St. *Catherine's* if poffible, in order to fave the hull of the fhip, and the guns and ftores on board her; but the crew inftantly left off pumping, and being enraged at the hardfhips they had fuffered, and the numbers they had loft, (there being at that time no lefs than thirty dead bodies lying on the deck) they all with one voice cried out *on fhore, on fhore,* and obliged the Captain to run the fhip in directly for the land, where, the 5th day after, fhe funk with her ftores, and all her furniture on board her, but the remainder of the crew, whom hunger and fatigue had fpared, to the number of four hundred, got fafe on fhore.

From this account of the adventures and cataftrophe of the *Guipufcoa,* we may form fome conjecture of the manner, in which the *Hermiona* was loft, and of the diftreffes endured by the three remaining fhips of the fquadron, which got into the river of *Plate.* Thefe laft being in great want of mafts, yards, rigging, and all kind of naval ftores, and having no fupply at *Buenos Ayres,* nor in any other of their fettlements, *Pizarro* difpatched an advice boat with

E   a letter a letter of credit to *Rio Janeiro*, to purchafe what was wanting from the *Portuguefe*. He, at the fame time, fent an exprefs acrofs the continent to *San Jago* in *Chili*, to be thence forwarded to the Viceroy of *Peru*, informing him of the difafters that had befallen his fquadron, and defiring a remittance of 200,000 dollars from the royal chefts at *Lima*, to enable him to victual and refit his remaining fhips, that he might be again in a condition to attempt the paffage to the *South-Seas*, as foon as the feafon of the year fhould be more favourable. It is mentioned by the *Spaniards* as a moft extraordinary circumftance, that the *Indian* charged with this exprefs (though it was then the depth of winter, when the *Cordilleras* are efteemed impaffable on account of the fnow) was only thirteen days in his journey from *Buenos Ayres* to St. *Jago* in *Chili*; though thefe places are diftant three hundred *Spanifh* leagues, near forty of which are amongft the fnows and precipices of the *Cordilleras*.

The return to this difpatch of *Pizarro*'s from the Viceroy of *Peru* was no ways favourable; inftead of 200,000 dollars, the fum demanded, the Viceroy remitted him only 100,000, telling him, that it was with great difficulty he was able to procure him even that: Though the inhabitants at *Lima*, who confidered the prefence of *Pizarro* as abfolutely neceffary to their fecurity, were much difcontented at this procedure, and did not fail to affert, that it was not the want of money, but the interefted views of fome of the Viceroy's confidents, that prevented *Pizarro* from having the whole fum he had afked for.

The advice-boat fent to *Rio Janeiro* alfo executed her commiffion, but imperfectly; for though fhe brought back a confiderable quantity of pitch, tar and cordage, yet fhe could not procure either mafts or yards: and as an additional misfortune, *Pizarro* was difappointed of fome mafts he expected from *Paraguay*; for a carpenter, whom he entrufted with a large fum of money, and had fent there to cut mafts, inftead of profecuting the bufinefs he was employed in, had married in the country, and refufed to return. However, by removing the mafts of the *Efperanza* into the *Afia*,

and

and making ufe of what fpare mafts and yards they had on board, they made a fhift to refit the *Afia* and the St. *Eftevan*. And in the *October* following, *Pizarro* was preparing to put to fea with thefe two fhips, in order to attempt the paffage round Cape *Horn* a fecond time; but the St. *Eftevan*, in coming down the river *Plate*, ran on a fhoal, and beat off her rudder, on which, and other damages fhe received, fhe was condemned and broke up, and *Pizarro* in the *Afia* proceeded to fea without her. Having now the fummer before him, and the winds favourable, no doubt was made of his having a fortunate and fpeedy paffage; but being off Cape *Horn*, and going right before the wind in very moderate weather, though in a fwelling fea, by fome mifconduct of the officer of the watch the fhip rolled away her mafts, and was a fecond time obliged to put back to the river of *Plate* in great diftrefs.

The *Afia* having confiderably fuffered in this fecond unfortunate expedition, the *Efperanza*, which had been left behind at *Monte Vedio*, was ordered to be refitted, the command of her being given to *Mindinuetta*, who was Captain of the *Guipufcoa*, when fhe was loft. He, in the *November* of the fucceeding year, that is, in *November* 1742, failed from the river of *Plate* for the *South-Seas*, and arrived fafe on the coaft of *Chili*; where his Commodore *Pizarro* paffing over land from *Buenos Ayres* met him. There were great animofities and contefts between thefe two Gentlemen at their meeting, occafioned principally by the claim of *Pizarro* to command the *Efperanza*, which *Mindinuetta* had brought round: For *Mindinuetta* refufed to deliver her up to him; infifting, that as he came into the *South-Seas* alone, and under no fuperior, it was not now in the power of *Pizarro* to refume that authority, which he had once parted with. However, the Prefident of *Chili* interpofing, and declaring for *Pizarro*, *Mindinuetta*, after a long and obftinate ftruggle, was obliged to fubmit.

But *Pizarro* had not yet compleated the feries of his adventures; for when he and *Mindinuetta* came back by land from *Chili* to *Buenos Ayres*, in the year 1745, they found at *Monte Vedio* the

Afia,

Afia, which near three years before they had left there. This fhip they refolved, if poffible, to carry to *Europe*, and with this view they refitted her in the beft manner they could : But their great difficulty was to procure a fufficient number of hands to navigate her, for all the remaining failors of the fquadron to be met with in the neighbourhood of *Buenos Ayres*, did not amount to a hundred men. They endeavoured to fupply this defect by preffing many of the inhabitants of *Buenos Ayres*, and putting on board befides all the *Englifh* prifoners then in their cuftody, together with a number of *Portuguefe* fmugglers, which they had taken at different times, and fome of the *Indians* of the country. Among thefe laft there was a Chief and ten of his followers, which had been furprized by a party of *Spanifh* foldiers about three months before. The name of this Chief was *Orellana*, he belonged to a very powerful Tribe, which had committed great ravages in the neighbourhood of *Buenos Ayres*. With this motly crew (all of them, except the *European Spaniards*, extremely averfe to the voyage) *Pizarro* fet fail from *Monte Vedio* in the river of *Plate*, about the beginning of *November* 1745, and the native *Spaniards* being no ftrangers to the diffatisfaction of their forced men, treated both thofe, the *Englifh* prifoners and the *Indians*, with great infolence and barbarity ; but more particularly the *Indians*, for it was common for the meaneft officers in the fhip to beat them moft cruelly on the flighteft pretences, and oftentimes only to exert their fuperiority. *Orellana* and his followers, though in appearance fufficiently patient and fubmiffive, meditated a fevere revenge for all thefe inhumanities. As he converfed very well in *Spanifh*, (thefe *Indians* having in time of peace a great intercourfe with *Buenos Ayres*) he affected to talk with fuch of the *Englifh* as underftood that language, and feemed very defirous of being informed how many *Englifhmen* there were on board, and which they were. As he knew that the *Englifh* were as much enemies to the *Spaniards* as himfelf, he had doubtlefs an intention of difclofing his purpofes to them, and making them partners in the fcheme he had projected for revenging his wrongs, and recovering his liberty ; but

having

having founded them at a diftance, and not finding them fo precipitate and vindictive as he expected, he proceeded no further with them, but refolved to truft alone to the refolution of his ten faithful followers. Thefe, it fhould feem, readily engaged to obferve his directions, and to execute whatever commands he gave them; and having agreed on the meafures neceffary to be taken, they firft furnifhed themfelves with *Dutch* knives fharp at the point, which being the common knives ufed in the fhip, they found no difficulty in procuring: Befides this, they employed their leifure in fecretly cutting out thongs from raw hides, of which there were great numbers on board, and in fixing to each end of thefe thongs the double-headed fhot of the fmall quarter-deck guns; this, when fwung round their heads, according to the practice of their country, was a moft mifchievous weapon, in the ufe of which the *Indians* about *Buenos Ayres* are trained from their infancy, and confequently are extremely expert. Thefe particulars being in good forwardnefs, the execution of their fcheme was perhaps precipitated by a particular outrage committed on *Orellana* himfelf. For one of the Officers, who was a very brutal fellow, ordered *Orellana* aloft, which being what he was incapable of performing, the Officer, under pretence of his difobedience, beat him with fuch violence, that he left him bleeding on the deck, and ftupified for fome time with his bruifes and wounds. This ufage undoubtedly heightened his thirft for revenge, and made him eager and impatient, till the means of executing it were in his power; fo that within a day or two after this incident, he and his followers opened their defperate refolves in the enfuing manner.

It was about nine in the evening, when many of the principal Officers were on quarter-deck, indulging in the frefhnefs of the night air; the wafte of the fhip was filled with live cattle, and the forecaftle was manned with its cuftomary watch. *Orellana* and his companions, under cover of the night, having prepared their weapons, and thrown off their trouzers and the more cumbrous part of their drefs, came all together on the quarter-deck, and drew towards

the

the door of the great cabbin. The Boatfwain immediately repri-
manded them, and ordered them to be gone. On this *Orellana*
fpoke to his followers in his native language, when four of them
drew off, two towards each gangway, and the Chief and the fix
remaining *Indians* feemed to be flowly quitting the quarter-deck.
When the detached *Indians* had taken poffeffion of the gangway,
Orellana placed his hands hollow to his mouth, and bellowed out
the war-cry ufed by thofe favages, which is faid to be the harfheft
and moft terrifying found known in nature. This hideous yell was
the fignal for beginning the maffacre : For on this they all drew
their knives, and brandifhed their prepared double-headed fhot, and
the fix with their Chief, which remained on the quarter-deck, immediately fell on the *Spaniards*, who were intermingled with them,
and laid near forty of them at their feet, of which above twenty
were killed on the fpot, and the reft difabled. Many of the Officers, in the beginning of the tumult, pufhed into the great cabbin,
where they put out the lights, and barricadoed the door. And of
the others, who had avoided the firft fury of the *Indians*, fome endeavoured to efcape along the gangways into the forecaftle, but the
Indians, placed there on purpofe, ftabbed the greateft part of them,
as they attempted to pafs by, or forced them off the gangways into
the wafte. Others threw themfelves voluntarily over the barricadoes into the wafte, and thought themfelves happy to lie concealed
amongft the cattle ; but the greateft part efcaped up the main
fhrouds, and fheltered themfelves either in the tops or rigging.
And though the *Indians* attacked only the quarter-deck, yet the
watch in the forecaftle finding their communication cut off, and being terrified by the wounds of the few, who not being killed on the
fpot, had ftrength fufficient to force their paffage along the gangways, and not knowing either who their enemies were, or what
were their numbers, they likewife gave all over for loft, and in
great confufion ran up into the rigging of the fore-maft and
bowfprit.

Thus

Thus thefe eleven *Indians*, with a refolution perhaps without example, poffeffed themfelves almoft in an inftant of the quarter-deck
of a fhip mounting fixty-fix guns, with a crew of near five hundred
men, and continued in peaceable poffeffion of this poft a confiderable time. For the Officers in the great cabbin, (amongft whom
were *Pizarro* and *Mindinuetta*) the crew between decks, and thofe
who had efcaped into the tops and rigging, were only anxious for
their own fafety, and were for a long time incapable of forming
any project for fuppreffing the infurrection, and recovering the poffeffion of the fhip. It is true, the yells of the *Indians*, the groans
of the wounded, and the confufed clamours of the crew, all heightned by the obfcurity of the night, had at firft greatly magnified
their danger, and had filled them with the imaginary terrors, which
darknefs, diforder, and an ignorance of the real ftrength of an enemy never fail to produce. For as the *Spaniards* were fenfible of
the difaffection of their preft hands, and were alfo confcious of their
barbarity to their prifoners, they imagined, the confpiracy was general, and confidered their own deftruction as infallible; fo that, it
is faid, fome of them had once taken the refolution of leaping into
the fea, but were prevented by their companions.

However, when the *Indians* had entirely cleared the quarter-deck,
the tumult in a great meafure fubfided; for thofe, who had efcaped, were kept filent by their fears, and the *Indians* were incapable of purfuing them to renew the diforder. *Orellana*, when he faw
himfelf mafter of the quarter-deck, broke open the arm-cheft,
which, on a flight fufpicion of mutiny, had been ordered there a
few days before, as to a place of the greateft fecurity. Here he
took it for granted, he fhould find cutlaffes fufficient for himfelf and
his companions, in the ufe of which weapon they were all extremely fkilful, and with thefe, it was imagined, they propofed to
have forced the great cabbin: But on opening the cheft, there appeared nothing but fire-arms, which to them were of no ufe. There
were indeed cutlaffes in the cheft, but they were hid by the fire-
arms being laid over them. This was a fenfible difappointment to
them,

them, and by this time *Pizarro* and his companions in the great
cabbin were capable of converfing aloud, through the cabbin windows and port-holes, with thofe in the gun-room and between
decks, and from hence they learnt, that the *Englifh* (whom they
principally fufpected) were all fafe below, and had not intermedled
in this mutiny ; and by other particulars they at laft difcovered, that
none were concerned in it but *Orellana* and his people. On this
Pizarro and the Officers refolved to attack them on the quarter-
deck, before any of the difcontented on board fhould fo far recover their firft furprize, as to reflect on the facility and certainty of
feizing the fhip by a junction with the *Indians* in the prefent emergency. With this view *Pizarro* got together what arms were in
the cabbin, and diftributed them to thofe who were with him :
But there were no other fire-arms to be met with but piftols, and
for thefe they had neither powder nor ball. However, having now
fettled a correfpondence with the gun-room, they lowered down a
bucket out of the cabbin-window, into which the gunner, out of
one of the gun-room ports, put a quantity of piftol cartridges.
When they had thus procured ammunition, and had loaded their
piftols, they fet the cabbin-door partly open, and fired fome fhot
amongft the *Indians* on the quarter-deck, at firft without effect.
But at laft *Mindinuetta*, whom we have often mentioned, had the
good fortune to fhoot *Orellana* dead on the fpot ; on which his
faithful companions abandoning all thoughts of farther refiftance,
inftantly leaped into the fea, where they every man perifhed. Thus
was this infurrection quelled, and the poffeffion of the quarter-
deck regained, after it had been full two hours in the power of this
great and daring Chief, and his gallant and unhappy countrymen.

Pizarro having efcaped this imminent peril fteered for *Europe*,
and arrived fafe on the coaft of *Galicia* in the beginning of the year
1746, after having been abfent between four and five years, and
having, by his attendance on our expedition, diminifhed the naval
power of *Spain* by above three thoufand hands, (the flower of their
failors) and by four confiderable fhips of war and a Patache. For
we

we have feen, that the *Hermiona* foundered at fea ; the *Guipufcoa* was ftranded, and funk on the coaft of *Brazil*; the St. *Eftevan* was condemned, and broke up in the river of *Plate* ; and the *Efperanza* being left in the *South-Seas*, is doubtlefs by this time incapable of returning to *Spain.* So that the *Afia* only, with lefs than one hundred hands, may be confidered as all the remains of that fquadron, with which *Pizarro* firft put to fea. And whoever attends to the very large proportion, which this fquadron bore to the whole navy of *Spain*, will, I believe, confefs, that had our undertaking been attended with no other advantages than that of ruining fo great a part of the fea-force of fo dangerous an enemy, this alone would be a fufficient equivalent for our equipment, and an inconteftible proof of the fervice, which the Nation has thence received. Having thus concluded this fummary of *Pizarro*'s adventures, I fhall now return again to the narration of our own tranfactions.

F

CHAP.

C H A P. IV.

From *Madera* to St. *Catherine's*.

I HAVE already mentioned, that on the 3d of *November* we weighed from *Madera*, after orders had been given to the Captains to rendezvous at St. *Jago*, one of the *Cape de Verd* Iflands, in cafe the fquadron was feparated. But the next day, when we were got to fea, the Commodore confidering that the feafon was far advanced, and that touching at St. *Jago* would create a new delay, he for this reafon thought proper to alter his rendezvous, and to appoint the Ifland of St. *Catherine's*, on the coaft of *Brazil*, to be the firft place to which the fhips of the fquadron were to repair in cafe of feparation.

In our paffage to the Ifland of St. *Catherine's*, we found the direction of the trade-winds to differ confiderably from what we had reafon to expect, both from the general hiftories given of thefe winds, and the experience of former Navigators. For the learned Dr. *Halley*, in his account of the trade winds, which take place in the *Ethiopic* and *Atlantic* Ocean, tells us, that from the latitude of 28° N, to the latitude of 10° N, there is generally a frefh gale of N. E. wind, which towards the *African* fide rarely comes to the eaftward of E. N. E, or paffes to the northward of N. N. E: But on the *American* fide, the wind is fomewhat more eafterly, though moft commonly even there it is a point or two to the northward of the Eaft : That from 10° N. to 4° N, the calms and tornadoes take place ; and from 4° N. to 30° S, the winds are generally and perpetually between the South and the Eaft. This account we expected to have verified by our own experience ; but we found confiderable variations from it, both in refpect to the fteadinefs of the winds, and the quarter from whence they blew. For though we met with a N. E. wind about the latitude of 28° N, yet from the

latitude

latitude of 25° to the latitude of 18° N, the wind was never once to the northward of the East, but on the contrary, almost constantly to the southward of it. However, from thence to the latitude of 6° : 20′ N, we had it usually to the northward of the East, though not entirely, it having for a short time changed to E. S. E. From hence, to about 4° 46′ N, the weather was very unsettled; sometimes the wind was N. E. then changed to S. E, and sometimes we had a dead calm, attended with small rain and lightning. After this, the wind continued almost invariably between the S. and E, to the latitude of 7° : 30′ S ; and then again as invariably between the N. and E, to the latitude of 15° : 30′ S ; then E. and S. E, to 21° : 37′ S. But after this, even to the latitude of 27° : 44′ S, the wind was never once between the S. and the E, though we had it at times in all the other quarters of the compass. But this last circumstance may be in some measure accounted for, from our approach to the main continent of the *Brazils*. I mention not these particulars with a view of cavilling at the received accounts of these trade-winds, which I doubt not are in general sufficiently accurate ; but I thought it a matter worthy of public notice, that such deviations from the established rules do sometimes take place. This observation may not only be of service to Navigators, by putting them on their guard against these hitherto unexpected irregularities, but may perhaps contribute to the solution of that great question about the causes of trade-winds, and monsoons, a question, which, in my opinion, has not been hitherto discussed with that clearness and accuracy, which its importance (whether it be considered as a naval or philosophical inquiry) seems to demand.

On the 16th of *November*, one of our Victuallers made a signal to speak with the Commodore, and we shortned sail for her to come up with us. The Master came on board, and acquainted Mr. *Anson*, that he had complied with the terms of his charter-party, and desired to be unloaded and dismissed. Mr. *Anson*, on consulting the Captains of the squadron, found all the ships had still such quantities of provision between their decks, and were withal so deep,

that

that they could not without great difficulty take in their feveral pro-
portions of brandy from the *Induftry Pink*, one of the Victualiers
only : And confequently he was obliged to continue the other of
them, the *Anna Pink*, in the fervice of attending the fquadron. And
the next day the Commodore made a fignal for the fhips to bring
to, and to take on board their fhares of the brandy from the *Induftry Pink* ; and in this, the long boats of the fquadron were employed the three following days, that is, till the 19th in the evening, when the *Pink* being unloaded, fhe parted company with us,
being bound for *Barbadoes*, there to take in a freight for *England*.
Moft of the Officers of the fquadron took the opportunity of writing
to their friends at home by this fhip; but fhe was afterwards, as I have
been fince informed, unhappily taken by the *Spaniards*.

On the 20th of *November*, the Captains of the fquadron reprefented to the Commodore, that their fhips companies were very
fickly, and that it was their own opinion as well as their furgeons, that
it would tend to the prefervation of the men to let in more air between decks ; but that their fhips were fo deep, they could not
poffibly open their lower ports. On this reprefentation, the Commodore ordered fix air fcuttles to be cut in each fhip, in fuch places
where they would leaft weaken it.

And on this occafion I cannot but obferve, how much it is the duty
of all thofe, who either by office or authority, have any influence
in the direction of our naval affairs, to attend to this important article, the prefervation of the lives and health of our feamen. If
it could be fuppofed, that the motives of humanity were infufficient
for this purpofe, yet policy, and a regard to the fuccefs of our arms,
and the intereft and honour of each particular Commander, fhould
naturally lead us to a careful and impartial examination of every
probable method propofed for maintaining a fhip's crew in health
and vigour. But hath this been always done? Have the late invented plain and obvious methods of keeping our fhips fweet and
clean, by a conftant fupply of frefh air, been confidered with that
candour and temper, which the great benefits promifed hereby

I   ought

ought naturally to have infpired? On the contrary, have not thefe falutary fchemes been often treated with neglect and contempt? And have not fome of thofe who have been entrufted with experimenting their effects, been guilty of the moft indefenfible partiality, in the accounts they have given of thefe trials? Indeed, it muft be confeffed, that many diftinguifhed perfons, both in the direction and command of our fleets, have exerted themfelves on thefe occafions with a judicious and difpaffionate examination, becoming the interefting nature of the inquiry; but the wonder is, that any could be found irrational enough to act a contrary part, in defpight of the ftrongeft dictates of prudence and humanity. I muft however own, that I do not believe this conduct to have arifen from motives fo favage, as the firft reflection thereon does naturally fuggeft: But I rather impute it to an obftinate, and in fome degree, fuperftitious attachment to fuch practices as have been long eftablifhed, and to a fettled contempt and hatred of all kinds of innovations, efpecially fuch as are projected by landmen and perfons refiding on fhore. But let us return from this, I hope not, impertinent digreffion.

We croffed the equinoctial with a fine frefh gale at S. E, on *Friday* the 28th of *November*, at four in the morning, being then in the longitude of 27°: 59′ W. from *London*. And on the 2d of *December*, in the morning, we faw a fail in the N. W. quarter, and made the *Gloucefter*'s and *Tryal*'s fignals to chafe; and half an hour after, we let out our reefs and chafed with the fquadron; and about noon a fignal was made for the *Wager* to take our remaining Victualler, the *Anna Pink*, in tow. But at feven in the evening, finding we did not near the chace, and that the *Wager* was very far a-ftern, we fhortened fail, and made a fignal for the cruizers to join the fquadron. The next day but one we again difcovered a fail, which, on a nearer approach, we judged to be the fame veffel. We chafed her the whole day, and though we rather gained upon her, yet night came on before we could overtake her, and obliged us to give over the chace, to collect our fcattered fquadron. We were

much

much chagrined at the efcape of this veffel, as we then apprehended her to be an advice-boat fent from *Old Spain* to *Buenos Ayres*, with notice of our expedition. But we have fince learnt, that we were deceived in this conjecture, and that it was our *Eaft-India* Company's Packet bound to St. *Helena*.

On the 10th of *December*, being by our accounts in the latitude of 20° S, and 36° : 30′ longitude Weft from *London*, the *Tryal* fired a gun to denote foundings. We immediately founded, and found fixty fathom water, the bottom coarfe ground with broken fhells. The *Tryal* being a-head of us, had at one time thirty-feven fathom, which afterwards increafed to 90 : And then fhe found no bottom, which happened to us too at our fecond trial, though we founded with a hundred and fifty fathom of line. This is the fhoal which is laid down in moft charts by the name of the *Abrollos* ; and it appeared we were upon the very edge of it ; perhaps farther in, it may be extremely dangerous. We were then, by our different accounts, from ninety to fixty leagues Eaft of the coaft of *Brazil*. The next day but one we fpoke with a *Portuguefe* Brigantine from *Rio Janeiro*, bound to *Bahia del todos Santos*, who informed us, that we were thirty four leagues from Cape St. *Thomas*, and forty leagues from Cape *Frio*, which laft bore from us W. S. W. By our accounts we were near eighty leagues from Cape *Frio* ; and though, on the information of this Brigantine, we altered our courfe, and ftood more to the fouthward, yet by our coming in with the land afterwards, we were fully convinced that our reckoning was much correcter than our *Portuguefe* intelligence. We found a confiderable current fetting to the fouthward, after we had paffed the latitude of 16° S. And the fame took place all along the coaft of *Brazil*, and even to the fouthward of the river of *Plate*, it amounting fometimes to thirty miles in twenty-four hours, and once to above forty miles.

If this current is occafioned (as it is moft probable) by the running off of the water, accumulated on the coaft of *Brazil* by the conftant fweeping of the eaftern trade-wind over the *Ethiopic*

Ocean,

Ocean, then it is moſt natural to ſuppoſe, that its general courſe is determined by the bearings of the adjacent ſhore. Perhaps too, in almoſt every other inſtance of currents, the ſame may hold true, as I believe no examples occur of conſiderable currents being obſerved at any great diſtance from land. If this then could be laid down for a general principle, it would be always eaſy to correct the reckoning by the obſerved latitude. But it were much to be wiſhed, for the general intereſts of navigation, that the actual ſettings of the different currents which are known to take place in various parts of the world, were examined more frequently and accurately than hitherto appears to have been done.

We now began to grow impatient for a ſight of land, both for the recovery of our ſick, and for the refreſhment and ſecurity of thoſe who as yet continued healthier. When we departed from St. *Helens*, we were in ſo good a condition, that we loſt but two men on board the *Centurion*, in our long paſſage to *Madera*. But in this preſent run between *Madera* and St. *Catherine's* we have been very ſickly, ſo that many died, and great numbers were confined to their hammocks, both in our own ſhip and in the reſt of the ſquadron, and ſeveral of theſe paſt all hopes of recovery. The diſorders they in general labour under are ſuch as are common to the hot climates, and what moſt ſhips bound to the ſouthward experience in a greater or leſs degree. Theſe are thoſe kind of fevers, which they uſually call Calentures: A diſeaſe, which was not only terrible in its firſt inſtance, but even the remains of it often proved fatal to thoſe who conſidered themſelves as recovered from it. For it always left them in a very weak and helpleſs condition, and uſually afflicted with fluxes and tenaſmus's. And by our continuance at ſea all our complaints were every day increaſing, ſo that it was with great joy that we diſcovered the coaſt of *Brazil* on the 18th of *December*, at ſeven in the morning.

The coaſt of *Brazil* appeared high and mountainous land, extending from the W. to W. S. W, and when we firſt ſaw it, it was

about

about feventeen leagues diftant. At noon we perceived a low double land, bearing W. S. W. about ten leagues diftant, which we
took to be the Ifland of St. *Catherine's.* That afternoon and the
next morning, the wind being N. N. W, we gained very little to
windward, and were apprehenfive of being driven to the leeward of
the Ifland; but a little before noon, the next day, the wind came about
the fouthward, and enabled us to fteer in between the North point
of St. *Catherine's,* and the neighbouring Ifland of *Alvoredo.* As
we ftood in for the land, we had regular foundings gradually decreafing, from thirty-fix to twelve fathom, all muddy ground. In
this laft depth of water we let go our anchor at five o'clock in the
evening of the 18th, the North Weft point of the Ifland of St. *Catherine's* bearing S.S.W, diftant three miles; and the Ifland *Alvoredo*
N. N. E, diftant two leagues. Here we found the tide to fet S. S. E.
and N. N. W, at the rate of two knots, the tide of flood coming
from the fouthward. We could from our fhips obferve two fortifications at a confiderable diftance within us, which feemed defigned
to prevent the paffage of an enemy between the Ifland of St. *Catherine's* and the main. And we could foon perceive that our fquadron had alarmed the coaft, for we faw the two forts hoift their
colours, and fire feveral guns, which we fuppofed to be intended
for affembling the inhabitants. To prevent any confufion, the
Commodore immediately fent a boat with an Officer on fhore, to
compliment the Governor, and to defire a Pilot to carry us into the
road. The Governor returned a very civil anfwer, and ordered us
a Pilot. On the morning of the 20th we weighed and ftood in,
and towards noon the Pilot came on board us, who, the fame afternoon, brought us to an anchor in five fathom and an half, in a
large commodious bay on the continent fide, called by the *French,
Bon Port.* In ftanding from our laft anchorage to this place, we
every where found an ouzy bottom, with a depth of water firft re-
gularly decreafing to five fathom, and then increafing to feven, after which we had fix and five fathom alternately. The next morning

ing we weighed again with the fquadron, in order to run above the two fortifications we have mentioned, which are called the caftles of *Santa Cruiz* and St. *Juan*. And now the foundings between the Ifland and the Main were four, five and fix fathom, with muddy ground. As we paffed by the caftle of *Santa Cruiz* we faluted it with eleven guns, and were anfwered by an equal number; and at one in the afternoon, the fquadron came to an anchor in five fathom and a half, the Governor's Ifland bearing N. N. W, St. *Juan's* Caftle N. E. $\frac{1}{2}$ E, and the Ifland of St. *Antonio* South. In this pofition we moored at the Ifland of St. *Catherine's* on *Sunday* the 21ft of *December*, the whole fquadron being, as I have already mentioned, fickly, and in great want of refrefhments: Both which inconveniencies we hoped to have foon removed at this fettlement, celebrated by former Navigators for its healthinefs and its provifions, and for the freedom, indulgence, and friendly affiftance there given to the fhips of all *European* Nations, in amity with the Crown of *Portugal*.

G   C H A P.

C H A P. V.

Proceedings at St. *Catherine's*, and a defcription of the place, with a fhort account of *Brazil*.

OUR firft care, after having moored our fhips, was to fend our fick men on fhore, each fhip being ordered by the Commodore to erect two tents for that purpofe: One of them for the reception of the difeafed, and the other for the accommodation of the furgeon and his affiftants. We fent about eighty fick from the *Centurion*, and the other fhips I believe fent nearly as many, in proportion to the number of their hands. As foon as we had performed this neceffary duty, we fcraped our decks, and gave our fhip a thorough cleanfing ; then fmoked it between decks, and after all wafhed every part well with vinegar. Thefe operations were extremely neceffary for correcting the noifome ftench on board, and deftroying the vermin ; for from the number of our men, and the heat of the climate, both thefe nuifances had increafed upon us to a very loathfome degree, and befides being moft intolerably offenfive, they were doubtlefs in fome fort productive of the ficknefs we had laboured under for a confiderable time, before our arrival at this Ifland.

Our next employment was wooding and watering our fquadron, caulking our fhips fides and decks, overhaling our rigging, and fecuring our mafts againft the tempeftuous weather we were, in all probability, to meet with in our paffage round Cape *Horn*, in fo advanced and inconvenient a feafon. But before I engage in the particulars of thefe tranfactions, it will not be improper to give fome account of the prefent ftate of this Ifland of St. *Catherine's*, and of the neighbouring country ; both as the circumftances of this place are now greatly changed from what they were in the time of former writers, and as thefe changes laid us under many more difficulties

and

A View of the N.E. End of the Island

Plate I.

ST CATHERINES *on the Coast of* BRASIL.

A view of the north entrance

Plate II.

c   d

harbour of St CATHERINES.

and perplexities than we had reaſon to expect, or than other *Britiſh* ſhips, hereafter bound to the *South-Seas*, may perhaps think it prudent to ſtruggle with.

This Iſland is eſteemed by the natives to be no where above two leagues in breadth, though about nine in length ; it lies in 49° : 45′ of Weſt longitude from *London*, and extends from the South latitude of 27° 35′, to that of 28°. Although it be of a conſiderable height, yet it is ſcarce diſcernible at the diſtance of ten leagues, being then obſcured under the continent of *Brazil*, whoſe mountains are exceeding high ; but on a nearer approach it is eaſy to be diſtinguiſhed, and may be readily known by a number of ſmall Iſlands lying at each end, and ſcattered along the Eaſt ſide of it. In the annexed plate there is exhibited a very exact view of the N. E. end of the Iſland, where (*a*) is its N. E. point, as it appears when it bears N. W. And (*b*) is the ſmall Iſland of *Alvoredo*, bearing N. N. W, at the diſtance of 7 leagues. The beſt entrance to the harbour is between the point (*a*) and the Iſland of *Alvoredo*, where ſhips may paſs under the guidance of their lead, without the leaſt apprehenſions of danger. The view of this North entrance of the harbour is repreſented in the ſecond plate, where (*a*) is the N. W. end of St. *Catherine's* Iſland, (*b*) *Parrot* Iſland, (*c*) a battery on St. *Catherine's*, and (*d*) a battery on a ſmall Iſland near the continent. *Frezier* has given a draught of this Iſland of St. *Catherine's*, and of the neighbouring coaſt, and the minuter iſles adjacent ; but he has by miſtake called the Iſland of *Alvoredo* the Iſle *de Gal*, whereas the true Iſle *de Gal* lies ſeven or eight miles to the North-weſtward of it, and is much ſmaller. He has alſo called an Iſland, to the ſouthward of St. *Catherine's*, *Alvoredo*, and has omitted the Iſland *Maſaqura* ; in other reſpects his plan is ſufficiently exact.

The North entrance of the harbour is in breadth about five miles, and the diſtance from thence to the Iſland of St. *Antonio* is eight miles, and the courſe from the entrance to St. *Antonio* is S. S. W. ½ W. About the middle of the Iſland the harbour is contracted by two points of land to a narrow channel, no more than a quarter of a

mile

mile broad; and to defend this paſſage, a battery was erecting on the point of land on the Iſland ſide. But this ſeems to be a very uſe-leſs work, as the channel has no more than two fathom water, and conſequently is navigable only for barks and boats, and therefore ſeems to be a paſſage that an enemy could have no inducement to at-tempt, eſpecially as the common paſſage at the North end of the Iſland is ſo broad and ſafe, that no ſquadron can be prevented from coming in by any of their fortifications, when the ſea-breeze is made. However, the Brigadier Don *Joſe Sylva de Paz*, the Governor of this ſettlement, is eſteemed an expert Engineer, and he doubtleſs under-ſtands one branch of his buſineſs very well, which is the advan-tages which new works bring to thoſe who are entruſted with the care of erecting them : For beſides the battery mentioned above, there are three other forts carrying on for the defence of the har-bour, none of which are yet compleated. The firſt of theſe, cal-led St. *Juan*, is built on a point of St. *Catherine's* near *Parrot* Iſland; the ſecond, in form of a half moon, is on the Iſland of St. *Anto-nio*; and the third, which ſeems to be the chief, and has ſome ap-pearance of a regular fortification, is on an Iſland near the conti-nent, where the Governor reſides.

The ſoil of the Iſland is truly luxuriant, producing fruits of moſt kinds ſpotaneouſly; and the ground is covered over with one conti-nued foreſt of trees of a perpetual verdure, which from the exube-rance of the ſoil, are ſo entangled with briars, thorns, and under-wood, as to form a thicket abſolutely impenetrable, except by ſome narrow pathways which the inhabitants have made for their own convenience. Theſe, with a few ſpots cleared for plantations along the ſhore facing the continent, are the only uncovered parts of the Iſland. The woods are extremely fragrant, from the many aromatick trees and ſhrubs with which they abound; and the fruits and vegetables of all climates thrive here, almoſt without cul-ture, and are to be procured in great plenty; ſo that here is no want of pine-apples, peaches, grapes, oranges, lemons, citrons, melons, apricots, nor plantains. There are beſides great abundance of two

other

other productions of no fmall confideration for a fea-ftore, I mean onions and potatoes. The provifions of other kinds are however inferior to their vegetables : There are fmall wild cattle to be purchafed, fomewhat like buffaloes, but thefe are very indifferent food, their flefh being of a loofe contexture, and generally of a difagreable flavour, which is probably owing to the wild calabafh on which they feed. There are likewife great plenty of pheafants, but they are much inferiour in tafte to thofe we have in *England*. The other provifions of the place are monkeys, parrots, and fifh of various forts, which abound in the harbour, and are all exceeding good, and are eafily catched, for there are a great number of fmall fandy bays very convenient for haling the *Seyne*.

The water both on the Ifland and the oppofite continent is excellent, and preferves at fea as well as that of the *Thames*. For after it has been in the cafk a day or two it begins to purge itfelf, and ftinks moft intolerably, and is foon covered over with a green fcum : But this, in a few days, fubfides to the bottom, and leaves the water as clear as chryftal, and perfectly fweet. The *French* (who, during their *South-Sea* trade in Queen *Anne*'s reign firft brought this place into repute) ufually wooded and watered in *Bon Port*, on the continent fide, where they likewife anchored with great fafety in fix fathom water; and this is doubtlefs the moft commodious road for fuch fhips as intend to make only a fhort ftay. But we watered on the St. *Catherine*'s fide, at a plantation oppofite to the Ifland of St. *Antonio*.

Thefe are the advantages of this Ifland of St *Catherine's*; but there are many inconveniencies attending it, partly from its climate, but more from its new regulations, and the late form of government eftablifhed there. With regard to the climate, it muft be remembred, that the woods and hills which furround the harbour, prevent a free circulation of the air. And the vigorous vegetation which conftantly takes place there, furnifhes fuch a prodigious quantity of vapour, that all the night and a great part of the morning a thick fog covers the whole country, and continues till either the

fun

sun gathers strength to dissipate it, or it is dispersed by a brisk
sea-breeze. This renders the place close and humid, and probably
occasioned the many fevers and fluxes we were there afflicted
with. To these exceptions I must not omit to add, that all the
day we were pestered with great numbers of muscatos, which are
not much unlike the gnats in *England*, but more venemous in their
stings. And at sun-set, when the muscatos retired, they were succeeded by an infinity of sand-flies, which, though scarce discerni-
ble to the naked eye, make a mighty buzzing, and wherever they
bite raise a small bump in the flesh, which is soon attended with a
painful itching, like that arising from the bite of an *English* harvest
bug.

But as the only light in which this place deserves our consideration, is its favourable situation for supplying and refreshing our
cruizers intended for the *South-Seas :* In this view its greatest inconveniencies remain still to be related ; and to do this more di-
stinctly, it will not be amiss to consider the changes which it has
lately undergone, both in its inhabitants, its police, and its governor.

In the time of *Frezier* and *Shelvocke*, this place served only as a
retreat to vagabonds and outlaws, who fled thither from all parts of
Brazil. They did indeed acknowledge a subjection to the Crown.
of *Portugal*, and had a person among them whom they called their
Captain, who was considered in some sort as their Governor: But
both their allegiance to their King, and their obedience to their
Captain, seemed to be little more than verbal. For as they had
plenty of provisions but no money, they were in a condition to
support themselves without the assistance of any neighbouring settlements, and had not amongst them the means of tempting any
adjacent Governor to busy his authority about them. In this situation they were extremely hospitable and friendly to such foreign
ships as came amongst them. For these ships wanting only provisions, of which the natives had great store ; and the natives wanting clothes, (for they often despised money, and refused to take it)

which

which the ſhips furniſhed them with in exchange for their proviſions, both ſides found their account in this traffic ; and their Captain or Governor had neither power nor intereſt to reſtrain it or to tax it. But of late (for reaſons which ſhall be hereafter mentioned) theſe honeſt vagabonds have been obliged to receive amongſt them a new colony, and to ſubmit to new laws and government. Inſtead of their former ragged bare legged Captain (whom however they took care to keep innocent) they have now the honour to be governed by Don *Joſe Sylva de Paz*, a Brigadier of the armies of *Portugal*. This Gentleman has with him a garriſon of ſoldiers, and has conſequently a more extenſive and a better ſupported power than any of his predeceſſors, and as he wears better clothes, and lives more ſplendidly, and has beſides a much better knowledge of the importance of money than they could ever pretend to: So he puts in practice certain methods of procuring it, with which they were utterly unacquainted. But it may be much doubted, if the inhabitants conſider theſe methods as tending to promote either their intereſts, or that of their Sovereign the King of *Portugal*. This is certain, that his behaviour cannot but be extremely embarraſſing to ſuch *Britiſh* ſhips as touch there in their way to the *South-Seas*. For one of his practices was placing centinels at all the avenues, to prevent the people from ſelling us any refreſhments, except at ſuch exorbitant rates as we could not afford to give. His pretence for this extraordinary ſtretch of power was, that he was obliged to preſerve their proviſions for upwards of an hundred families, which they daily expected to reinforce their colony. Hence he appears to be no novice in his profeſſion, by his readineſs at inventing a plauſible pretence for his intereſted management. However, this, though ſufficiently provoking, was far from being the moſt exceptionable part of his conduct. For by the neighbourhood of the river *Plate*, a conſiderable ſmuggling traffic is carried on between the *Portugueſe* and the *Spaniards*, eſpecially in the exchanging gold for ſilver, by which both Princes are defrauded of their fifths, and in this prohibited commerce *Don Joſe* was ſo deeply

I   engaged,

engaged, that in order to ingratiate himself with his *Spanish* corref-pondents (for no other reason can be given for his procedure) he treacherously dispatched an express to *Buenos Ayres* in the river of *Plate*, where *Pizarro* then lay, with an account of our arrival, and of the strength of our squadron; particularly the number of ships, guns and men, and every circumstance which he could suppose our ene-my desirous of being acquainted with. And the same perfidy every *British* cruizer may expect, who touches at St. *Catherine's*, while it is under the Government of Don *Jose Sylva de Paz*.

Thus much, with what we shall be necessitated to relate in the course of our own proceedings, may suffice as to the present state of St. *Catherine's*, and the character of its Governor. But as the reader may be desirous of knowing to what causes the late new mo-delling of this settlement is owing; to satisfy him in this particular, it will be necessary to give a short account of the adjacent continent of *Brazil*, and of the wonderful discoveries which have been made there within this last forty years, which, from a country of but mean estimation, has rendered it now perhaps the most considerable colony on the face of the globe.

This country was first discovered by *Americus Vesputio* a *Floren-tine*, who had the good fortune to be honoured with giving his name to the immense continent, some time before found out by *Columbus* : He being in the service of the *Portuguese*, it was settled and planted by that Nation, and with the other dominions of *Portu-gal*, devolved to the Crown of *Spain*, when that Kingdom became subject to it. During the long war between *Spain* and the State of *Holland*, the *Dutch* possessed themselves of the northermost part of *Brazil*, and were masters of it for some years. But when the *Portuguese* revolted from the *Spanish* Government, this country took part in the revolt, and soon repossessed themselves of the places the *Dutch* had taken; since which time it has continued without inter-ruption under the Crown of *Portugal*, being, till the beginning of the present century, only productive of sugar, and tobacco, and a few other commodities of very little account.

I   **But**

But this country, which for many years was only confidered for the produce of its plantations, has been lately difcovered to abound with the two minerals, which mankind hold in the greateft efteem, and which they exert their utmoft art and induftry in acquiring, I mean, gold and diamonds. Gold was firft found in the mountains, which lie adjacent to the city of *Rio Janeiro*. The occafion of its difcovery is varioufly related, but the moft common account is, that the *Indians*, lying on the back of the *Portuguefe* fettlements, were obferved by the foldiers employed in an expedition againft them to make ufe of this metal for their fifh hooks; and their manner of procuring it being enquired into, it appeared that great quantities of it were annually wafhed from the hills, and left amongft the fand and gravel, which remained in the vallies after the running off, or evaporation of the water. It is now little more than forty years fince any quantities of gold worth notice have been imported to *Europe* from *Brazil*; but fince that time the annual imports from thence have been continually augmented by the difcovery of places in other provinces, where it is to be met with as plentifully as at firft about *Rio Janeiro*. And it is now faid, that there is a fmall flender vein of it fpread through all the country, at about twenty-four feet from the furface, but that this vein is too thin and poor to anfwer the expence of digging; however where the rivers or rains have had any courfe for a confiderable time, there gold is always to be collected, the water having feparated the metal from the earth, and depofited it in the fands, thereby faving the expences of digging: So that it is efteemed an infallible gain to be able to divert a ftream from its channel, and to ranfack its bed. From this account of gathering this metal, it fhould follow, that there are properly no gold mines in *Brazil*; and this the Governor of *Rio Grande* (who being at St. *Catherine's*, frequently vifited Mr. *Anfon*) did moft confidently affirm, affuring us, that the gold was all collected either from rivers, or from the beds of torrents after floods. It is indeed afferted, that in the mountains, large rocks are found abounding with this metal; and I myfelf have feen the fragment of one of thefe rocks with a confiderable lump of gold in-

H   tangled

tangled in it; but even in this cafe, the workmen break off the rocks, and do not properly mine into them; and the great expence in fubfifting among thefe mountains, and afterwards in feparating the metal from the ftone, makes this method of procuring gold to be but rarely put in practice.

The examining the bottoms of rivers, and the gullies of torrents, and the wafhing the gold found therein from the fand and dirt, with which it is always mixed, are works performed by flaves, who are principally Negroes, kept in great numbers by the *Portuguefe* for thefe purpofes. The regulation of the duty of thefe flaves is fingular: For they are each of them obliged to furnifh their mafter with the eighth part of an ounce of gold *per diem*; and if they are either fo fortunate or induftrious as to collect a greater quantity, the furplus is confidered as their own property, and they have the liberty of difpofing of it as they think fit. So that it is faid fome Negroes who have accidentally fallen upon rich wafhing places have themfelves purchafed flaves, and have lived afterwards in great fplendor, their original mafter having no other demand on them than the daily fupply of the forementioned eighth; which as the *Portuguefe* ounce is fomewhat lighter than our troy ounce, may a-mount to about nine fhillings fterling.

The quantity of gold thus collected in the *Brazils*, and returned annually to *Lisbon*, may be in fome degree eftimated from the a-mount of the King's fifth. This hath of late been efteemed one year with another to be one hundred and fifty arroves of 32 *l. Por-tuguefe* weight, each of which, at 4 *l.* the troy ounce, makes very near 300,000 *l.* fterling; and confequently the capital, of which this is the fifth, is about a million and a half fterling. And the annual return of gold to *Lifbon* cannot be lefs than this, though it be dif-ficult to determine how much it exceeds it; perhaps we may not be very much miftaken in our conjecture, if we fuppofe the gold ex-changed for filver with the *Spaniards* at *Buenos Ayres*, and what is brought privily to *Europe*, and efcapes the duty, amounts to near half a million more, which will make the whole annual produce of the *Brafilian* gold near two millions fterling; a prodigious fum to be found in a country, which a few years since was not known to furnish a single grain.

I have already mentioned, that besides gold, this country does likewise produce diamonds. The discovery of these valuable stones is much more recent than that of gold, it being as yet scarce twenty years since the first were brought to *Europe*. They are found in the same manner as the gold, in the gullies of torrents and beds of rivers, but only in particular places, and not so universally spread through the country. They were often found in washing the gold before they were known to be diamonds, and were consequently thrown away with the sand and gravel separated from it. And it is very well remembered, that numbers of very large stones, which would have made the fortunes of the possessors, have passed unregarded through the hands of those, who now with impatience support the mortifying reflection. However, about twenty years since, a person acquainted with the appearance of rough diamonds, conceived that these pebbles, as they were then esteemed, were of the same kind : But it is said, that there was a considerable interval between the first starting of this opinion, and the confirmation of it by proper trials and examination, it proving difficult to persuade the inhabitants, that what they had been long accustomed to despise, could be of the importance represented by the discovery ; and I have been informed, that in this interval, a Governor of one of their places procured a good number of these stones, which he pretended to make use of at cards to mark with, instead of counters. But to proceed : It was at last confirmed by skilful Jewellers in *Europe*, consulted on this occasion, that the stones thus found in *Brazil* were truly diamonds, many of which were not inferiour either in lustre, or any other quality to those of the *East-Indies*. On this determination the *Portuguese*, in the neighbourhood of those places where they had first been observed, set themselves to search for them with great assiduity. And they were not without great hopes of discovering considerable masses of them, as they found large

rocks

rocks of chriftal in many of the mountains, from whence the ftreams came which wafhed down the diamonds.

But it was foon reprefented to the King of *Portugal*, that if fuch plenty of diamonds fhould be met with as their fanguine conjectures feemed to indicate, this would fo debafe their value, and diminifh their eftimation, that befides ruining all the *Europeans* who had any quantity of *Indian* diamonds in their poffeffion, it would render the difcovery itfelf of no importance, and would prevent his Majefty from receiving any advantages from it. And on thefe confiderations his Majefty has thought proper to reftrain the general fearch of diamonds, and has erected a Diamond Company for that purpofe, with an exclufive charter. This Company, in confideration of a fum paid by them to the King, have the property of all diamonds found in *Brazil*: But to hinder their collecting too large quantities, and thereby debafing their value, they are prohibited from employing above eight hundred flaves in fearching after them. And to prevent any of his other fubjects from acting the fame part, and likewife to fecure the Company from being defrauded by the interfering of interlopers in their trade, he has depopulated a large town, and a confiderable diftrict round it, and has obliged the inhabitants, who are faid to amount to fix thoufand, to remove to another part of the country; for this town being in the neighbourhood of the diamonds, it was thought impoffible to prevent fuch a number of people, who were on the fpot, from frequently fmuggling.

In confequence of thefe important difcoveries in *Brazil*, new laws, new governments, and new regulations have been eftablifhed in many parts of the country. For not long fince, a confiderable tract, poffeffed by a fet of inhabitants, who from their principal fettlement were called *Paulifts*, was almoft independent of the Crown of *Portugal*, to which they fcarcely acknowledged more than a nominal allegiance. Thefe are faid to be defcendants of thofe *Portuguefe*, who retired from the northern part of *Brazil*,

when

when it was invaded and poffeffed by the *Dutch*. And being for a long time neglected and obliged to provide for their own fecurity and defence, the neceffity of their affairs produced a kind of government amongft them, which they found fufficient for the confined manner of life to which they were inured. And therefore rejecting and defpifing the authority and mandate of the Court of *Lifbon*, they were often engaged in a ftate of downright rebellion: And the mountains furrounding their country, and the difficulty of clearing the few paffages that open into it, generally put it in their power to make their own terms before they fubmitted. But as gold was found to abound in this country of the *Paulifts*, the prefent King of *Portugal* (during whofe reign almoft the whole difcoveries I have mentioned were begun and compleated) thought it incumbent on him to reduce this province, which now became of great confequence, to the fame dependency and obedience with the reft of the country, which, I am told, he has at laft, though with great difficulty, happily effected. And the fame motives which induced his Majefty to undertake the reduction of the *Paulifts*, has alfo occafioned the changes I have mentioned, to have taken place at the Ifland of St. *Catherine's*. For the Governor of *Rio Grande*, of whom I have already fpoken, affured us, that in the neighbourhood of this Ifland there were confiderable rivers which were found to be extremely rich, and that this was the reafon that a garrifon, a military Governor, and a new colony was fettled there. And as the harbour at this Ifland is by much the fecureft and the moft capacious of any on the coaft, it is not improbable, if the riches of the neighbourhood anfwer their expectation, but it may become in time the principal fettlement in *Brazil*, and the moft confiderable port in all South *America*.

Thus much I have thought neceffary to infert, in relation to the prefent ftate of *Brazil*, and of the Ifland of St. *Catherine's*. For as this laft place has been generally recommended as the moft eligible port for our cruifers to refrefh at, which are bound to the *South-Seas*, I believed it to be my duty to inftruct my countrymen, in the

hitherto

hitherto unfufpected inconveniencies which attend that place. And as the *Brafilian* gold and diamonds are fubjects, about which, from their novelty, very few particulars have been hitherto publifhed, I conceived this account I had collected of them, would appear to the reader to be neither a trifling nor a ufelefs digreffion. Thefe fub-jects being thus difpatched, I fhall now return to the feries of our own proceedings.

When we firft arrived at St. *Catherine's*, we were employed in refrefhing our fick on fhore, in wooding and watering the fquadron, cleanfing our fhips, and examining and fecuring our mafts and rig-ging, as I have already obferved in the foregoing chapter. At the fame time Mr. *Anfon* gave directions, that the fhips companies fhould be fupplied with frefh meat, and that they fhould be victu-alled with whole allowance of all the kinds of provifion. In con-fequence of thefe orders, we had frefh beef fent on board us con-tinually for our daily expence, and what was wanting to make up our allowance we received from our Victualler the *Anna Pink*, in order to preferve the provifions on board our fquadron entire for our future fervice. The feafon of the year growing each day lefs favou-rable for our paffage round Cape *Horn*, Mr. *Anfon* was very defi-rous of leaving this place affoon as poffible; and we were at firft in hopes that our whole bufinefs would be done, and we fhould be in a readinefs to fail in about a fortnight from our arrival: But, on examining the *Tryal's* mafts, we, to our no fmall vexation, found inevitable employment for twice that time. For, on a furvey, it was found that the main-maft was fprung at the upper woulding, though it was thought capable of being fecured by a couple of fifhes; but the fore-maft was reported to be unfit for fervice, and thereupon the Carpenters were fent into the woods, to endeavour to find a ftick proper for a fore-maft. But after a fearch of four days, they returned without having been able to meet with any tree fit for the purpofe. This obliged them to come to a fecond con-fultation about the old fore-maft, when it was agreed to endeavour to fecure it by cafing it with three fifhes: And in this work the

Carpenters

Carpenters were employed, till within a day or two of our failing. In the mean time, the Commodore thinking it neceſſary to have a clean veſſel on our arrival in the *South-Seas*, ordered the *Tryal* to be hove down, as this would not occaſion any loſs of time, but might be compleated while the Carpenters were refitting her maſts, which was done on ſhore.

On the 27th of *December* we diſcovered a ſail in the offing, and not knowing but ſhe might be a *Spaniard*, the eighteen oared-boat was manned and armed, and ſent under the command of our ſecond Lieutenant, to examine her, before ſhe arrived within the protection of the forts. She proved to be a *Portugueſe* Brigantine from *Rio Grande*. And though our Officer, as it appeared on inquiry, had behaved with the utmoſt civility to the Maſter, and had refuſed to accept a calf, which the Maſter would have forced on him as a preſent: Yet the Governor took great offence at our ſending our boat; and talked of it in a high ſtrain, as a violation of the peace ſubſiſting between the Crowns of *Great-Britain* and *Portugal*. We at firſt imputed this ridiculous bluſtering to no deeper a cauſe, than Don *Joſe*'s inſolence; but as we found he proceeded ſo far as to charge our Officer with behaving rudely, and opening letters, and particularly with an attempt to take out of the veſſel, by violence, the very calf which we knew he had refuſed to receive as a preſent, (a circumſtance which we were ſatisfied the Governor was well acquainted with) we had hence reaſon to ſuſpect, that he purpoſely ſought this quarrel, and had more important motives for engaging in it, than the mere captious biaſs of his temper. What theſe motives were it was not ſo eaſy for us to determine at that time; but as we afterwards found by letters, which fell into our hands in the *South-Seas*, that he had diſpatched an expreſs to *Buenos Ayres*, where *Pizarro* then lay, with an account of our ſquadron's arrival at St. *Catherine's*, together with the moſt ample and circumſtantial intelligence of our force and condition, we thence conjectured that Don *Joſe* had raiſed this groundleſs clamour, only to prevent our viſiting the Brigantine when ſhe ſhould

put

put to fea again, leaft we might there find proofs of his perfidious behaviour, and perhaps at the fame time difcover the fecret of his fmuggling correfpondence with his neighbouring Governors, and the *Spaniards* at *Buenos Ayres.* But to proceed,

It was near a month before the *Tryal* was refitted ; for not only her lower mafts were defective, as hath been already mentioned, but her main top-maft and fore-yard were likewife decayed and rotten. While this work was carrying on, the other fhips of the fquadron fixed new ftanding rigging, and fet up a fufficient number of preventer fhrouds to each maft, to fecure them in the moft effectual manner. And in order to render the fhips ftiffer, and to enable them to carry more fail abroad, and to prevent their labouring in hard gales of wind, each Captain had orders given him to ftrike down fome of their great guns into the hold. Thefe precautions being complied with, and each fhip having taken in as much wood and water as there was room for, the *Tryal* was at laft compleated, and the whole fquadron was ready for the fea : On which the tents on fhore were ftruck, and all the fick were received on board. And here we had a melancholy proof how much the healthinefs of this place had been over-rated by former writers, for we found that though the *Centurion* alone had buried no lefs than twenty-eight men fince our arrival, yet the number of her fick was in the fame interval increafed from eighty to ninety-fix. And now our crews being embarked, and every thing prepared for our departure, the Commodore made a fignal for all Captains, and delivered them their orders, containing the fucceffive places of rendezvous from hence to the coaft of *China.* And then, on the next day, being the 18th of *January*, the fignal was made for weighing, and the fquadron put to fea, leaving without regret this Ifland of St. *Catherine's*; where we had been fo extremely difappointed in our refrefhments, in our accommodations, and in the humane and friendly offices which we had been taught to expect in a place, which hath been fo much celebrated for its hofpitality, freedom, and conveniency.

CHAP.

C H A P. VI.

The run from St. Catherine's to port St. Julian, with some account of that port, and of the country to the southward of the river of Plate.

IN leaving St. *Catherine's*, we left the last amicable port we proposed to touch at, and were now proceeding to an hostile, or at best, a desart and inhospitable coast. And as we were to expect a more boisterous climate to the southward than any we had yet experienced, not only our danger of separation would by this means be much greater than it had been hitherto, but other accidents of a more pernicious nature were likewise to be apprehended, and as much as possible to be provided against. And therefore Mr. *Anson*, in appointing the various stations at which the ships of the squadron were to rendezvous, had considered, that it was possible his own ship might be disabled from getting round Cape *Horn*, or might be lost, and had given proper directions, that even in that case the expedition should not be abandoned. For the orders delivered to the Captains, the day before we sailed from St. *Catherine's*, were, that in case of separation, which they were with the utmost care to endeavour to avoid, the first place of rendezvous should be the bay of port St. *Julian*; describing the place from Sir *John Narborough's* account of it: There they were to supply themselves with as much salt as they could take in, both for their own use, and for the use of the squadron; and if, after a stay there of ten days, they were not joined by the Commodore, they were then to proceed through *Streights le Maire* round Cape *Horn*, into the *South-Seas*, where the next place of rendezvous was to be the Island of *Nostra Senora del Socoro*, in the latitude of 45° South, and longitude from the *Lizard* 71°: 12′ West. They were to bring

I   this

this Iſland to bear E. N. E, and to cruize from five to twelve leagues diſtance from it, as long as their ſtore of wood and water would permit, both which they were to expend with the utmoſt frugality. And when they were under an abſolute neceſſity of a freſh ſupply, they were to ſtand in, and endeavour to find out an anchoring place ; and in caſe they could not, and the weather made it dangerous· to ſupply their ſhips by ſtanding off and on, they were then to make the beſt of their way to the Iſland of *Juan Fernandes*, in the latitude of 33° : 37' South. And as ſoon as they had there recruited their wood and water, they were to continue cruizing off the anchoring place of that Iſland for fifty-ſix days; in which time, if they were not joined by the Commodore, they might conclude that ſome accident had befallen him, and they were forthwith to put themſelves under the command of the ſenior Officer, who was to uſe his utmoſt endeavours to annoy the enemy both by ſea and land. That with theſe views their new Commodore was to continue in thoſe ſeas as long as his proviſions laſted, or as long as they were recruited by what he ſhould take from the enemy, reſerving only· a ſufficient quantity to carry him and the ſhips under his command to *Macao*, at the entrance of the river *Tigris* near *Canton* on the coaſt of *China*, where having ſupplied himſelf with a new ſtock of proviſions, he was thence, without delay, to make the beſt of his way to *England*. And as it was found impoſſible as yet to unload our Victualler the *Ann Pink*, the Commodore gave the Maſter of her the ſame rendezvous, and the ſame orders to put himſelf under the command of the remaining ſenior Officer.

Under theſe orders the ſquadron ſailed from St. *Catherine's* on *Sunday* the 18th of *January*, as hath been already mentioned in the preceding chapter. The next day we had very ſqually weather, attended with rain, lightning and thunder, but it ſoon became fair again with light breezes, and continued thus till *Wedneſday* evening, when it blew freſh again; and encreaſing all night, by eight the next morning it became a moſt violent ſtorm, and we had with it ſo thick a fog, that it was impoſſible to ſee at the diſtance

of

of two fhips length, fo that the whole fquadron difappeared. On this, a fignal was made, by firing guns, to bring to with the larboard tacks, the wind being then due Eaft. We ourfelves immediately handed the top-fails, bunted the main-fail, and lay to under a reefed mizen till noon, when the fog difperfed, and we foon difcovered all the fhips of the fquadron except the *Pearl*, who did not join us till near a month afterwards. The *Tryal* Sloop was a great way to leeward, having loft her main-maft in this fquall, and having been obliged, for fear of bilging, to cut away the raft. We bore down with the fquadron to her relief, and the *Gloucefter* was ordered to take her in tow, for the weather did not entirely abate till the day after, and even then, a great fwell continued from the eaftward, in confequence of the preceding ftorm.

After this accident we ftood to the fouthward with little interruption, and here we experienced the fame fetting of the current, which we had obferved before our arrival at St. *Catherine's*; that is, we generally found ourfelves to the fouthward of our reckoning, by about twenty miles each day. This error continued, with a little variation, till we had paffed the latitude of the river of *Plate*; and even then, we found that the fame current, however difficult to be accounted for, did yet undoubtedly take place; for we were not fatisfied in deducing it from the error in our reckoning, but we actually tried it more than once, when a calm made it practicable.

When we had paffed the latitude of the river of *Plate*, we had foundings all along the coaft of *Patagonia*. Thefe foundings, when well afcertained, being of great ufe in determining the pofition of the fhip, and we having tried them more frequently, in greater depths, and with more attention, than I believe had been done before us, I fhall recite our obfervations as fuccinctly as I can, referring to the chart hereafter inferted in the ninth chapter of this book, for a general view of the whole. In the latitude of 36°: 52' we had fixty fathom of water, with a bottom of fine black and grey fand; from thence, to 39°: 55', we varied our depths from fifty to eighty fathom, though we had conftantly the fame bottom as before; be-

tween

tween the laſt mentioned latitude, and 43° : 16', we had only fine grey ſand, with the ſame variation of depths, except that we once or twice leſſened our water to forty fathom. After this, we continued in forty fathom for about half a degree, having a bottom of coarſe ſand and broken ſhells, at which time we were in ſight of land, and not above ſeven leagues from it : As we edged from the land we met with variety of ſoundings ; firſt black ſand, then muddy, and ſoon after rough ground with ſtones ; but then encreaſing our water to forty-eight fathom, we had a muddy bottom to the latitude of 46° : 10. We then returned again into thirty-ſix fathom, and kept ſhoaling our water, till at length we came into twelve fathom, having conſtantly ſmall ſtones and pebbles at the bottom. Part of this time we had a view of Cape *Blanco,* which lies in about the latitude of 46° : 52', and longitude Weſt from *London* 66° : 43'. This is the moſt remarkable land upon the coaſt : Two very exact views of it are exhibited in the annexed plate, where (*b*) repreſents the Cape itſelf ; theſe draughts will fully enable future Voyagers to diſtinguiſh it. Steering from hence S. by E. nearly, we, in a run of about thirty leagues, deepned out water to fifty fathom, without once altering the bottom ; and then drawing towards the ſhore with a S. W. courſe, varying rather to the weſtward, we had every where a ſandy bottom, till our coming into thirty fathom, where we had again a ſight of land diſtant from us, about eight leagues, lying in the latitude of 48° : 31. We made this land on the 17th of *February,* and at five in the afternoon we came to an anchor upon the ſame bottom, in the latitude of 48° : 58', the ſouthermoſt land then in view bearing S. S W, the northermoſt N. ½ E, a ſmall Iſland N. W, and the weſtermoſt hummock W. S. W. In this ſtation we found the tide to ſet S. by W ; and weighing again at five the next morning, we, an hour afterwards, diſcovered a ſail, upon which the *Severn* and *Glouceſter* were both directed to give chace ; but we ſoon perceived it to be the *Pearl,* which ſeparated from us a few days after we left St. *Catherine's,* and on this we made a ſignal for the *Severn* to rejoin the ſquadron,

leaving

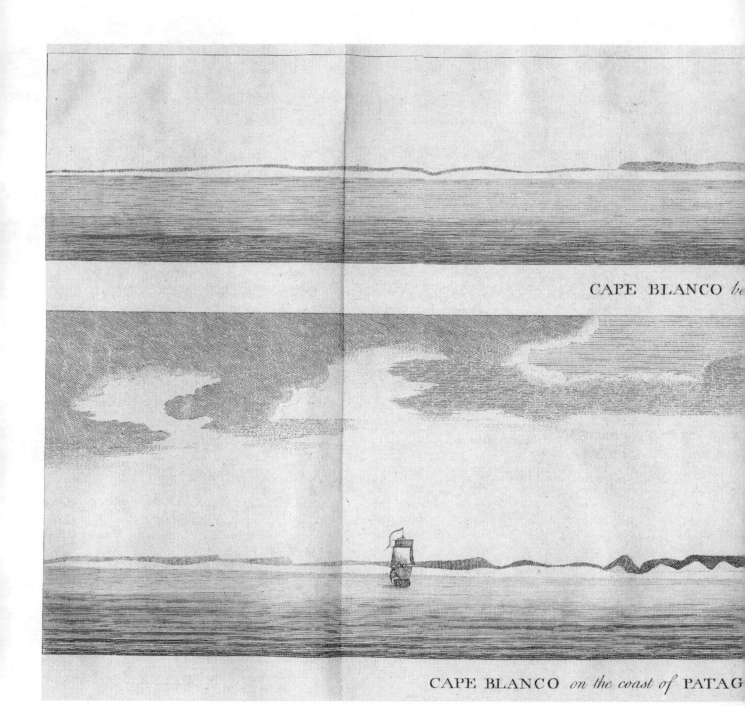

CAPE BLANCO *be*

CAPE BLANCO *on the coast of* PATAG

Plate III.

S W distant 5 leagues.

A: *Bearing S^t W ½ W distant 4 leagues.*

leaving the *Gloucester* alone in the pursuit. And now we were surprized to see, that on the *Gloucester*'s approach, the people on board the *Pearl* increased their sail, and stood from her. However, the *Gloucester* came up with them, but found them with their hammocks in their nettings, and every thing ready for an engagement. At two in the afternoon the *Pearl* joined us, and running up under our stern, Lieutenant *Salt* hailed the Commodore, and acquainted him that Captain *Kidd* died on the 31st of *January*. He likewise informed him, that he had seen five large ships the 10th instant, which he for some time imagined to be our squadron: That he suffered the commanding ship, which wore a red broad pennant, exactly resembling that of the Commodore, at the main top-mast head, to come within gun-shot of him before he discovered his mistake; but then finding it not to be the *Centurion*, he haled close upon the wind, and crowded from them with all his sail, and standing cross a ripling, where they hesitated to follow him, he happily escaped. He made them to be five *Spanish* men of war, one of them exceedingly like the *Gloucester*, which was the occasion of his apprehensions when the *Gloucester* chased him. By their appearance he thought they consisted of two ships of seventy guns, two of fifty, and one of forty guns. The whole squadron continued in chace of him all that day, but at night finding they could not get near him, they gave over the chace, and directed their course to the southward.

And now had it not been for the necessity we were under of refitting the *Tryal*, this piece of intelligence would have prevented our making any stay at St. *Julian's*; but as it was impossible for that sloop to proceed round the Cape in her present condition, some stay there was inevitable, and therefore the same evening we came to an anchor again in twenty five fathom water, the bottom a mixture of mud and sand, and the high hummock bearing S. W. by W. And weighing at nine in the morning, we soon after sent the two Cutters belonging to the *Centurion* and *Severn* in shore, to discover the harbour of St. *Julian*, while the ships kept standing along the

coast,

coaſt, at about the diſtance of a league from the land. At ſix o'clock we anchored in the bay of St. *Julian*, in nineteen fathom, the bottom muddy ground with ſand, the northermoſt land in ſight bearing N. and by E, the ſouthermoſt S. ½ E, and the high hummock, to which Sir *John Narborough* formerly gave the name of *Wood's Mount*, W. S. W. Soon after, the Cutter returned on board having diſcovered the harbour, which did not appear to us in our ſituation, the northermoſt point ſhutting in upon the ſouthermoſt, and in appearance cloſing the entrance. To facilitate the knowledge of this coaſt to future Navigators, there are two views annexed; the firſt of the land of *Patagonia*, to the northward of port St. *Julian*, where (*w*) is *Wood's Mount*, and the bay of St. *Julian* lies round the point (*c*). The ſecond view is of the bay itſelf; and here again (*w*) is *Wood's Mount*, (*a*) is cape St. *Julian*, and (*b*) the port or river's mouth.

Being come to an anchor in this bay of St. *Julian*, principally with a view of refitting the *Tryal*, the Carpenters were immediately employed in that buſineſs, and continued ſo during our whole ſtay at the place. The *Tryal*'s main maſt having been carried away about twelve feet below the cap, they contrived to make the remaining part of the maſt ſerve again; and the *Wager* was ordered to ſupply her with a ſpare main top-maſt, which the Carpenters converted into a new fore-maſt. And I cannot help obſerving, that this accident to the *Tryal*'s maſt, which gave us ſo much uneaſineſs at that time, on account of the delay it occaſioned, was, in all probability, the means of preſerving the ſloop, and all her crew. For before this, her maſts, how well ſoever proportioned to a better climate, were much too lofty for theſe high ſouthern latitudes: So that had they weathered the preceding ſtorm, it would have been impoſſible for them to have ſtood againſt thoſe ſeas and tempeſts we afterwards encountered in paſſing round Cape *Horn*, and the loſs of maſts in that boiſterous climate, would ſcarcely have been attended with leſs than the loſs of the veſſel, and of every man on board her; ſince it would have been impracticable for the other ſhips to have

given

A view of the land of PATAGONI.

little to the northward of PORT St IULIAN.

I.S.Müller sculp.

A view of the bay of St. JULIAN *when* MOUNT WOOD *bears*

Plate V.

S. W. ½ S. and the PORT or rivers mouth S. W. distant ten miles.

given them any relief, during the continuance of thofe impetuous ftorms.

Whilft we ftayed at this place, the Commodore appointed the Honourable Captain *Murray* to fucceed to the *Pearl*, and Captain *Cheap* to the *Wager*, and he promoted Mr. *Charles Saunders*, his firft Lieutenant, to the command of the *Tryal* Sloop. But Capt. *Saunders* lying dangeroufly ill of a fever on board the *Centurion*, and it being the opinion of the furgeons, that the removing him on board his own fhip, in his prefent condition, might tend to the hazard of his life; Mr. *Anfon* gave an order to Mr. *Saumarez*, firft Lieutenant of the *Centurion*, to act as Mafter and Commander of the *Tryal*, during the illnefs of Captain *Saunders*.

Here the Commodore too, in order to eafe the expedition of all unneceffary expence, held a farther confultation with his Captains about unloading and difcharging the *Anna Pink*; but they reprefented to him, that they were fo far from being in a condition of taking any part of her loading on board, that. they had ftill great quantities of provifions in the way of their guns between decks, and that their fhips were withal fo very deep, that they were not fit for action without being cleared. This put the Commodore under a neceffity of retaining the *Pink* in the fervice; and as it was apprehended we fhould certainly meet with the *Spanifh* fquadron, in paffing the Cape, Mr. *Anfon* thought it advifeable to give orders to the Captains, to put all their provifions, which were in the way of their guns, on board the *Anna Pink*, and to remount fuch of their guns as had formerly, for the eafe of their fhips, been ordered into the hold.

This bay of St. *Julian*, where we are now at anchor, being a convenient rendezvous, in cafe of feparation, for all cruifers bound to the fouthward, and the whole coaft of *Patagonia*, from the river of *Plate* to the Streights of *Magellan*, lying nearly parallel to their ufual route, a fhort account of the fingularity of this country, with a particular defcription of port St. *Julian*, may perhaps be neither unacceptable to the curious, nor unworthy the attention of future

Navigators,

Navigators, as some of them, by unforeseen accidents, may be obliged to run in with the land, and to make some stay on this coast, in which case the knowledge of the country, its produce and inhabitants, cannot but be of the utmost consequence to them.

To begin then with the tract of country usually stiled *Patagonia*. This is the name often given to the southermost part of South *America*, which is unpossessed by the *Spaniards*, extending from their settlements to the Streights of *Magellan*. On the east side, this country is extremely remarkable, for a peculiarity not to be paralleled in any other known part of the globe; for though the whole territory to the northward of the river of *Plate* is full of wood, and stored with immense quantities of large timber trees, yet to the southward of the river no trees of any kind are to be met with, except a few peach-trees, first planted and cultivated by the *Spaniards* in the neighbourhood of *Buenos Ayres*: So that on the whole eastern coast of *Patagonia*, extending near four hundred leagues in length, and reaching as far back as any discoveries have yet been made, no other wood has been found than a few insignificant shrubs. Sir *John Narborough* in particular, who was sent out, by King *Charles* the second, expresly to examine this country, and the Streights of *Magellan*, and who, in pursuance of his orders, wintered upon this coast in port St. *Julian* and port *Desire*, in the year 1670; Sir *John Narborough*, I say, tells us, that he never saw a stick of wood in the country, large enough to make the handle of an hatchet.

But though this country be so destitute of wood, it abounds with pasture. For the land appears in general to be made up of downs of a light dry gravelly soil, and produces great quantities of long coarse grass, which grows in tufts interspersed with large barren spots of gravel between them. This grass, in many places, feeds immense herds of cattle: For the *Spaniards* at *Buenos Ayres*, having brought over a few black cattle from *Europe* at their first settlement, they have thriven prodigiously by the plenty of herbage which they found here, and are now encreased to that degree, and are extended

so

ſo far into the country, that they are not conſidered as private property; but many thouſands at a time are ſlaughtered every year by
the Hunters, only for their hides and tallow. The manner of killing theſe cattle, being a practice peculiar to that part of the world,
merits a more circumſtantial deſcription. The Hunters employed
on this occaſion being all of them mounted on horſeback, (and
both the *Spaniards* and *Indians* in that part of the world are uſually
moſt excellent horſemen) they arm themſelves with a kind of a
ſpear, which, at its end, inſtead of a blade fixed in the ſame
line with the wood in the uſual manner, has its blade fixed acroſs;
with this inſtrument they ride at a beaſt, and ſurround him. The
Hunter that comes behind him hamſtrings him; and as after this
operation the beaſt ſoon tumbles, without being able to raiſe himſelf
again, they leave him on the ground, and purſue others, whom
they ſerve in the ſame manner. Sometimes there is a ſecond party,
who attend the Hunters, to ſkin the cattle as they fall: But it is ſaid,
that at other times the Hunters chuſe to let them languiſh in tor-
ment till the next day, from an opinion that the anguiſh, which the
animal in the mean time endures, may burſt the lymphaticks, and
thereby facilitate the ſeparation of the ſkin from the carcaſs: And
though their Prieſts have loudly condemned this moſt barbarous
practice, and have gone ſo far, if my memory does not fail me,
as to excommunicate thoſe who follow it, yet all their efforts to put
an entire ſtop to it have hitherto proved ineffectual.

Beſides the numbers of cattle which are every year ſlaughtered for
their hides and tallow, in the manner already deſcribed, it is often
neceſſary for the purpoſes of agriculture, and likewiſe with other
views, to take them alive, and without wounding them: This is
performed with a moſt wonderful and almoſt incredible dexterity,
and principally by the uſe of a machine, which the *Engliſh*, who
have reſided at *Buenos Ayres*, generally denominate a laſh. It is
made of a thong of ſeveral fathoms in length, and very ſtrong,
with a running nooſe at one end of it: This the Hunters (who in
this caſe are alſo mounted on horſeback) take in their right hands,

it

it being firft properly coiled up, and having its end oppofite to the noofe faftened to the faddle; and thus prepared they ride at a herd of cattle. When they arrive within a certain diftance of a beaft, they throw their thong at him with fuch exactnefs, that they never fail of fixing the noofe about his horns. The beaft, when he finds himfelf entangled, generally runs, but the horfe, being fwifter, attends him, and prevents the thong from being too much ftrained, till a fecond Hunter, who follows the game, throws another noofe about one of its hind legs; and this being done, both horfes (they being trained for this purpofe) inftantly turn different ways, in order to ftrain the two thongs in contrary directions, on which the beaft, by their oppofite pulls, is prefently overthrown, and then the horfes ftop, keeping the thongs-ftill upon the ftretch: Being thus on the ground, and incapable of refiftance, (for he is extended between the two horfes) the Hunters alight, and fecure him in fuch a manner, that they afterwards eafily convey him to whatever place they pleafe. In the fame manner they noofe horfes, and, as it is faid, even tygers; and however ftrange this laft circumftance may appear, there are not wanting perfons of credit who affert it. Indeed, it muft be owned, that the addrefs both of the *Spaniards* and *Indians* in that part of the world, in the ufe of this lafh or noofe, and the certainty with which they throw it, and fix it on any intended part of the beaft at a confiderable diftance, are matters only to be believed, from the repeated and concurrent teftimony of all who have frequented that country, and might reafonably be queftioned, did it rely on a fingle report, or had it been ever contradicted or denied by any one who had refided at *Buenos Ayres*.

The cattle which are killed in the manner I have already obferved, are flaughtered only for their hides and tallow, to which fometimes are added their tongues, and the reft of their flefh is left to putrify, or to be devoured by the birds and wild beafts; but the greateft part of this carion falls to the fhare of the wild dogs, of which there are immenfe numbers to be found in that country. They are fuppofed to have been originally produced by *Spanifh*

dogs

dogs from *Buenos Ayres*, who, allured by the great quantity of carion, and the facility they had by that means of fubfifting, left their
Mafters, and ran wild amongft the cattle; for they are plainly of
the breed of the *European* dogs, an animal not originally found in
America. But though thefe dogs are faid to be fome thoufands in
a company, they hitherto neither diminifh nor prevent the increafe
of the cattle, not daring to attack them, by reafon of the numbers
which conftantly feed together; but contenting themfelves with the
carion left them by the Hunters, and perhaps now and then with a
few ftragglers, who, by accidents, are feparated from the herd they
belong to.

Befides the wild cattle which have fpread themfelves in fuch vaft
herds from *Buenos Ayres* towards the fouthward, the fame country
is in like manner furnifhed with horfes. Thefe too were firft
brought from *Spain*, and are alfo prodigioufly encreafed, and run
wild to a much greater diftance than the black cattle : And though
many of them are excellent, yet their number makes them of very
little value; the beft of them being often fold, in a country where
money is plenty and commodities very dear, for not more than a
dollar a-piece. It is not as yet certain how far to the fouthward
thefe herds of wild cattle and horfes have extended themfelves; but
there is fome reafon to conjecture, that ftragglers of both kinds are
to be met with very near the Streights of *Magellan*; and they will
in time doubtlefs fill the fouthern part of this Continent with their
breed, which cannot fail of proving of confiderable advantage to
fuch fhips as may touch upon the coaft; for the horfes themfelves
are faid to be very good eating, and as fuch, to be preferred by fome
of the *Indians* even before the black cattle. But whatever plenty of
this kind may be hereafter found here, there is one material refrefhment which this eaftern fide of *Patagonia* feems to be very defective in, and that is frefh water; for the land being generally of
a nitrous and faline nature, the ponds and ftreams are frequently brackifh. However, as good water has been found there, though in fmall

K 2

quantities,

quantities, it is not improbable, but on a further fearch, this inconvenience may be removed.

Befides the cattle and horfes which I have mentioned, there are in all parts of this country a good number of *Vicunnas* or *Peruvian* fheep; but thefe, by reafon of their fhynefs and fwiftnefs, are killed with difficulty. On the eaftern coaft too, there abounds immenfe quantities of feals, and a vaft variety of fea-fowl, amongft which the moft remarkable are the *Penguins*; they are in fize and fhape like a goofe, but inftead of wings they have fhort ftumps like fins, which are of no ufe to them except in the water; their bills are narrow, like that of an *Albitrofs*, and they ftand and walk in an erect pofture. From this, and their white bellies, Sir *John Narborough* has whimfically likened them to little children ftanding up in white aprons.

The inhabitants of this eaftern coaft (to which I have all along hitherto confined my relation) appear to be but few, and have rarely been feen more than two or three at a time, by any fhips that have touched here. We, during our ftay at the port of St. *Julian,* faw none. However, towards *Buenos Ayres* they are fufficiently numerous, and oftentimes very troublefome to the *Spaniards*; but there the greater breadth and variety of the country, and a milder climate, yield them a better protection; for in that place the Continent is between three and four hundred leagues in breadth, whereas at port St. *Julian* it is little more than a hundred : So that I conceive the fame *Indians*, that frequent the weftern coaft of *Patagonia* and the Streights of *Magellan*, often ramble to this fide. As the *Indians* near *Buenos Ayres* exceed thefe fouthern *Indians* in number, fo they greatly furpafs them in activity and fpirit, and feem in their manners to be nearly allied to thofe gallant *Chilian Indians*, who have long fet the whole *Spanifh* power at defiance, have often ravaged their country, and remain to this hour independent. For the *Indians* about *Buenos Ayres* have learnt to be excellent horfemen, and are extreamly expert in the management of all cutting weapons, though ignorant of the ufe of fire-arms, which

the

A

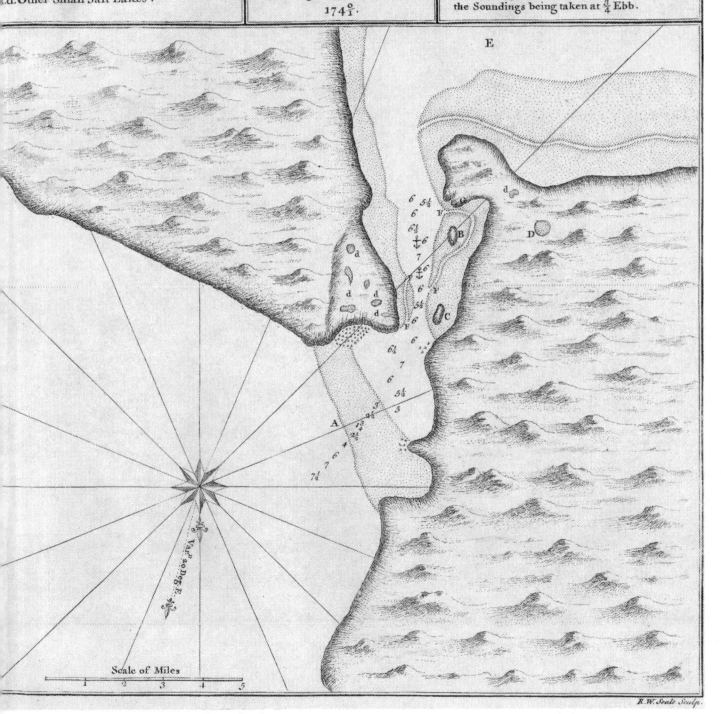

PLAN of
St. JULIAN'S HARBOUR
on the Coast of
PATAGONIA
Laying in the Lat.d of 49:30.S.º
& W.t Long.d from London 70:44.
174 0/1.

Scale of Miles

R.W. Seale Sculp.

A Prospect of St. JULIAN'S RIVER, *as it*

Plate VII.

J. Mason Sculp

...ears looking up the River at Low Water.

A prospect of PORT S.ᵗ IULIAN, *as it appears at low water, looking down t*

Plate VIII.

r, and extending from the island of TRUE IUSTICE *to the rivers mouth.*

the *Spaniards* are very folicitous to keep out of their hands. And of the vigour and refolution of thefe *Indians,* the behaviour of *O-rellana* and his followers, whom we have formerly mentioned, is a memorable inftance. Indeed were we difpofed to aim at the utter fubverfion of the *Spanifh* power in *America,* no means feem more probable to effect it, than due encouragement and affiftance given to thefe *Indians* and thofe of *Chili.*

Thus much may fuffice in relation to the eaftern coaft of *Patagonia.* The weftern coaft is of lefs extent; and by reafon of the *Andes* which fkirt it, and ftretch quite down to the water, is a very rocky and dangerous fhore. However, I fhall be hereafter neceffitated to make further mention of it, and therefore fhall not enlarge thereon at this time, but fhall conclude this account with a fhort defcription of the harbour of St. *Julian,* the general form of which may be conceived from the annexed fketch. But it muft be remembered, that the bar, which is there marked at the entrance, is often fhifting and has many holes in it. The tide flows here N. and S, and at full and change, rifes four fathom.

We, on our firft arrival here, fent an Officer on fhore to the falt-pond, marked (D) in the plan, in order to procure a quantity of falt for the ufe of the Squadron, Sir *John Narborough* having obferved, when he was here, that the falt produced in that place was very white and good, and that in *February* there was enough of it to fill a thoufand fhips; but our Officer returned with a fample which was very bad, and he told us, that even of this there was but little to be got; I fuppofe the weather had been more rainy than ordinary, and had deftroyed it. To give the reader a better idea of this port, and of the adjacent country, to which the whole coaft I have defcribed bears a great refemblance, I have inferted two very accurate views, one of them reprefenting the appearance of the country, when looking up the river; the other, being a view taken from the fame fpot, but the obferver is now fuppofed to turn round oppofite to his former fituation, and confequently this is a reprefentation of the appearance of the country down the river, betwixt the ftation of the obferver, and the river's mouth.   CHAP

C H A P. VII.

Departure from the bay of St. *Julian*, and the paffage from thence to Streights *Le Maire*.

THE *Tryal* being nearly refitted, which was our principal occupation at this bay of St. *Julian*, and the fole occafion of our ftay, the Commodore thought it neceffary, as we were now directly bound for the *South-Seas* and the enemy's coafts, to regulate the plan of his future operations : And therefore, on the 24th of *February*, a fignal was made for all Captains, and a Council of war was held on board the *Centurion*, at which were prefent the Honourable *Edward Legg*, Captain *Matthew Mitchel*, the Honourable *George Murray*, Captain *David Cheap*, together with Colonel *Mordaunt Cracherode*, Commander of the land-forces. At this Council Mr. *Anfon* propofed, that their firft attempt, after their arrival in the *South-Seas*, fhould be the attack of the town and harbour of *Baldivia*, the principal frontier of the diftrict of *Chili*; Mr. *Anfon* informing them, at the fame time, that it was an article contained in his Majefty's inftructions to him, to endeavour to fecure fome port in the *South-Seas*, where the fhips of the fquadron might be careened and refitted. To this propofition made by the Commodore, the Council unanimoufly and readily agreed ; and in confequence of this refolution, new inftructions were given to the Captains of the fquadron, by which, though they were ftill directed, in cafe of feparation, to make the beft of their way to the Ifland of *Nueftra Senora del Socoro*, yet (notwithftanding the orders they had formerly given them at St. *Catherine's*) they were to cruize off that Ifland only ten days ; from whence, if not joined by the Commodore, they were to proceed, and cruize off the harbour of *Baldivia*, making the land between the latitudes of 40°, and 40°: 30 ; and taking care to keep to the fouthward of the port ; and, if

I   in in fourteen days they were not joined by the reft of the fquadron, they were then to quit this ftation, and to direct their courfe to the Ifland of *Juan Fernandes*, after which they were to regulate their further proceedings by their former orders. The fame directions were alfo given to the Mafter of the *Anna Pink*, and he was particularly inftructed to be very careful in anfwering the fignals made by any fhip of the fquadron, and likewife to deftroy his papers and orders, if he fhould be fo unfortunate, as to fall into the hands of the enemy. And as the feparation of the fquadron might prove of the utmoft prejudice to his Majefty's fervice, each Captain was ordered to give it in charge to the refpective Officers of the watch, not to keep their fhip at a greater diftance from the *Centurion* than two miles, as they would anfwer it at their peril ; and if any Captain fhould find his fhip beyond the diftance fpecified, he was to acquaint the Commodore with the name of the Officer, who had thus neglected his duty.

Thefe neceffary regulations being eftablifhed, and the *Tryal* Sloop compleated, the fquadron weighed on *Friday* the 27th of *February*, at feven in the morning, and ftood to the fea; the *Gloucefter* indeed found a difficulty in purchafing her anchor, and was left a confiderable way a-ftern, fo that in the night we fired feveral guns as a fignal to her Captain to make fail, but he did not come up to us till the next morning, when we found that they had been obliged to cut their cable, and leave their beft bower behind them. At ten in the morning, the day after our departure, *Wood's Mount*, the highland over St. *Julian*, bore from us N. by W. diftant ten leagues, and we had fifty-two fathom of water. And now ftanding to the fouthward, we had great expectation of falling in with *Pizarro*'s fquadron; for, during our ftay at port St. *Julian*, there had generally been hard gales between the W. N. W. and S. W, fo that we had reafon to conclude the *Spaniards* had gained no ground upon us in that interval. And it was the profpect of meeting with them, that had occafioned our Commodore to be fo very folicitous to prevent the feparation of our fhips : For had we

been

been folely intent on getting round Cape *Horn* in the fhorteft time, the propereft method for this purpofe would have been, to have ordered each fhip to have made the beft of her way to the rendezvous, without waiting for the reft.

From our departure from St. *Julian* to the 4th of *March*, we had little wind, with thick hazy weather, and fome rain ; and our foundings were generally from forty to fifty fathom, with a bottom of black and grey fand, fometimes intermixed with pebble ftones. On the 4th of *March* we were in fight of cape *Virgin Mary*, and not more than fix or feven leagues diftant from it : This is the northern cape of the Streights of *Magellan*, it lies in the latitude of 52° : 21′ South, and longitude from *London* 71° : 44′ Weft, and feems to be a low flat land, ending in a point. And for a direction to fuch fhips as may, by particular reafons, be induced hereafter to pafs through thofe Streights into the *South-Seas*, I have annexed a very accurate draught of its appearance, where (*a*) reprefents the Cape itfelf. Off this Cape our depth of water was from thirty-five to forty-eight fathom. The afternoon of this day was very bright and clear, with fmall breezes of wind, inclinable to a calm, and moft of the Captains took the opportunity of this favourable weather to pay a vifit to the Commodore ; but while they were in company together, they were all greatly alarmed by a fudden flame, which burft out on board the *Gloucefter*, and which was fucceeded by a cloud of fmoak. However, they were foon relieved from their apprehenfions, by receiving information, that the blaft was occafioned by a fpark of fire from the forge, lighting on fome gunpowder and other combuftibles, which an Officer on board was preparing for ufe, in cafe we fhould fall in with the *Spanifh* fleet ; and that it had been extinguifhed, without any damage to the fhip.

We here found what was conftantly verified by all our obfervations in thefe high latitudes, that fair weather was always of an exceeding fhort duration, and that when it was remarkably fine, it was a certain prefage of a fucceeding ftorm, for the calm and

funfhine

Cape VIRGIN MARY *at the north entran*

MAGELLANS STREIGHTS.

A view of part of the N. E. side of TERRA DE

Plate X.

JEGO *and the entrance of streights* LE MAIRE.

sunshine of our afternoon ended in a most turbulent night, the
wind freshning from the S. W. as the night came on, and en-
creasing its violence continually till nine in the morning the next
day, when it blew so hard, that we were obliged to bring to with
the squadron, and to continue under a reefed mizen till eleven at
night, having in that time from forty-three to fifty-seven fathom
water, with black sand and gravel; and by an observation we had
at noon, we concluded a current had set us twelve miles to the
southward of our reckoning. Towards midnight, the wind abating,
we made sail again; and steering South, we discovered in the
morning for the first time the land, called *Terra del Fuego*, stretch-
ing from the S. by W, to the S. E. $\frac{1}{2}$ E. This indeed afforded
us but a very uncomfortable prospect, it appearing of a stupendous
height, covered every where with snow. And though the dreari-
ness of this scene can be but imperfectly represented by any Drawing, yet the annexed plate contains so exact a delineation of the
form of the country, that it may greatly assist the reader in fram-
ing some idea of this uncouth and rugged coast. In this Drawing
(*a*) is the opening of Streights *Le Maire*, (*b*) Cape St. *Diego*,
(1) (2) (3) the three hills, called the three brothers, and (4) *Montegorda*, an highland which lies up in the country, and appears over
the three brothers. We steered along this shore all day, having
soundings from forty to fifty fathom, with stones and gravel. And
as we intended to pass through Streights *Le Maire* next day, we
lay to at night, that we might not overshoot them, and took this
opportunity to prepare ourselves for the tempestuous climate we
were soon to be engaged in; with which view, we employed
ourselves good part of the night in bending an entire new suit of
sails to the yards. At four the next morning, being the 7th of
March, we made sail, and at eight saw the land; and soon after
we began to open the Streights, at which time Cape St. *James* bore
from us E. S. E, Cape St. *Vincent* S. E. $\frac{1}{2}$ E, the middlemost of
the three brothers S. and by W, *Montegorda* South, and Cape St.
Bartholomew, which is the southermost point of *Staten-land*, E. S. E.

L   The

The appearance of the Streights in this situation, is reprefented in the annexed plate, where (*a*) is part of *Staten-land,* (*b*) Cape St. *Bartholomew,* (*c*) part of *Terra del Fuego,* (*d*) port *Maurice,* and (*e*) fuppofed to be *Valentine*'s bay, or the bay of good fuccefs. And here I muft obferve, that *Frezier* has given us a very correct profpect of the part of *Terra del Fuego,* which borders on the Streights, but has omitted that of *Staten-land,* which forms the oppofite fhore: Hence we found it difficult to determine exactly where the Streights lay, till they began to open to our view; and for want of this, if we had not happened to have coafted a confiderable way along fhore, we might have miffed the Streights, and have got to the eaftward of *Staten-land* before we knew it. This is an accident that has happened to many fhips, particularly, as *Frezier* mentions, to the *Incarnation* and *Concord*; who intending to pafs through Streights *Le Maire,* were deceived by three hills on *Staten land* like the three brothers, and fome creeks refembling thofe of *Terra del Fuego,* and thereby over-fhot the Streights. To prevent thefe accidents for the future, there is inferted the Weft profpect of *Staten-land,* where (*a*) is Cape St. *Diego,* on *Terra del Fuego,* (*b*) Cape St. *Bartholomew,* on *Staten-land.* This Drawing will hereafter render it impoffible for any fhips to be deceived in the manner abovementioned, or to find any difficulty in diftinguifhing the points of land by which the Streights are formed.

And on occafion of this profpect of *Staten-land* here inferted, I cannot but remark, that though *Terra del Fuego* had an afpect extremely barren and defolate, yet this Ifland of *Staten-land* far furpaffes it, in the wildnefs and horror of its appearance: It feeming to be entirely compofed of inacceffible rocks, without the leaft mixture of earth or mold between them. Thefe rocks terminate in a vaft number of ragged points, which fpire up to a prodigious height, and are all of them covered with everlafting fnow; the points themfelves are on every fide furrounded with frightful precipices, and often overhang in a moft aftonifhing manner; and the hills which bear them, are generally feparated from each other by

narrow

A View of Streights LE MAIRE *between*

Plate XI.

J. Moson Sculp.

ERRA DEL FUEGO *and* STATEN LAND.

The west prospect

Plate XII.

STATEN ISLAND.

narrow clefts, which appear as if the country had been rent by earthquakes ; for thefe chafms are nearly perpendicular, and extend through the fubftance of the main rocks, almoft to their very bottoms : So that nothing can be imagined more favage and gloomy, than the whole afpect of this coaft.   But to proceed,

I have above-mentioned, that on the 7th of *March*, in the morning, we opened Streights *Le Maire*, and foon after, or about ten o'clock, the *Pearl* and the *Tryal* being ordered to keep a-head of the fquadron, we entered them with fair weather and a brifk gale, and were hurried through by the rapidity of the tide in about two hours, though they are between feven and eight leagues in length.   As thefe Streights are often confidered as the boundary between the *Atlantick* and *Pacifick* Oceans, and as we prefumed we had nothing now before us but an open fea, till we arrived on thofe opulent coafts where all our hopes and wifhes centered, we could not help flattering ourfelves, that the greateft difficulty of our paffage was now at an end, and that our moft fanguine dreams were upon the point of being realifed ; and hence we indulged our imaginations in thofe romantick fchemes, which the fancied poffeffion of the *Chilian* gold and *Peruvian* filver might be conceived to infpire.   Thefe joyous ideas were heightened by the brightnefs of the fky, and the ferenity of the weather, which was indeed moft remarkably pleafing ; for tho' the winter was now advancing apace, yet the morning of this day, in its brilliancy and mildnefs, gave place to none we had feen fince our departure from *England*.   Thus animated by thefe delufions, we travers'd thefe memorable Streights, ignorant of the dreadful calamities that were then impending, and juft ready to break upon us ; ignorant that the time drew near, when the fquadron would be feparated never to unite again, and that this day of our paffage was the laft chearful day that the greateft part of us would ever live to enjoy.

CHAP.

CHAP. VIII.

From Streights *Le Maire* to Cape *Noir*.

WE had fcarcely reached the fouthern extremity of the
Streights of *Le Maire*, when our flattering hopes were inftantly loft in the apprehenfions of immediate deftruction:
For before the fternmoft fhips of the fquadron were clear of the
Streights, the ferenity of the fky was fuddenly changed, and gave
us all the prefages of an impending ftorm ; and immediately the
wind fhifted to the fouthward, and blew in fuch violent fqualls,
that we were obliged to hand our top-fails, and reef our main-fail :
The tide too, which had hitherto favoured us, now turned againft
us, and drove us to the eaftward with prodigious rapidity, fo that
we were in great anxiety for the *Wager* and the *Anna Pink*, the
two fternmoft veffels, fearing they would be dafhed to pieces againft
the fhore of *Staten-land*; nor were our apprehenfions without
foundation, for it was with the utmoft difficulty they efcaped. And
now the whole fquadron, inftead of purfuing their intended courfe to
the S. W, were driven to the eaftward by the united force of the
ftorm, and of the currents ; fo that next day in the morning we
found ourfelves near feven leagues to the eaftward of *Staten-land*,
which then bore from us N. W. The violence of the current,
which had fet us with fo much precipitation to the eaftward, together with the force and conftancy of the wefterly winds, foon
taught us to confider the doubling of Cape *Horn* as an enterprize,
that might prove too mighty for our efforts, though fome amongft
us had lately treated the difficulties which former voyagers were faid
to have met with in this undertaking, as little better than chimerical, and had fuppofed them to arife rather from timidity and unfkilfulnefs, than from the real embarraffments of the winds and

feas ;

feas; but we were now feverely convinced, that thefe cenfures were rafh and ill-grounded : For the diftreffes with which we ftruggled, during the three fucceeding months, will not eafily be paralleled in the relation of any former naval expedition. This will, I doubt not, be readily allowed by thofe who fhall carefully perufe the enfuing narration.

From the ftorm which came on before we had well got clear of Streights *Le Maire*, we had a continual fucceffion of fuch tempeftuous weather, as furprized the oldeft and moft experienced Mariners on board, and obliged them to confefs, that what they had hitherto called ftorms were inconfiderable gales, compared with the violence of thefe winds, which raifed fuch fhort, and at the fame time fuch mountainous waves, as greatly furpaffed in danger all feas known in any other part of the globe: And it was not without great reafon, that this unufual appearance filled us with continual terror ; for had any one of thefe waves broke fairly over us, it muft, in all probability, have fent us to the bottom. Nor did we efcape with terror only ; for the fhip rolling inceffantly gunwale to, gave us fuch quick and violent motions, that the men were in perpetual danger of being dafhed to pieces againft the decks, or fides of the fhip. And though we were extremely careful to fecure ourfelves from thefe fhocks, by grafping fome fixed body, yet many of our people were forced from their hold ; fome of whom were killed, and others greatly injured ; in particular, one of our beft feamen was canted over-board and drowned, another diflocated his neck, a third was thrown into the main-hold and broke his thigh, and one of our Boatfwain's Mates broke his collar-bone twice ; not to mention many other accidents of the fame kind. Thefe tempefts, fo dreadful in themfelves, though unattended by any other unfavourable circumftance, were yet rendered more mifchievous to us by their inequality, and the deceitful intervals which they at fome-times afforded ; for though we were oftentimes obliged to lie to for days together under a reefed mizen, and were fometimes reduced to lie at the mercy of the waves under our bare poles, yet

now and then we ventured to make fail with our courfes double
reefed; and the weather proving more tolerable, would perhaps encourage us to fet our top-fails; after which, the wind, without any
previous notice, would return upon us with redoubled force, and
would in an inftant tear our fails from the yards. And that no circumftance might be wanting which could aggrandize our diftrefs,
thefe blafts generally brought with them a great quantity of fnow
and fleet, which cafed our rigging, and froze our fails, thereby
rendring them and our cordage brittle, and apt to fnap upon the
flighteft ftrain, adding great difficulty and labour to the working of
the fhip, benumbing the limbs of our people, and making them
incapable of exerting themfelves with their ufual activity, and even
difabling many of them, by mortifying their toes and fingers. It
were indeed endlefs to enumerate the various difafters of different
kinds which befel us; and I fhall only mention the moft material,
which will fufficiently evince the calamitous condition of the whole
fquadron, during the courfe of this navigation.

It was on the 7th of *March*, as hath been already obferved, that
we paffed Streights *Le Maire*, and were immediately afterwards
driven to the eaftward by a violent ftorm, and the force of the current which fet that way. For the four or five fucceeding days we
had hard gales of wind from the fame quarter, with a moft prodigious fwell; fo that though we ftood, during all that time, towards
the S. W, yet we had no reafon to imagine, we had made any
way to the weftward. In this interval we had frequent fqualls of
rain and fnow, and fhipped great quantities of water; after which,
for three or four days, though the feas ran mountains high, yet
the weather was rather more moderate: But, on the 18th, we had
again ftrong gales of wind with extreme cold, and at midnight the
main top-fail fplit, and one of the ftraps of the main dead eyes
broke. From hence, to the 23d, the weather was more favourable,
though often intermixed with rain and fleet, and fome hard gales;
but as the waves did not fubfide, the fhip, by labouring in this lofty
fea, was now grown fo loofe in her upper works, that fhe let in the

water

water at every feam, fo that every part within board was conftantly expofed to the fea-water, and fcarcely any of the Officers ever lay in dry beds. Indeed it was very rare, that two nights ever paffed without many of them being driven from their beds, by the deluge of water that came upon them.

On the 23d, we had a moft violent ftorm of wind, hail, and rain, with a very great fea; and though we handed the main top-fail before the height of the fquall, yet we found the yard fprung; and foon after the foot-rope of the main-fail breaking, the main-fail itfelf fplit inftantly to rags, and, in fpite of our endeavours to fave it, much the greater part of it was blown over-board. On this, the Commodore made the fignal for the fquadron to bring to; and the ftorm at length flattening to a calm, we had an opportunity of getting down our main top-fail yard to put the Carpenters at work upon it, and of repairing our rigging; after which, having bent a new mainfail, we got under fail again with a moderate breeze; but in lefs than twenty-four hours we were attacked by another ftorm ftill more furious than the former; for it proved a perfect hurricane, and reduced us to the neceffity of lying to under our bare poles. As our fhip kept the wind better than any of the reft, we were obliged, in the afternoon, to wear fhip, in order to join the fquadron to the leeward, which otherwife we fhould have been in danger of lofing in the night: And as we dared not venture any fail abroad, we were obliged to make ufe of an expedient, which anfwered our purpofe; this was putting the helm a weather, and manning the fore-fhrouds: But though this method proved fuc-cefsful for the end intended, yet in the execution of it, one of our ableft feaman was canted over-board; and notwithftanding the pro-digious agitation of the waves, we perceived that he fwam very ftrong, and it was with the utmoft concern that we found ourfelves incapable of affifting him; and we were the more grieved at his unhappy fate, fince we loft fight of him ftruggling with the waves, and conceived from the manner in which he fwam, that he might

continue

continue fenfible for a confiderable time longer, of the horror at
tending his irretrievable fituation.

Before this laft mentioned ftorm was quite abated, we found two
of our main-fhrouds and one mizen-fhroud broke, all which we
knotted, and fet up immediately; and from hence we had an interval of three or four days lefs tempeftuous than ufual, but accom-
panied with a thick fog, in which we were obliged to fire
guns almoft every half hour, to keep our fquadron together. On
the 31ft, we were alarmed by a gun fired from the *Gloucefter*, and
a fignal made by her to fpeak with the Commodore; we immediately bore down to her, and were prepared to hear of fome terrible difafter; but we were apprized of it before we joined her, for we
faw that her main-yard was broke in the flings. This was a grie-
vous misfortune to us all at this juncture; as it was obvious it would
prove an hindrance to our failing, and would detain us the longer in
thefe inhofpitable latitudes. But our future fuccefs and fafety was
not to be promoted by repining, but by refolution and activity; and
therefore, that this unlucky incident might delay us as little as
poffible, the Commodore ordered feveral Carpenters to be put on
board the *Gloucefter* from the other fhips of the fquadron, in order
to repair her damage with the utmoft expedition. And the Captain of the *Tryal* complaining at the fame time, that his pumps
were fo bad, and the floop made fo great a quantity of water, that
he was fcarcely able to keep her free, the Commodore ordered him
a pump ready fitted from his own fhip. It was very fortunate for
the *Gloucefter* and the *Tryal*, that the weather proved more favourable this day than for many days, both before and after; fince by
this means they were enabled to receive the affiftance which feemed
effential to their prefervation, and which they could fcarcely have
had at any other time, as it would have been extremely hazardous
to have ventured a boat on board.

The next day, that is, on the 1ft of *April*, the weather returned again to its cuftomary bias, the fky looked dark and gloomy,

and

and the wind began to freſhen and to blow in ſqualls; however, it was not yet ſo boiſterous, as to prevent our carrying our top-ſails cloſe reefed; but its appearance was ſuch, as plainly prognoſticated that a ſtill ſeverer tempeſt was at hand : And accordingly, on the 3d of *April,* there came on a ſtorm, which both in its violence and continuation (for it laſted three days) exceeded all that we had hitherto encountered. In its firſt onſet we received a furious ſhock from a ſea which broke upon our larboard quarter, where it ſtove in the quarter gallery, and ruſhed into the ſhip like a deluge ; our rigging too ſuffered extremely, for one of the ſtraps of the main dead-eyes was broke, as was alſo a main-ſhroud and puttock-ſhroud, ſo that to eaſe the ſtreſs upon the maſts and ſhrouds, we lowered both our main and fore-yards, and furled all our ſails, and in this poſture we lay to for three days, when the ſtorm ſomewhat abating, we ventured to make ſail under our courſes only ; but even this we could not do long, for, the next day, which was the 7th, we had another hard gale of wind, with lightning and rain, which obliged us to lie to again till night. It was wonderful, that notwithſtanding the hard weather we had endured, no extraordinary accident had happened to any of the ſquadron ſince the breaking of the *Glouceſter's* main-yard: But this wonder ſoon ceaſed ; for at three the next morning, ſeveral guns were fired to leeward as ſignals of diſtreſs. And the Commodore making a ſignal for the ſquadron to bring to, we, at day-break, ſaw the *Wager* a conſiderable way to leeward of any of the other ſhips; and we ſoon perceived that ſhe had loſt her mizen-maſt, and main top-ſail yard. We immediately bore down to her, and found this diſaſter had ariſen from the badneſs of her iron work ; for all the chain-plates to windward had given way, upon the ſhip's fetching a deep roll. This proved the more unfortunate to the *Wager,* as her Carpenter had been on board the *Glouceſter* ever ſince the 31ſt of *March,* and the weather was now too ſevere to permit him to return : Nor was the *Wager* the only ſhip of the ſquadron that had ſuffered in the late tempeſt ; for, the next day, a ſignal of diſtreſs was made

M   by

by the *Anna Pink*, and, upon fpeaking with the Mafter, we learnt that they had broke their fore-ftay and the gammon of the bow-fprit, and were in no fmall danger of having all the mafts come by the board ; fo that we were obliged to bear away until they had made all faft, after which we haled upon a wind again.

And now, after all our folicitude, and the numerous ills of every kind, to which we had been inceffantly expofed for near forty days, we had great confolation in the flattering hopes we entertained, that our fatigues were drawing to a period, and that we fhould foon arrive in a more hofpitable climate, where we fhould be amply re-payed for all our paft fufferings. For, towards the latter end of *March*, we were advanced, by our reckoning, near 10° to the weftward of the weftermoft point of *Terra del Fuego*, and this al-lowance being double what former Navigators have thought ne-ceffary to be taken, in order to compenfate the drift of the eaftern current, we efteemed ourfelves to be well advanced within the limits of the fouthern Ocean, and had therefore been ever fince ftanding to the northward with as much expedition, as the turbu-lence of the weather, and our frequent difafters permitted. And, on the 13th of *April*, we were but a degree in latitude to the fouthward of the Weft entrance of the Streights of *Magellan*; fo that we fully expected, in a very few days, to have experienced the celebrated tranquility of the *Pacifick* Ocean.

But thefe were delufions which only ferved to render our dif-appointment more terrible ; for the next morning, between one and two, as we were ftanding to the northward, and the weather, which had till then been hazy, accidentally cleared up, the *Pink* made a fignal for feeing land right a head ; and it being but two miles diftant, we were all under the moft dreadful apprehenfions of running on fhore ; which, had either the wind blown from its ufual quarter with its wonted vigour, or had not the moon fud-denly fhone out, not a fhip amongft us could poffibly have avoided : But the wind, which fome few hours before blew in fqualls from the S. W, having fortunately fhifted to W. N. W, we were enabled

to

to ſtand to the ſouthward, and to clear ourſelves of this unexpected danger; ſo that by noon we had gained an offing of near
twenty leagues.

By the latitude of this land we fell in with, it was agreed to be
a part of *Terra del Fuego*, near the ſouthern outlet deſcribed in
Frezier's Chart of the Streights of *Magellan*, and was ſuppoſed to
be that point called by him Cape *Noir*. It was indeed moſt wonderful, that the currents ſhould have driven us to the eaſtward with
ſuch ſtrength; for the whole ſquadron eſteemed themſelves upwards of ten degrees more weſterly than this land, ſo that in running
down, by our account, about nineteen degrees of longitude, we
had not really advanced above half that diſtance. And now, inſtead of having our labours and anxieties relieved by approaching a
warmer climate and more tranquil ſeas, we were to ſteer again to
the ſouthward, and were again to combat thoſe weſtern blaſts,
which had ſo often terrified us; and this too, when we were weakned by our men falling ſick, and dying apace, and when our ſpirits,
dejected by a long continuance at ſea, and by our late diſappointment, were much leſs capable of ſupporting us in the various difficulties, which we could not but expect in this new undertaking.
Add to all this too, the diſcouragement we received by the diminution of the ſtrength of the ſquadron; for three days before this, we
loſt ſight of the *Severn* and the *Pearl* in the morning; and though
we ſpread our ſhips, and beat about for them ſome time, yet we never ſaw them more; whence we had apprehenſions that they too
might have fallen in with this land in the night, and by being leſs
favoured by the wind and the moon than we were, might have run
on ſhore and have periſhed. Full of theſe dejected thoughts and
gloomy preſages, we ſtood away to the S. W, prepared by our late
diſaſter to ſuſpect, that how large ſoever an allowance we made in
our weſting for the drift of the eaſtern current, we might ſtill,
upon a ſecond trial, perhaps find it inſufficient.

CHAP.

C H A P. IX.

Obſervations and directions for facilitating the paſſage of our future Cruiſers round Cape *Horn.*

THE improper ſeaſon of the year in which we attempted to double Cape *Horn,* and to which is to be imputed the diſappointment (recited in the foregoing chapter) in falling in with *Terra del Fuego,* when we reckoned ourſelves at leaſt a hundred leagues to the weſtward of that whole coaſt, and conſequently well advanced into the *Pacifick* Ocean ; this unſeaſonable navigation, I ſay, to which we were neceſſitated by our too late departure from *England,* was the fatal ſource of all the misfortunes we afterwards encountered. For from hence proceeded the ſeparation of our ſhips, the deſtruction of our people, the ruin of our project on *Baldivia,* and of all our other views on the *Spaniſh* places, and the reduction of our ſquadron from the formidable condition in which it paſſed Streights *Le Maire,* to a couple of ſhattered half manned cruiſers and a ſloop, ſo far diſabled, that in many climates they ſcarcely durſt have put to ſea. To prevent therefore, as much as in me lies, all ſhips hereafter bound to the *South-Seas* from ſuffering the ſame calamities, I think it my duty to inſert in this place, ſuch directions and obſervations, as either my own experience and reflection, or the converſe of the moſt ſkilful Navigators on board the ſquadron could furniſh me with, in relation to the moſt eligible manner of doubling Cape *Horn,* whether in regard to the ſeaſon of the year, the courſe proper to be ſteered, or the places of refreſhment both on the Eaſt and Weſt-ſide of *South America.*

And firſt with regard to the proper place for refreſhment on the Eaſt-ſide of *South America.* For this purpoſe the Iſland of St. *Catherine's* has been uſually recommended by former writers, and on

their

their faith we put in there, as has been formerly mentioned: But the treatment we met with, and the small store of refreshments we could procure there, are sufficient reasons to render all ships for the future cautious, how they trust themselves in the government of *Don Jose Silva de Paz*; for they may certainly depend on having their strength, condition and designs betrayed to the *Spaniards*, as far as the knowledge, the Governor can procure of these particulars, will give leave. And as this treacherous conduct is inspired by the views of private gain, in the illicit commerce carried on to the river of *Plate*, rather than by any national affection which the *Portuguese* bear the *Spaniards*, the same perfidy may perhaps be expected from most of the Governors of the *Brazil* coast; since these smuggling engagements are doubtless very extensive and general. And though the Governors should themselves detest so faithless a procedure, yet as ships are perpetually passing from some or other of the *Brazil* ports to the river of *Plate*, the *Spaniards* could scarcely fail of receiving, by this means, casual intelligence of any *British* ships upon the coast; which, however imperfect such intelligence might be, would prove of dangerous import to the views and interests of those cruisers who were thus discovered.

For the *Spanish* trade in the *South-Seas* running all in one track from North to South, with very little deviation to the eastward or westward, it is in the power of two or three cruisers, properly stationed in different parts of this track, to possess themselves of every ship that puts to sea: But this is only so long as they can continue concealed from the neighbouring coast; for the instant an enemy is known to be in those seas, all navigation is stopped, and consequently all captures are at an end; since the *Spaniards*, well apprized of these advantages of the enemy, send expresses along the coast, and lay a general embargo on all their trade; a measure, which they prudentially foresee will not only prevent their vessels being taken, but will soon lay any cruisers, who have not strength sufficient to attempt their places, under a necessity of returning home. Hence then appears the great importance of concealing all

expeditions

expeditions of this kind; and hence too it follows, how extremely prejudicial that intelligence may prove, which is given by the *Portuguese* Governors to the *Spaniards*, in relation to the defigns of fhips touching at the ports of *Brazil*.

However, notwithftanding the inconveniencies we have mentioned of touching on the coaft of *Brazil*, it will oftentimes happen, that fhips bound round Cape *Horn* will be obliged to call there for a fupply of wood and water, and other refrefhments. In this cafe St. *Catherine's* is the laft place I would recommend, both as the proper animals for a live ftock at fea, as hogs, fheep and fowls cannot be procured there, (for want of which we found ourfelves greatly diftreffed, by being reduced to live almoft entirely on falt provifions) but alfo becaufe from its being nearer the river of *Plate* than many of their other fettlements, the inducements and conveniencies of betraying us are much ftronger. The place I would recommend is *Rio Janeiro*, where two of our fquadron put in after they were feparated from us in paffing Cape *Horn*; for here, as I have been informed by one of the Gentlemen on board thofe fhips, any quantity of hogs and poultry may be procured, and this place being more diftant from the river of *Plate*, the difficulty of intelligence is fomewhat inhanced, and confequently the chance of continuing there undifcovered, in fome degree augmented. Other meafures, which may effectually obviate all thefe embarraffments, will be confidered more at large hereafter.

And now I proced to the confideration of the proper courfe to be fteered for doubling Cape *Horn*. And here, I think, I am fufficiently authorifed by our own fatal experience, and by a careful comparifon and examination of the journals of former Navigators, to give this piece of advice, which in prudence I think ought never to be departed from: That is, that all fhips bound to the *South-Seas*, inftead of paffing through Streights *le Maire*, fhould conftantly pafs to the eaftward of *Staten-land*, and fhould be invariably bent on running to the fouthward, as far as the latitude of 61 or 62 degrees, before they endeavour to ftand to the weftward; and that when they are

got

got into that latitude, they fhould then make fure of fufficien wefting, before they once think of fteering to the northward.

But as directions diametrically oppofite to thefe have been formerly given by other writers, it is incumbent on me to produce my reafons for each part of this maxim. And firft, as to the paffing to the eaftward of *Staten-land.* Thofe who have attended to the rifque we ran in paffing Streights *Le Maire,* the danger we were in of being driven upon *Staten-land* by the current, when, though we happily efcaped being put on fhore, we were yet carried to the eaftward of that Ifland : Thofe who reflect on this, and on the like accidents which have happened to other fhips, will furely not efteem it prudent to pafs through Streights *Le Maire,* and run the rifque of fhipwreck, and after all find themfelves no farther to the weftward (the only reafon hitherto given for this practice) than they might have been in the fame time, by a fecure navigation in an open fea.

And next, as to the directions I have given for running into the latitude of 61 or 62 South, before any endeavour is made to ftand to the weftward. The reafons for this precept are, that in all probability the violence of the currents will be hereby avoided, and the weather will prove lefs tempeftuous and uncertain. This laft circumftance we ourfelves experienced moft remarkably ; for after we had unexpectedly fallen in with the land, as has been mentioned in the preceding chapter, we ftood away to the fouthward to run clear of it, and were no fooner advanced into fixty degrees or upwards, but we met with much better weather, and fmoother water than in any other part of the whole paffage : The air indeed was very cold and fharp, and we had ftrong gales, but they were fteady and uniform, and we had at the fame time funfhine and a clear fky ; whereas in the lower latitudes, the winds every now and then intermitted, as it were, to recover new ftrength, and then returned fuddenly in the moft violent gufts, threatening at each blaft the lofs of our mafts, which muft have ended in our certain deftruction. And that the currents in this high latitude would be of much lefs ef-

ficacy

ficacy than nearer the land, feems to be evinced from thefe confiderations, that all currents run with greater violence near the fhore than at fea, and that at greater diftances from fhore they are fcarcely perceptible : Indeed the reafon of this feems fufficiently obvious, if we confider, that conftant currents are, in all probability, produced by conftant winds, the wind driving before it, though with a flow and imperceptible motion, a large body of water, which being accumulated upon any coaft that it meets with, this fuperfluous water muft efcape along the fhore by the endeavours of its furface, to reduce itfelf to the fame level with the reft of the Ocean. And it is reafonable to fuppofe, that thofe violent gufts of wind which we experienced near the fhore, fo very different from what we found in the latitude of fixty degrees and upwards, may be owing to a fimilar caufe ; for a wefterly wind almoft perpetually prevails in the fouthern part of the *Pacifick* Ocean: And this current of air being interrupted by thofe immenfe hills called the *Andes*, and by the mountains on *Terra del Fuego*, which together bar up the whole country to the fouthward as far as Cape *Horn*, a part of it only can efcape over the tops of thofe prodigious precipices, and the reft muft naturally follow the direction of the coaft, and muft range down the land to the fouthward, and fweep with an impetuous and irregular blaft round Cape *Horn*, and the fouthermoft part of *Terra del Fuego*. However, not to rely on thefe fpeculations, we may, I believe, eftablifh, as inconteftable, thefe matters of fact, that both the rapidity of the currents, and the violence of the weftern gales, are lefs fenfible in the latitude of 61 or 62 degrees, than nearer the fhore of *Terra del Fuego*.

But though I am fatisfied both from our own experience, and the relations of other Navigators, of the importance of the precept I here infift on, that of running into the latitude of 61 or 62 degrees, before any endeavours are made to ftand to the weftward ; yet I would advife no fhips hereafter to truft fo far to this management, as to neglect another moft effential maxim, which is the making this paffage in the heighth of fummer, that is, in the months of *December*

and

and *January*; and the more diftant the time of paffing is taken from this feafon, the more difaftrous it may be reafonably expected to prove. Indeed, if the mere violence of the weftern winds be confidered, the time of our paffage, which was about the Equinox, was perhaps the moft unfavourable feafon; but then it muft be confidered, that in the depth of winter there are many other inconveniencies to be apprehended in this navigation, which are almoft infuperable: For the feverity of the cold, and the fhortnefs of the days, would render it impracticable at that feafon to run fo far to the fouthward as is here recommended; and the fame reafons would greatly augment the alarms of failing in the neighbourhood of an unknown fhore, dreadful in its appearance in the midft of fummer, and would make a winter navigation on this coaft to be, of all others, the moft difmaying and terrible. As I would therefore advife all fhips to make their paffage in *December* and *January*, if poffible, fo I would warn them never to attempt the feas to the fouthward of Cape *Horn*, after the month of *March*.

And now as to the remaining confideration, that is, the propereft port for cruifers to refrefh at on their firft arrival in the *South-Seas*. On this head there is fcarcely any choice, the Ifland of *Juan Fernandes* being the only place that can be prudently recommended for this purpofe. For though there are many ports on the weftern fide of *Patagonia*, between the Streights of *Magellan* and the *Spanifh* fettlements (a plan of one of which I fhall infert in the courfe of this work) where fhips might ride in great fafety, might recruit their wood and water, and might procure fome few refrefhments; yet that coaft is in itfelf fo terrible, from the rocks and breakers it abounds with, and from the violence of the weftern winds, which blow conftantly full upon it, that it is by no means advifeable to fall in with that land, at leaft till the roads, channels and anchorage in each part of it are accurately furveyed, and both the dangers and fhelter it abounds with are more diftinctly known.

N   Thus

Thus having given the beſt directions in my power for the ſucceſs of future cruiſers bound to the *South-Seas,* it might be expected that I ſhould again reſume the thread of my narration. But as both in the preceding and ſubſequent parts of this work, I have thought it my duty not only to recite all ſuch facts, and to inculcate ſuch maxims as had the leaſt appearance of proving beneficial to future Navigators, but alſo occaſionally to recommend ſuch meaſures to the Public, as I conceive are adapted to promote the ſame laudable purpoſe, I cannot deſiſt from the preſent ſubject, without beſeeching thoſe to whom the conduct of our naval affairs is committed, to endeavour to remove the many perplexities and embarraſſments with which the navigation to the *South-Seas* is, at preſent, neceſſarily encumbered. An effort of this kind could not fail of proving highly honourable to themſelves, and extremely beneficial to their country. For it is to me ſufficiently evident, that whatever advantages navigation ſhall receive, either by the invention of methods that ſhall render its practice leſs hazardous, or by the more accurate delineation of the coaſts, roads and ports already known, or by the diſcovery of new nations, or new ſpecies of commerce; it is evident, I ſay, to me, that by whatever means navigation is promoted the conveniencies hence ariſing muſt ultimately redound to the emolument of *Great-Britain.* Since as our fleets are at preſent ſuperior to thoſe of the whole world united, it muſt be a matchleſs degree of ſupineneſs or mean-ſpiritedneſs, if we permitted any of the advantages which new diſcoveries, or a more extended navigation may produce to mankind, to be raviſhed from us.

As therefore it appears that all our future expeditions to the *South-Seas* muſt run a conſiderable riſque of proving abortive, whilſt we are under the neceſſity of touching at *Brazil* in our paſſage thither, an expedient that might relieve us from this difficulty, would ſurely be a ſubject worthy of the attention of the Public; and this ſeems capable of being effected, by the diſcovery of ſome place more to the ſouthward, where ſhips might refreſh and ſupply them

ſelves

felves with the neceſſary ſea-ſtock for their voyage round Cape *Horn*. And we have in reality the imperfect knowledge of two places, which might perhaps, on examination, prove extremely convenient for this purpoſe; the firſt of them is *Pepys*'s Iſland, in the latitude of 47° South, and laid down by Dr. *Halley*, about eighty leagues to the eaſtward of Cape *Blanco*, on the coaſt of *Patagonia*; the ſecond, is *Falkland*'s Iſles, in the latitude of 51° ½ nearly South of *Pepys*'s Iſland. The firſt of theſe was diſcovered by Captain *Cowley*, in his Voyage round the World in the year 1686; who repreſents it as a commodious place for ſhips to wood and water at, and ſays, it is provided with a very good and capacious harbour, where a thouſand ſail of ſhips might ride at anchor in great ſafety; that it abounds with fowls, and as the ſhore is either rocks or ſands, it ſeems to promiſe great plenty of fiſh. The ſecond place, or *Falkland*'s Iſles, have been ſeen by many ſhips both *French* and *Engliſh*, being the land laid down by *Frezier*, in his Chart of the extremity of South *America*, under the title of the *New Iſlands*. *Woods Rogers*, who run along the N. E. coaſt of theſe Iſles in the year 1708, tells us, that they extended about two degrees in length, and appeared with gentle deſcents from hill to hill, and ſeemed to be good ground, with woods and harbours. Either of theſe places, as they are Iſlands at a conſiderable diſtance from the Continent, may be ſuppoſed, from their latitude, to lie in a climate ſufficiently temperate. It is true, they are too little known to be at preſent recommended for proper places of refreſhment for ſhips bound to the ſouthward: But if the Admiralty ſhould think it adviſeable to order them to be ſurveyed, which may be done at a very ſmall expence, by a veſſel fitted out on purpoſe; and if, on this examination, one or both of theſe places ſhould appear proper for the purpoſe intended, it is ſcarcely to be conceived, of what prodigious import a convenient ſtation might prove, ſituated ſo far to the ſouthward, and ſo near Cape *Horn*. The Duke and Ducheſs of *Briſtol* were but thirty-five days from their loſing ſight of *Falkland*'s Iſles to their arrival at *Juan Fernandes* in the *South-Seas*: And as the re-

turning

turning back is much facilitated by the weftern winds, I doubt not
but a voyage might be made from *Falkland's* Ifles to *Juan Fernandes* and back again, in little more than two months. This, even in
time of peace, might be of great confequence to this Nation; and,
in time of war, would make us mafters of thofe feas.

And as all difcoveries of this kind, though extremely honourable to thofe who direct and promote them, may yet be carried on
at an inconfiderable expence, fince fmall veffels are much the propereft to be employed in this fervice, it were to be wifhed, that
the whole coaft of *Patagonia*, *Terra del Fuego*, and *Staten-land*,
were carefully furveyed, and the numerous channels, roads and harbours with which they abound, accurately examined; this might
open to us facilities of paffing into the *Pacifick* Ocean, which as yet
we may be unacquainted with, and would render all that fouthern
navigation infinitely fecurer than at prefent; and particularly, an exact draught of the Weft coaft of *Patagonia*, from the Streights of
Magellan to the *Spanifh* fettlements, might perhaps furnifh us with
better and more convenient ports for refrefhment, and better fituated for the purpofes either of war or commerce, and above a fortnight's fail nearer to *Falkland's* Iflands, than the Ifland of *Juan Fernandes*. The difcovery of this coaft hath formerly been thought of
fuch confequence, by reafon of its neighbourhood to the *Araucos*
and other *Chilian Indians*, who are generally at war, or at leaft on
ill terms with their *Spanifh* neighbours, that Sir *John Narborough*
was purpofely fitted out in the reign of King *Charles* II, to furvey
the Streights of *Magellan*, the neighbouring coaft of *Patagonia*,
and the *Spanifh* ports on that frontier, with directions, if poffible,
to procure fome intercourfe with the *Chilian Indians*, and to eftablifh a commerce and a lafting correfpondence with them. His
Majefty's views in employing Sir *John Narborough* in this expedition, were not folely the advantage he might hope to receive, from
the alliance of thofe favages, in reftraining and intimidating the
Crown of *Spain*; but he conceived, that, independent of thofe motives, the immediate traffick with thefe *Indians* might prove extremely

tremely advantagious to the *Englifh* Nation. For it is well known, that at the firft difcovery of *Chili* by the *Spaniards*, it abounded with vaft quantities of gold, much beyond what it has at any time produced, fince it has been in their poffeffion. And hence it has been generally believed, that the richeft mines are prudently concealed by the *Indians*, as well knowing that the difcovery of them to the *Spaniards* would only excite in them a greater thirft for conqueft and tyranny, and render their own independence precarious. But with refpect to their commerce with the *Englifh*, thefe reafons would no longer influence them ; fince it would be in our power to furnifh them with arms and ammunition of all kinds, of which they are extremely defirous, together with many other conveniencies which their intercourfe with the *Spaniards* has taught them to relifh. They would then, in all probability, open their mines, and gladly embrace a traffick of fuch mutual convenience to both Nations; for then their gold, inftead of proving the means of enflaving them, would procure them weapons to affert their liberty, to chaftife their tyrants, and to fecure themfelves for ever from the *Spanifh* ycke ; whilft with our affiftance, and under our protection, they might become a confiderable people, and might fecure to us that wealth, which formerly by the Houfe of *Auftria*, and lately by the Houfe of *Bourbon*, has been moft mifchievoufly lavifhed in the purfuit of univerfal Monarchy.

It is true, that Sir *John Narborough* did not fucceed in opening this commerce, which in appearance promifed fo many advantages to this Nation. However, his difappointment was merely accidental, and his tranfactions upon that coaft (befides the many valuable improvements he furnifhed to geography and navigation) are rather an encouragement for future trials of this kind, than any objection againft them ; his principal misfortune being the lofing company of a fmall bark which attended him, and having fome of his people trapanned at *Baldivia*. However, it appeared, by the precautions and fears of the *Spaniards*, that they were fully convinced of the

practicability

practicability of the fcheme he was fent to execute, and extremely alarmed with the apprehenfion of its confequences.

It is faid, that his Majefty King *Charles* the Second was fo far prepoffeffed with the hopes of the advantages redounding from this expedition, and fo eager to be informed of the event of it, that having intelligence of Sir *John Narborough*'s paffing through the *Downs* on his return, he had not patience to attend his arrival at Court, but went himfelf in his barge to *Gravefend* to meet him.

To facilitate as much as poffible any attempts of this kind, which may be hereafter undertaken, I have, in the annexed plate, given a chart of that part of the world, as far as it is hitherto known, which I flatter myfelf is in fome refpects much correcter than any hitherto publifhed. To evince which, it may be neceffary to mention what materials I have principally made ufe of, and what changes I have introduced different from other authors.

The two moft celebrated charts hitherto publifhed of the fouthermoft part of South *America*, are thofe of Dr. *Halley*, in his general chart of the magnetic variation, and of *Frezier* in his voyage to the *South-Seas*. But befides thefe, there is a chart of the Streights of *Magellan*, and of fome part of the adjacent coaft, by Sir *John Narborough* abovementioned, which is doubtlefs infinitely exacter in that part than *Frezier*, and in fome refpects fuperior to *Halley*, particularly in what relates to the longitudes of the different parts of thofe Streights. The coaft from Cape *Blanco* to *Terra del Fuego*, and thence to Streights *Le Maire*, we were in fome meafure capable of correcting by our own obfervations, as we ranged that fhore generally in fight of land. The pofition of the land, to the northward of the Streights of *Magellan*, on the Weft fide, is doubtlefs laid down in our chart but very imperfectly ; and yet I believe it to be much nearer the truth than what has hitherto been done : As it is drawn from the information of fome of the *Wager*'s crew, who were fhipwrecked on that fhore, and afterwards coafted it down ;

and

and as it agrees pretty nearly with the defcription of fome *Spanifh* manufcripts I have feen.

The Channel dividing *Terra del Fuego* is drawn from *Frezier*; but in the *Spanifh* manufcripts there are feveral Channels delineated, and I have reafon to fuppofe, that whenever this country is thoroughly examined, this circumftance will prove true, and *Terra del Fuego* will be found to confift of feveral Iflands.

And having mentioned *Frezier* fo often, I muft not omit warning all future Navigators, againft relying on the longitude of Streights *Le Maire*, or of any part of that coaft, laid down in his chart; the whole being from 8 to 10 degrees too far to the eaftward, if any faith can be given to the concurrent evidences of a great number of journals, verified in fome particulars by aftronomical obfervation. For inftance, Sir *John Narborough* lays down Cape *Virgin Mary* in 65° : 42 of Weft longitude from the *Lizard*, that is in 71° : 20 from *London*. And the fhips of our fquadron, who took their departure from St. *Catherine's* (where the longitude was rectified by an obfervation of the eclipfe of the moon) found Cape *Virgin Mary* to be from 70° : 46', to 71° : 30' from *London*, according to their different reckonings : And there were no circumftances in our run that could render it confiderably erroneous, fo that it cannot be efteemed in lefs than 71 degrees of Weft longitude; whereas *Frezier* lays it down in lefs than 66 degrees from *Paris*, that is little more than 63 degrees from *London*, which is doubtlefs 8 degrees fhort of its true quantity. Again, our fquadron found Cape *Virgin Mary* and Cape St. *Bartholomew* on the eaftern fide of Streights *Le Maire* to be only 2° : 8' different in longitude, which in *Frezier* are diftant near 4 degrees; fo that not only the longitude of Cape St. *Bartholomew* is laid down in him near 10 degrees too little, but the whole coaft, from the Streights of *Magellan* to Streights *Le Maire*, is enlarged to near double its real extent.

But to have done with *Frezier*, whofe errors, the importance of the fubject and not a fondnefs for cavilling, has obliged me to remark, (though his treatment of Dr. *Halley* might, on the prefent occasion, authorise much severer usage) I must, in the next place, particularize wherein the chart I have here inserted differs from that of our learned countryman.

It is well known that this Gentleman was sent abroad by the Public, to make such geographical and astronomical observations, as might facilitate the future practice of navigation, and particularly to determine the variation of the compass in such places as he should touch at, and if possible, to ascertain its general laws and affections.

These things Dr. *Halley*, to his immortal reputation and the honour of our Nation, in good measure accomplished, particularly with regard to the variation of the compass, a subject, of all others, the most interesting to those employed in the art of navigation. He likewise corrected the position of the coast of *Brazil*, which had been very erroneously laid down by all former Hydrographers; and by a judicious comparison of the observations of others, has happily succeeded in settling the geography of many parts of the globe, where he had not himself been. So that the chart he published, with the variation of the needle marked thereon, being the result of his labours on this subject, was allowed by all *Europe* to be far compleater in its geography than any that had then appeared, and at the same time most surprizingly exact in the quantity of variation assigned to the different parts of the globe; a subject so very intricate and perplexing, that all general determinations about it had till then appeared impossible.

But as the only means he had of correcting those coasts where he did not touch himself was the observations of others; where those observations were wanting, or were inaccurate, it was no imputation on his skill, that his determinations were defective. And this, upon the best comparison I have been able to make, is the case with regard to that part of his chart, which contains the South part of South *America*. For though the coast of *Brazil*, and the opposite coast of *Peru* on the *South-Seas* are laid down, I presume, with the greatest accuracy, yet from about the river of *Plate* on the East side, and its opposite point on the West, the coast gradually declines

too

too much to the weftward, fo as at the Streights of *Magellan.* to be, as I conceive, about fifty leagues removed from its true pofition : At leaft, this is the refult of the obfervations of our fquadron, which agree extremely well with thofe of Sir *John Narborough.* I muft add, that Dr. *Halley* has, in the Philofophical Tranfactions, given the foundation on which he has proceeded, in fixing Port St. *Julian* in 76° ½ of Weft longitude : (which the concurrent journals of our fquadron place from 70° ¾ to 71° ½) This, he tells us, was an obfervation of an eclipfe of the moon, made at that place by Mr. *Wood,* then Sir *John Narborough*'s Lieutenant, and which is faid to have happened there at eight in the evening, on the 18th of *September,* 1670. But Capt. *Wood*'s journal of this whole voyage under Sir *John Narborough* is fince publifhed, together with this obfervation, in which he determines the longitude of Port St. *Julian* to be 73 degrees from *London,* and the time of the eclipfe to have been different from Dr. *Halley*'s account. But the numbers he has given are fo faultily printed, that nothing can be determined from them.

To what I have already mentioned with regard to the chart hereunto annexed, I fhall only add, that to render it more compleat, I have inferted therein the rout of our fquadron, and have delineated, in the paffage round Cape *Horn,* both the real tract which we defcribed, and the imaginary tract exhibited by our reckoning; whence the violence of the currents in that part of the world, and the enormous deviations which they produce, will appear by infpection. And that no material article might be omitted in this important affair, the foundings on the coaft of *Patagonia,* and the variation of the magnetic needle, are annexed to thofe parts of this tract, where, by our obfervations, we found them to be of the quantity there fpecified.

O   CHAP.

CHAP. X.

From Cape *Noir* to the Ifland of *Juan Fernandes*.

AFTER the mortifying difappointment of falling in with the coaft of *Terra del Fuego*, when we efteemed ourfelves ten degrees to the weftward of it ; after this difappointment, I fay, recited in the eighth chapter, we ftood away to the S. W. till the 22d of *April*, when we were in upwards of 60° of South latitude, and by our account near 6° to the weftward of Cape *Noir* ; and in this run, we had a feries of as favourable weather, as could well be expected in that part of the world, even in a better feafon : So that this interval, fetting the inquietude of our thoughts afide, was by far the moft eligible of any we enjoyed from Streights *Le Maire* to the Weft coaft of *America*. This moderate weather continued, with little variation, till the 24th ; but on the 24th, in the evening, the wind began to blow frefh, and foon encreafed to a prodigious ftorm ; and the weather being extremely thick, about midnight we loft fight of the other four fhips of the fquadron, which, notwithftanding the violence of the preceding ftorms, had hitherto kept in company with us. Nor was this our fole misfortune ; for, the next morning, endeavouring to hand the top-fails, the clew-lines and bunt-lines broke, and the fheets being half flown, every feam in the top-fails was foon fplit from top to bottom, and the main top-fail fhook fo ftrongly in the wind, that it carried away the top lanthorn, and endangered the head of the maft ; however, at length fome of the moft daring of our men ventured upon the yard, and cut the fail away clofe to the reefs, though with the utmoft hazard of their lives. At the fame time, the foretop-fail beat about the yard with fo much fury, that it was foon blown to pieces ; and that we might have full employment, the main-fail blew loofe, which

obliged

obliged us to lower down the yard to secure the sail, and the fore-yard being likewise lowered, we lay to under a mizen : And besides the loss of our top sails, we had much of our other rigging broke, and lost a main studding-sail-boom out of the chains.

On the 25th, about noon, the weather became more moderate, which enabled us to sway up our yards, and to repair, in the best manner we could, our shattered rigging ; but still we had no sight of the rest of our squadron, nor indeed were we joined by any of them again, till after our arrival at *Juan Fernandes*; nor did any two of them, as we have since learned, continue in company together : And this total separation was the more wonderful, as we had hitherto kept together for seven weeks, through all the reiterated tempests of this turbulent climate. It must indeed be owned, that this separation gave us room to expect, that we might make our passage in a shorter time, than if we had continued together, because we could now make the best of our way without being retarded by the misfortunes of the other ships ; but then we had the melancholy reflection, that we ourselves were hereby deprived of the assistance of others, and our safety would depend upon our single ship ; so that if a plank started, or any other accident of the same nature should take place, we must all irrecoverably perish ; or should we be driven on shore, we had the uncomfortable prospect of ending our days on some desolate coast, without any reasonable hope of ever getting away ; whereas with another ship in company, all these calamities are much less formidable, since in every kind of danger, there would be some probability that one ship at least might escape, and might be capable of preserving or relieving the crew of the other.

The remaining part of this month of *April* we had generally hard gales, although we had been every day, since the 22d, edging to the northward; however, on the last day of the month, we flattered ourselves with the hopes of soon terminating all our sufferings, for we that day found ourselves in the latitude of $52^\circ : 13'$, which being to the northward of the Streights of *Magellan*, we were af-

sured

fured that we had compleated our paffage, and had arrived in the confines of the fouthern Ocean; and this Ocean being nominated *Pacifick*, from the equability of the feafons which are faid to prevail there, and the facility and fecurity with which navigation is there carried on, we doubted not but we fhould be fpeedily cheared with the moderate gales, the fmooth water, and the temperate air, for which that tract of the globe has been fo renowned. And under the influence of thefe pleafing circumftances, we hoped to experience fome kind of compenfation; for the complicated miferies which had fo conftantly attended us for the laft eight weeks. But here we were again difappointed; for in the fucceeding month of *May*, our fufferings rofe to a much higher pitch than they had ever yet done, whether we confider the violence of the ftorms, the fhattering of our fails and rigging, or the diminifhing and weakening of our crew by deaths and ficknefs, and the probable profpect of our total deftruction. All this will be fufficiently evident, from the following circumftantial account of our diverfified misfortunes.

Soon after our paffing Streights *Le Maire*, the fcurvy began to make its appearance amongft us; and our long continuance at fea, the fatigue we underwent, and the various difappointments we met with, had occafioned its fpreading to fuch a degree, that at the latter end of *April* there were but few on board, who were not in fome degree afflicted with it, and in that month no lefs than forty-three died of it on board the *Centurion*. But though we thought that the diftemper had then rifen to an extraordinary height, and were willing to hope, that as we advanced to the northward its malignity would abate, yet we found, on the contrary, that in the month of *May* we loft near double that number: And as we did not get to land till the middle of *June*, the mortality went on increafing, and the difeafe extended itfelf fo prodigioufly, that after the lofs of above two hundred men, we could not at laft mufter more than fix fore-maft men in a watch capable of duty.

This

This difeafe fo frequently attending all long voyages, and fo particularly deftructive to us, is furely the moft fingular and unaccountable of any that affects the human body. For its fymptoms are inconftant and innumerable, and its progrefs and effects extremely irregular ; for fcarcely any two perfons have the fame complaints, and where there hath been found fome conformity in the fymptoms, the order of their appearance has been totally different. However, though it frequently puts on the form of many other difeafes, and is therefore not to be defcribed by any exclufive and infallible criterions ; yet there are fome fymptoms which are more general than the reft, and therefore, occurring the oftneft, deferve a more particular enumeration. Thefe common appearances are large difcoloured fpots difperfed over the whole furface of the body, fwelled legs, putrid gums, and above all, an extraordinary laffitude of the whole body, efpecially after any exercife, however inconfiderable ; and this laffitude at laft degenerates into a pronenefs to fwoon on the leaft exertion of ftrength, or even on the leaft motion.

This difeafe is likewife ufually attended with a ftrange dejection of the fpirits, and with fhiverings, tremblings, and a difpofition to be feized with the moft dreadful terrors on the flighteft accident. Indeed it was moft remarkable, in all our reiterated experience of this malady, that whatever difcouraged our people, or at any time damped their hopes, never failed to add new vigour to the diftemper ; for it ufually killed thofe who were in the laft ftages of it, and confined thofe to their hammocks, who were before capable of fome kind of duty ; fo that it feemed as if alacrity of mind, and fanguine thoughts, were no contemptible prefervatives from its fatal malignity.

But it is not eafy to compleat the long roll of the various concomitants of this difeafe ; for it often produced putrid fevers, pleurifies, the jaundice, and violent rheumatick pains, and fometimes it occafioned an obftinate coftivenefs, which was generally attended with a difficulty of breathing ; and this was efteemed the moft deadly of all the fcorbutick fymptoms : At other times the whole

body,

body, but more especially the legs, were subject to ulcers of the worst kind, attended with rotten bones, and such a luxuriancy of funguous flesh, as yielded to no remedy. But a most extraordinary circumstance, and what would be scarcely credible upon any single evidence, is, that the scars of wounds which had been for many years healed, were forced open again by this virulent distemper: Of this, there was a remarkable instance in one of the invalids on board the *Centurion*, who had been wounded above fifty years before at the battle of the *Boyne*; for though he was cured soon after, and had continued well for a great number of years past, yet on his being attacked by the scurvy, his wounds, in the progress of his disease, broke out afresh, and appeared as if they had never been healed: Nay, what is still more astonishing, the callous of a broken bone, which had been compleatly formed for a long time, was found to be hereby dissolved, and the fracture seemed as if it had never been consolidated. Indeed, the effects of this disease were in almost every instance wonderful; for many of our people, though confined to their hammocks, appeared to have no inconsiderable share of health, for they eat and drank heartily, were chearful, and talked with much seeming vigour, and with a loud strong tone of voice; and yet on their being the least moved, though it was only from one part of the ship to the other, and that in their hammocks, they have immediately expired; and others, who have confided in their seeming strength, and have resolved to get out of their hammocks, have died before they could well reach the deck; and it was no uncommon thing for those who were able to walk the deck, and to do some kind of duty, to drop down dead in an instant, on any endeavours to act with their utmost vigour, many of our people having perished in this manner during the course of this voyage.

With this terrible disease we struggled the greatest part of the time of our beating round Cape *Horn*; and though it did not then rage with its utmost violence, yet we buried no less than forty-three men on board the *Centurion*, in the month of *April*, as hath

been

been already obferved, but we ftill entertained hopes, that when we fhould have once fecured our paffage round the Cape, we fhould put a period to this, and all the other evils which had fo conftantly purfued us. But it was our misfortune to find, that the *Pacifick* Ocean was to us lefs hofpitable than the turbulent neighbourhood of *Terra del Fuego* and Cape *Horn*: For being arrived, on the 8th of *May*, off the Ifland of *Socoro*, which was the firft rendezvous appointed for the fquadron, and where we hoped to have met with fome of our companions, we cruized for them in that ftation feveral days. And here we were not only difappointed in our hopes of being joined by our friends, and were thereby induced to favour the gloomy fuggeftions of their having all perifhed ; but we were likewife perpetually alarmed with the fears of being driven on fhore upon this coaft, which appeared too craggy and irregular to give us the leaft hopes, that in fuch a cafe any of us could poffibly efcape immediate deftruction. For the land had indeed a moft tremendous afpect : The moft diftant part of it, and which appeared far within the country, being the mountains ufually called the *Andes* or *Cordilleras*, was extremely high, and covered with fnow; and the coaft itfelf feemed quite rocky and barren, and the water's edge fkirted with precipices. In fome places indeed there appeared feveral deep bays running into the land, but the entrance into them was generally blocked up by numbers of little Iflands ; and though it was not improbable but there might be convenient fhelter in fome of thofe bays, and proper channels leading thereto ; yet as we were utterly ignorant of the coaft, had we been driven afhore by the weftern winds which blew almoft conftantly there, we did not expect to have avoided the lofs of our fhips and of our lives.

And this continued peril, which lafted for above a fortnight, was greatly aggravated by the difficulties we found in working the fhip ; as the fcurvy had by this time deftroyed fo great a part of our hands, and had in fome degree affected almoft the whole crew. Nor did we, as we hoped, find the winds lefs violent, as we advanced to the northward; for we had often prodigious fqualls which fplit

our

our fails, greatly damaged our rigging, and endangered our mafts. Indeed, during the greateft part of the time we were upon this coaft, the wind blew fo hard, that in another fituation, where we had fufficient fea-room, we fhould certainly have lain to; but in the prefent exigency we were neceffitated to carry both our courfes and top-fails, in order to keep clear of this lee-fhore. In one of thefe fqualls, which was attended by feveral violent claps of thunder, a fudden flafh of fire darted along our decks, which, dividing, exploded with a report like that of feveral piftols, and wounded many of our men and officers as it paffed, marking them in different parts of the body: This flame was attended with a ftrong fulphurous ftench, and was doubtlefs of the fame nature with the larger and more violent blafts of lightning which then filled the air.

It were endlefs to recite minutely the various difafters, fatigues and terrors which we encountered on this coaft; all thefe went on encreafing till the 22d of *May*, at which time, the fury of all the ftorms which we had hitherto encountered, feemed to be combined, and to have confpired our deftruction. In this hurricane almoft all our fails were fplit, and great part of our ftanding rigging broken; and, about eight in the evening, a mountainous overgrown-fea took us upon our ftarboard-quarter, and gave us fo prodigious a fhock, that feveral of our fhrouds broke with the jerk, by which our mafts were greatly endangered; our ballaft and ftores too were fo ftrangely fhifted, that the fhip heeled afterwards two ftreaks to port. Indeed it was a moft tremendous blow, and we were thrown into the utmoft confternation from the apprehenfion of inftantly foundering; and though the wind abated in a few hours, yet, as we had no more fails left in a condition to bend to our yards, the fhip laboured very much in a hollow fea, rolling gunwale to, for want of fail to fteady her: So that we expected our mafts, which were now very flenderly fupported, to come by the board every moment. However, we exerted ourfelves the beft we could to ftirrup our fhrouds, to reeve new lanyards, and to mend our fails; but while thefe neceffary operations were carrying on, we ran great rifque of being

driven

driven on fhore on the Ifland of *Chiloe*, which was not far diftant from us; but in the midft of our peril the wind happily fhifted to the fouthward, and we fteered off the land with the main-fail only, the Mafter and myfelf undertaking the management of the helm, while every one elfe on board was bufied in fecuring the mafts, and bending the fails as faft as they could be repaired. This was the laft effort of that ftormy climate; for in a day or two after, we got clear of the land, and found the weather more moderate than we had yet experienced fince our paffing Streights *Le Maire*. And now having cruized in vain for more than a fortnight in queft of the other fhips of the fquadron, it was refolved to take the advantage of the prefent favourable feafon and the offing we had made from this terrible coaft, and to make the beft of our way for the Ifland of *Juan Fernandes*. For though our next rendezvous was appointed off the harbour of *Baldivia*, yet as we had hitherto feen none of our companions at this firft rendezvous, it was not to be fuppofed that any of them would be found at the fecond : Indeed we had the greateft reafon to fufpect, that all but ourfelves had perifhed. Befides, we were by this time reduced to fo low a condition, that inftead of attempting to attack the places of the enemy, our utmoft hopes could only fuggeft to us the poffibility of faving the fhip, and fome part of the remaining enfeebled crew, by our fpeedy arrival at *Juan Fernandes*; for this was the only road in that part of the world where there was any probability of our recovering our fick, or refitting our veffel, and confequently our getting thither was the only chance we had left to avoid perifhing at fea.

Our deplorable fituation then allowing no room for deliberation, we ftood for the Ifland of *Juan Fernandes*; and to fave time, which was now extremely precious, (our men dying four, five and fix in a day) and likewife to avoid being engaged again with a lee-fhore, we refolved, if poffible, to hit the Ifland upon a meridian. And, on the 28th of *May*, being nearly in the parallel upon which it is laid down, we had great expectations of feeing it : But not finding

P

it

it in the pofition in which the charts had taught us to expect it, we began to fear that we had got too far to the weftward; and therefore, though the Commodore himfelf was ftrongly perfuaded, that he faw it on the morning of the 28th, yet his Officers believing it to be only a cloud, to which opinion the hazinefs of the weather gave fome kind of countenance, it was, on a confultation, refolved to ftand to the eaftward, in the parallel of the Ifland; as it was certain, that by this courfe we fhould either fall in with the Ifland, if we were already to the weftward of it; or fhould at leaft make the main-land of *Chili*, from whence we might take a new departure, and affure ourfelves, by running to the weftward afterwards, of not miffing the Ifland a fecond time.

On the 30th of *May* we had a view of the Continent of *Chili*, diftant about twelve or thirteen leagues; the land made exceeding high and uneven, and appeared quite white; what we faw being doubtlefs a part of the *Cordilleras*, which are always covered with fnow. Though by this view of the land we afcertained our pofition, yet it gave us great uneafinefs to find that we had fo needlefsly altered our courfe, when we were, in all probability juft upon the point of making the Ifland; for the mortality amongft us was now encreafed to a moft dreadful degree, and thofe who remained alive were utterly difpirited by this new difappointment, and the profpect of their longer continuance at fea: Our water too began to grow fcarce; fo that a general dejection prevailed amongft us, which added much to the virulence of the difeafe, and deftroyed numbers of our beft men; and to all thefe calamities there was added this vexatious circumftance, that when, after having got a fight of the Main, we tacked and ftood to the weftward in queft of the Ifland, we were fo much delayed by calms and contrary winds, that it coft us nine days to regain the wefting, which, when we ftood to the eaftward, we ran down in two. In this defponding condition, with a crazy fhip, a great fcarcity of frefh water, and a crew fo univerfally difeafed, that there were not above ten fore-maft men in a watch capable of doing duty, and even fome of thefe lame, and

unable

unable to go aloft: Under thefe difheartning circumftances, I fay, we ftood to the weftward; and, on the 9th of *June*, at day-break, we at laft difcovered the long-wifhed for Ifland of *Juan Fernandes*. And with this difcovery I fhall clofe this chapter and the firft book, after obferving (which will furnifh a very ftrong image of our unparalleled diftreffes) that by our fufpecting ourfelves to be to the weftward of the Ifland on the 28th of *May*, and, in confequence of this, ftanding in for the Main, we loft between feventy and eighty of our men, whom we fhould doubtlefs have faved had we made the Ifland that day, which, had we kept on our courfe for a few hours longer, we could not have failed to have done.

END *of* BOOK I.

A VOYAGE

A VOYAGE ROUND THE WORLD, &c.

BOOK II.

CHAP. I.

The arrival of the Centurion *at the Island of* Juan Fernandes, *with a description of that Island.*

ON the 9th of *June*, at day break, as is mentioned in the preceding chapter, we first descried the Island of *Juan Fernandes*, bearing N. by E. $\frac{1}{2}$ E, at eleven or twelve leagues distance. And though, on this first view, it appeared to be a very mountainous place, extremely ragged and irregular; yet as it was land, and the land we sought for, it was to us a most agreeable sight: For at this place only we could hope to put a period to those terrible calamities we had so long struggled with, which had already swept away above half our crew, and which, had we continued a few days longer at sea, would inevitably have compleated our destruction. For we were by this time reduced to so helpless a condition, that out of two hundred and odd men which

remained

remained alive, we could not, taking all our watches together, muster hands enough to work the ship on an emergency, though we included the officers, their servants, and the boys.

The wind being northerly when we first made the Island, we kept plying all that day, and the next night, in order to get in with the land; and wearing the ship in the middle watch, we had a melancholy instance of the almost incredible debility of our people; for the Lieutenant could muster no more than two Quarter-masters, and six Fore-mast men capable of working; so that without the assistance of the officers, servants and the boys, it might have proved impossible for us to have reached the Island, after we had got sight of it; and even with this assistance they were two hours in trimming the sails: To so wretched a condition was a sixty gun ship reduced, which had passed Streights *Le Maire* but three months before, with between four and five hundred men, almost all of them in health and vigour.

However, on the 10th in the afternoon, we got under the lee of the Island, and kept ranging along it, at about two miles distance, in order to look out for the proper anchorage, which was described to be in a bay on the North side. And now being nearer in with the shore, we could discover that the broken craggy precipices, which had appeared so unpromising at a distance, were far from barren, being in most places covered with woods; and that between them there were every where interspersed the finest vallies, clothed with a most beautiful verdure, and watered with numerous streams and cascades, no valley, of any extent, being unprovided of its proper rill. The water too, as we afterwards found, was not inferiour to any we had ever tasted, and was constantly clear: So that the aspect of this country would, at all times, have been extremely delightful, but in our distressed situation, languishing as we were for the land and its vegetable productions, (an inclination constantly attending every stage of the sea-scurvy) it is scarcely credible with what eagerness and transport we viewed the shore, and with how much impatience we longed for the greens and other refreshments
which

which were then in fight, and particularly for the water, for of this we had been confined to a very fparing allowance for a confiderable time, and had then but five ton remaining on board. Thofe only who have endured a long feries of thirft, and who can readily re-cal the defire and agitation which the ideas alone of fprings and brooks have at that time raifed in them, can judge of the emotion with which we eyed a large cafcade of the moft tranfparent water, which poured itfelf from a rock near a hundred feet high into the fea, at a fmall diftance from the fhip. Even thofe amongft the difeafed, who were not in the very laft ftages of the diftemper, though they had been long confined to their hammocks, exerted the fmall remains of ftrength that was left them, and crawled up to the deck to feaft themfelves with this reviving profpect. Thus we coafted the fhore, fully employed in the contemplation of this di-verfified landfkip, which ftill improved upon us the farther we ad-vanced. But at laft the night clofed upon us, before we had fa-tisfied ourfelves which was the proper bay to anchor in ; and there-fore we refolved to keep in foundings all night, (we having then from fixty-four to feventy fathom) and to fend our boat next morning to difcover the road : However, the current fhifted in the night, and fet us fo near the land, that we were obliged to let go the beft bower in fifty fix fathom, not half a mile from the fhore. At four in the morning, the Cutter was difpatched with our third Lieutenant to find out the bay we were in fearch of, who returned again at noon with the boat laden with feals and grafs ; for though the Ifland abounded with better vegetables, yet the boats-crew, in their fhort ftay, had not met with them ; and they well knew that even grafs would prove a dainty, and indeed it was all foon and eagerly devoured. The feals too were confidered as frefh provifion ; but as yet were not much admired, tho' they grew afterwards into more repute : For what rendered them lefs valuable at this juncture, was the prodigious quantity of excellent fifh, which the people on board had taken, during the abfence of the boat.

The

The Cutter, in this expedition, had difcovered the bay where we intended to anchor, which we found was to the weftward of our prefent ftation ; and, the next morning, the weather proving favourable, we endeavoured to weigh, in order to proceed thither : But though, on this occafion, we muftered all the ftrength we could, obliging even the fick, who were fcarce able to keep on their leggs, to affift us ; yet the capftan was fo weakly manned, that it was near four hours before we hove the cable right up and down : After which, with our utmoft efforts, and with many furges and fome purchafes we made ufe of to encreafe our power, we found ourfelves incapable of ftarting the anchor from the ground. However, at noon, as a frefh gale blew towards the bay, we were induced to fet the fails, which fortunately tripped the anchor ; on which we fteered along fhore, till we came a-breaft of the point that forms the eaftern part of the bay. On the opening of the bay, the wind, that had befriended us thus far, fhifted and blew from thence in fqualls ; but by means of the head-way we had got, we loofed clofe in, till the anchor brought us up in fifty-fix fathom. Soon after we had thus got to our new birth, we difcovered a fail, which we made no doubt was one of our fquadron ; and on its nearer approach, we found it to be the *Tryal* Sloop. We immediately fent fome of our hands on board her, by whofe affiftance fhe was brought to an anchor between us and the land. We foon found that the Sloop had not been exempted from thofe calamities which we had fo feverely felt ; for her Commander, Captain *Saunders*, waiting on the Commodore, informed him, that out of his fmall complement, he had buried thirty-four of his men ; and thofe that remained were fo univerfally afflicted with the fcurvy, that only himfelf, his Lieutenant, and three of his men, were able to ftand by the fails. The *Tryal* came to an anchor within us, on the 12th, about noon, and we carried our hawfers on board her, in order to moor ourfelves nearer in fhore ; but the wind coming off the land in violent gufts, prevented our mooring in the birth we intended, efpecially as our principal attention was now employed on bufinefs

rather

rather of more importance; for we were now extremely occupied in fending on fhore materials to raife tents for the reception of the fick, who died apace on board, and doubtlefs the diftemper was confiderably augmented, by the ftench and filthinefs in which they lay; for the number of the difeafed was fo great, and fo few could be fpared from the neceffary duty of the fails to look after them, that it was impoffible to avoid a great relaxation in the article of cleanlinefs, which had rendered the fhip extremely loathfome between decks. But notwithftanding our defire of freeing the fick from their hateful fituation, and their own extreme impatience to get on fhore, we had not hands enough to prepare the tents for their reception before the 16th; but on that and the two following days we fent them all on fhore, amounting to a hundred and fixty-feven perfons, befides at leaft a dozen who died in the boats, on their being expofed to the frefh air. The greateft part of our fick were fo infirm, that we were obliged to carry them out of the fhip in their hammocks, and to convey them afterwards in the fame manner from the water-fide to their tents, over a ftony beach. This was a work of confiderable fatigue to the few who were healthy, and therefore the Commodore, with his accuftomed humanity, not only affifted herein with his own labour, but obliged his Officers, without diftinction, to give their helping hand. The extreme weaknefs of our fick may in fome meafure be collected from the numbers who died after they had got on fhore; for it had generally been found, that the land, and the refrefhments it produces, very foon recover moft ftages of the fea-fcurvy; and we flattered ourfelves, that thofe who had not perifhed on this firft expofure to the open air, but had lived to be placed in their tents, would have been fpeedily reftored to their health and vigour: But, to our great mortification, it was near twenty days after their landing, before the mortality was tolerably ceafed; and for the firft ten or twelve days, we buried rarely lefs than fix each day, and many of thofe, who furvived, recovered by very flow and infenfible degrees. Indeed, thofe who were well enough at their firft getting on fhore,

Q

to

to creep out of their tents, and crawl about, were foon relieved, and recovered their health and ftrength in a very fhort time; but in the reft, the difeafe feemed to have acquired a degree of inveteracy which was altogether without example.

Having proceeded thus far, and got our fick on fhore, I think it neceffary, before I enter into any longer detail of our tranfactions, to give a diftinct account of this Ifland of *Juan Fernandes*, its fituation, productions, and all its conveniencies. Thefe particulars we were well enabled to be minutely inftructed in, during our three months ftay there; and as it is the only commodious place in those feas, where *Britifh* cruifers can refrefh and recover their men after their paffage round Cape *Horn*, and where they may remain for fome time without alarming the *Spanifh* coaft, thefe its advantages well merit a circumftantial defcription. And indeed Mr. *Anfon* was particularly induftrious in directing the roads and coafts to be furveyed, and other obfervations to be made, knowing, from his own experience, of how great confequence thefe materials might prove to any *Britifh* veffels hereafter employed in thofe feas. For the uncertainty we were in of its pofition, and our ftanding in for the Main on the 28th of *May*, in order to fecure a fufficient eafting, when we were indeed extremely near it, coft us the lives of between feventy and eighty of our men, by our longer continuance at fea: From which fatal accident we might have been exempted, had we been furnifhed with fuch an account of its fituation, as we could fully have depended on.

The Ifland of *Juan Fernandes* lies in the latitude of 33°: 40' South, and is a hundred and ten leagues diftant from the Continent of *Chili*. It is faid to have received its name from a *Spaniard*, who formerly procured a grant of it, and refided there fome time with a view of fettling it, but afterwards abandoned it. On approaching it on its eaft fide, it appears, as reprefented in the annexed plate, where (A) is a fmall Ifland, called *Goat Ifland*, to the S. W. of it; (B) a rock, called *Monkey key*, almoft contiguous to it; (C) is the Eaft bay, (D) *Cumberland Bay*, where we moored, and which, as
will

The east prospect of the Island

Plate XIV.

AN FERNANDES *in the south sea.*

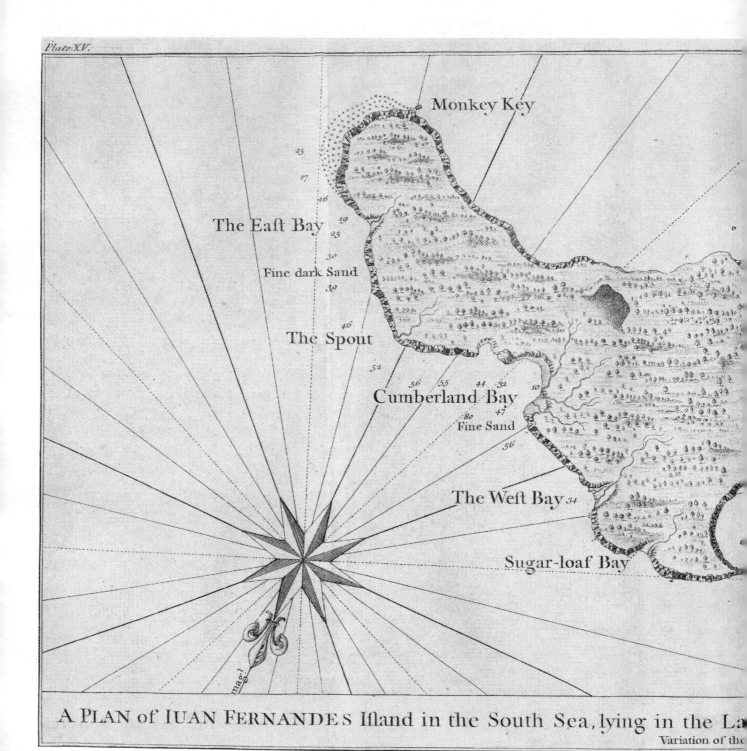

Plate XV.

Monkey Key

The East Bay

Fine dark Sand

The Spout

Cumberland Bay

Fine Sand

The West Bay

Sugar-loaf Bay

mag.l

A PLAN of IUAN FERNANDES Island in the South Sea, lying in the La

Variation of the

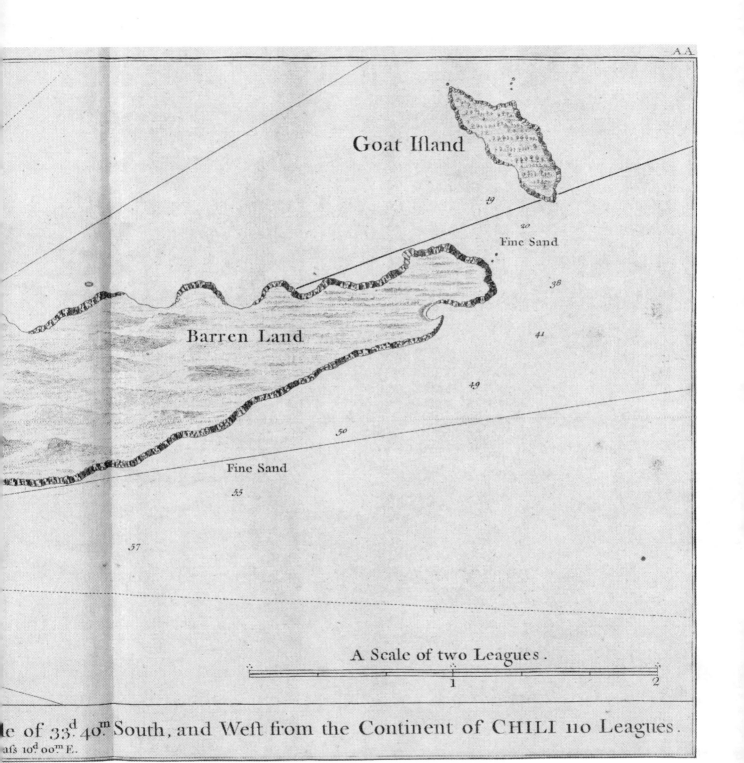

Goat Iſland

19

40

Fine Sand

38

Barren Land

41

49

50

Fine Sand

55

57

A Scale of two Leagues.

1   2

le of 33.ᵈ 40.ᵐ South, and Weſt from the Continent of CHILI 110 Leagues.

aſs 10ᵈ 00ᵐ E.

Plate XVI.

A SURVEY OF THE NORTH EAST SIDE OF I

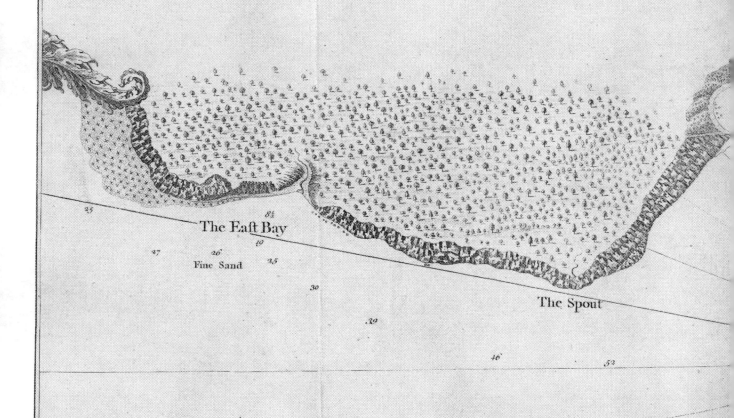

25

8¼

The East Bay

27   26   19

Fine Sand   25

30

39

The Spout

46   52

A Scale of two Miles.

1   2

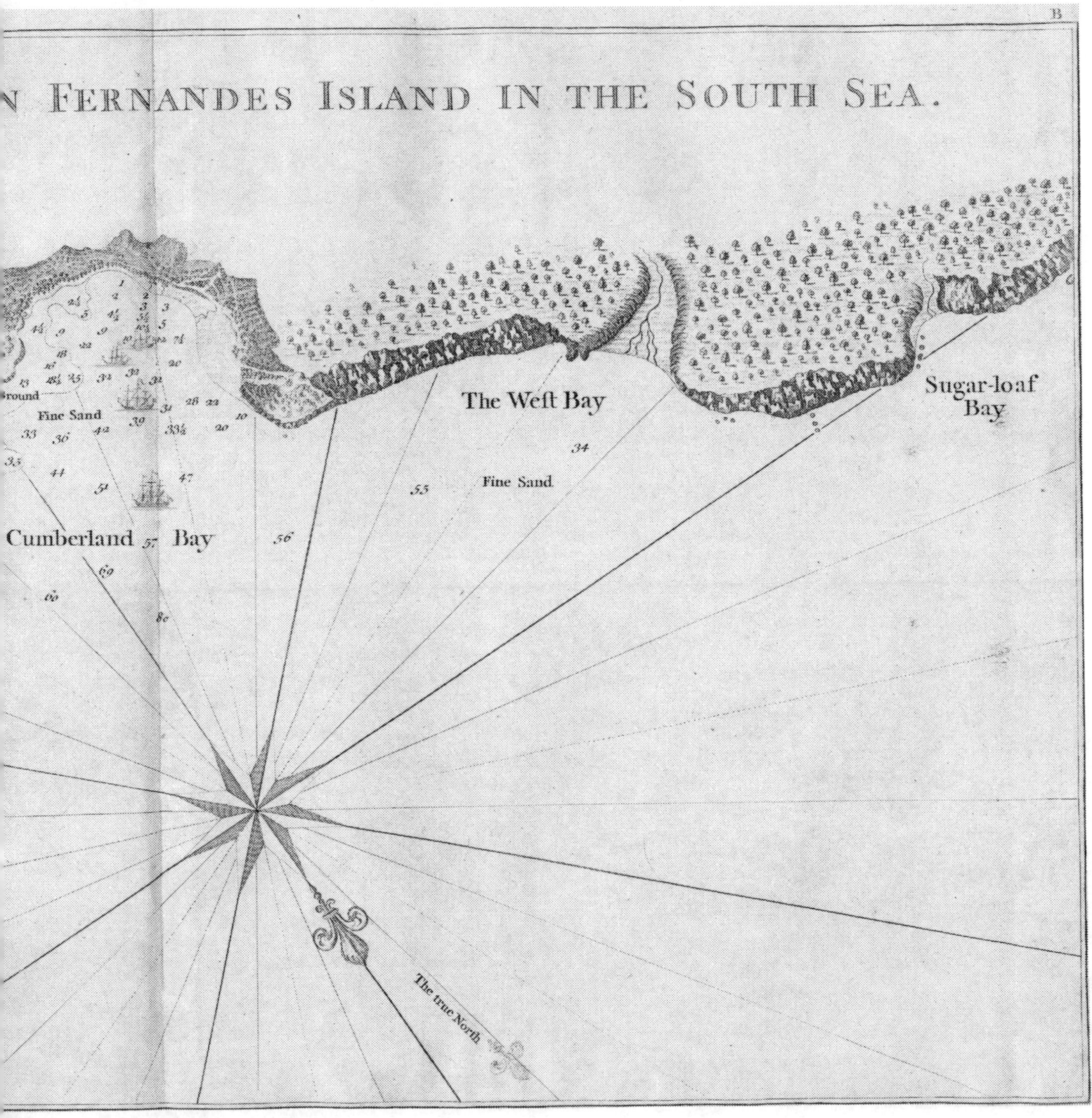

n FERNANDES ISLAND IN THE SOUTH SEA.

Sugar-loaf
Bay

The Weſt Bay

34

Fine Sand.

55

Fine Sand

Ground

Fine Sand

Cumberland *57* Bay

56

69

60

80

The true North

A view of CUMBERLAND BAY *at*

Island of IUAN FERNANDES.

will be obferved, is the beft road for fhipping, and (E) the Eaft
bay. The Ifland itfelf is of an irregular figure, as may be feen by
the very exact plan of it here inferted; its greateft extent being between four and five leagues, and its greateft breadth fomewhat
fhort of two leagues. The only fafe anchoring at this Ifland is on
the North fide, where are the three bays mentioned above, but the
middlemoft known by the name of *Cumberland Bay*, is the wideft
and deepeft, and in all refpects much the beft; the other two
bays, denominated the Eaft and Weft bays, are fcarcely more than
good landing places, where boats may conveniently put their cafk
on fhore. A plan of the N. E. fide of the Ifland, containing thefe
three bays, drawn by a large fcale, is here inferted, where it appears, that *Cumberland Bay* is pretty well fecured to the fouthward,
lying only expofed from the N. by W. to the E. by S; and as the
northerly winds feldom blow in that climate, and never with any
violence, the danger from that quarter is not worth attending to.
To diftinguifh this bay the better at fea, I have added a very exact
view of it, which will enable all future Navigators readily to
find it.

As the bay laft defcribed, or *Cumberland Bay*, is by far the moft
commodious road in the Ifland, fo it is advifeable for all fhips to anchor on the weftern fide of this bay, within little more than two
cables length of the beach. Here they may ride in forty fathom of
water, and be, in a great meafure, fheltered from a large heavy
fea, which comes rolling in whenever an eaftern or a weftern wind
blows. It is however expedient, in this cafe, to cackle or arm the
cables with an iron chain, or good rounding, for five or fix fathom
from the anchor, to fecure them from being rubbed by the foulnefs
of the ground.

I have before obferved, that a northerly wind, to which alone
this bay is expofed, very rarely blew during our ftay here; and as
it was then winter, it may be fuppofed, in other feafons, to be lefs
frequent. Indeed, in thofe few inftances when it was in that quarter, it did not blow with any great force: But this perhaps might

Q 2

be

be owing to the highlands on the fouthward of the bay, which checked its current, and thereby abated its violence; for we had reafon to fuppofe, that a few leagues off, it blew with confiderable force, fince it fometimes drove before it a prodigious fea, in which we rode fore-caftle in. But though the northern winds are never to be apprehended, yet the fouthern winds, which generally prevail here, frequently blow off the land in violent gufts and fqualls, which however rarely laft longer than two or three minutes. This feems to be owing to the obftruction of the fouthern gale, by the hills in the neighbourhood of the bay; for the wind being collected by this means, at laft forces its paffage through the narrow vallies, which, like fo many funnels, both facilitate its efcape, and increafe its violence. Thefe frequent and fudden gufts make it difficult for fhips to work in with the wind off fhore, or to keep a clear hawfe when anchored.

The northern part of this Ifland is compofed of high craggy hills, many of them inacceffible, though generally covered with trees. The foil of this part is loofe and fhallow, fo that very large trees on the hills foon perifh for want of root, and are eafily overturned; which occafioned the unfortunate death of one of our failors, who being upon the hills in fearch of goats, caught hold of a tree upon a declivity to affift him in his afcent, and this giving way, he immediately rolled down the hill, and though in his fall he faftened on another tree of confiderable bulk, yet that too gave way, and he fell amongft the rocks, and was dafhed to pieces. Mr *Brett* too met with an accident only by refting his back againft a tree, near as large about as himfelf, which ftood on a flope, for the tree giving way, he fell to a confiderable diftance, though without receiving any harm.

The fouthern, or rather the S. W. part of the Ifland, as diftin-guifhed in the plan, is widely different from the reft, being dry, ftony, and deftitute of trees, but very flat and low, compared with the hills on the northern part. This part of the Ifland is never fre-quented by fhips, being furrounded by a fteep fhore, and having

little

little or no frefh water ; and befides, it is expofed to the foutherly wind, which generally blows here the whole year round, and in the winter folftice very hard. The trees of which the woods on the northern fide of the Ifland are compofed, are moft of them aromaticks, and of many different forts : There are none of them of a fize to yield any confiderable timber, except the myrtle-trees, which are the largeft on the Ifland, and fupplied us with all the timber we made ufe of; but even thefe would not work to a greater length than forty feet. The top of the myrtle-tree is circular, and appears as uniform and regular, as if it had been clipped by art ; it bears on its bark an excrefcence like mofs, which in tafte and fmell refembles garlick, and was ufed by our people inftead of it. We found here too the piemento-tree and likewife the cabbage-tree, though in no great plenty

Our prifoners obferved, that the appearance of the hills in fome part of the Ifland refembled that of the mountains in *Chili*, where the gold is found : So that it is not impoffible but mines might be difcovered here. We obferved, in fome places, feveral hills of a peculiar fort of red earth, exceeding vermilion in colour, which perhaps, on examination, might prove ufeful for many purpofes.

Befides a great number of plants of various kinds which are to be met with upon the Ifland, but which we were not botanifts enough either to defcribe, or attend to, we found there almoft all the vegetables, which are ufually efteemed to be particularly adapted to the cure of thofe fcorbutick diforders, which are contracted by falt diet and long voyages. For here we had great quantities of watercreffes and purflain, with excellent wild forrel, and a vaft profufion of turnips and *Sicilian* radifhes : Thefe two laft, having fome refemblance to each other, were confounded by our people under the general name of turnips. We ufually preferred the tops of the turnips to the roots, which were often ftringy ; though fome of them were free from that exception, and remarkably good. Thefe vegetables, with the fifh and flefh we found here, and which I fhall more particularly defcribe hereafter, were not only extremely grateful

ful to our palates, after the long courfe of falt diet which we had been confined to, but were likewife of the moft falutary confequence to our fick in recovering and invigorating them, and of no mean fervice to us who were well, in deftroying the lurking feeds of the fcurvy, from which perhaps none of us were totally exempt, and in refrefhing and reftoring us to our wonted ftrength and activity.

Befides the vegetables I have mentioned, of which we made perpetual ufe, we found many acres of ground covered with oats and clover. There were alfo fome few cabbage-trees upon the Ifland, as obferved before; but as they generally grew on the precipices, and in dangerous fituations, and as it was neceffary to cut down a large tree for every fingle cabbage, this was a dainty that we were able but rarely to indulge in.

The excellence of the climate and the loofenefs of the foil render this place extremely proper for all kinds of vegetation; for if the ground be any where accidentally turned up, it is immediately overgrown with turnips and *Sicilian* radifhes; and therefore Mr. *Anfon* having with him garden feeds of all kinds, and ftones of different forts of fruits, he, for the better accommodation of his countrymen who fhould hereafter touch here, fowed both lettices, carrots, and other garden plants, and fett in the woods a great variety of plumb, apricock, and peach ftones: And thefe laft he has been informed have fince thriven to a very remarkable degree; for fome Gentlemen, who in their paffage from *Lima* to *Old Spain* were taken and brought to *England*, having procured leave to wait upon Mr. *Anfon*, to thank him for his generofity and humanity to his prifoners, fome of whom were their relations, they, in cafual difcourfe with him about his tranfactions in the *South-Seas*, particularly afked him, if he had not planted a great number of fruit-ftones on the Ifland of *Juan Fernandes*, for they told him, their late Navigators had difcovered there numbers of peach-trees and apricock-trees, which being fruits before unobferved in that place, they concluded them to be produced from kernels fett by him.

And

And this may in general fuffice as to the foil and vegetable productions of this place : But the face of the country, at leaft of
the North part of the Ifland, is fo extremely fingular, that I cannot
avoid giving it a particular confideration. I have already taken notice of the wild, inhofpitable air with which it firft appeared to us,
and the gradual improvement of this uncouth landfkip as we drew
nearer, till we were at laft captivated by the numerous beauties we
difcovered on the fhore. And I muft now add, that we found,
during the time of our refidence there, that the inland parts of the
Ifland did no ways fall fhort of the fanguine prepoffeffions which we
firft entertained in their favour.

For the woods which covered moft of the fteepeft hills, were
free from all bufhes and underwood, and afforded an eafy paffage
through every part of them ; and the irregularities of the hills and
precipices, in the northern part of the Ifland, neceffarily traced out
by their various combinations a great number of romantic vallies;
moft of which had a ftream of the cleareft water running through
them, that tumbled in cafcades from rock to rock, as the bottom
of the valley, by the courfe of the neighbouring hills, was at any
time broken into a fudden fharp defcent : Some particular fpots occurred in thefe vallies, where the fhade and fragrance of the contiguous woods, the loftinefs of the overhanging rocks, and the tran-
fparency and frequent falls of the neighbouring ftreams, prefented
fcenes of fuch elegance and dignity, as would perhaps with difficulty be rivalled in any other part of the globe. It is in this
place, perhaps, that the fimple productions of unaffifted nature
may be faid to excel all the fictitious defcriptions of the moft animated imagination. I fhall finifh this article with a fhort account
of that fpot where the Commodore pitched his tent, and which he
made choice of for his own refidence, though I defpair of con-
veying an adequate idea of its beauty. This piece of ground which
he chofe was a fmall lawn, that lay on a little afcent, at the diftance
of about half a mile from the fea. In the front of his tent there
was a large avenue cut through the woods to the fea-fide, which

floping

floping to the water with a gentle defcent, opened a profpect of the bay and the fhips at anchor. This lawn was fcreened behind by a tall wood of myrtle fweeping round it, in the form of a theatre, the ground on which the wood ftood, rifing with a much fharper afcent than the lawn itfelf, though not fo much, but that the hills and precipices within land towered up confiderably above the tops of the trees, and added to the grandeur of the view. There were, befides, two ftreams of chryftal water, which ran on the right and left of the tent, within an hundred yards diftance, and were fhaded by the trees which fkirted the lawn on either fide, and compleated the fymmetry of the whole. Some faint conceptions of the elegance of this fituation may perhaps be better deduced from the draught of it, inferted in the adjoining plate.

It remains now only that we fpeak of the animals and provifions which we met with at this place. Former writers have related, that this Ifland abounded with vaft numbers of goats, and their accounts are not to be queftioned, this place being the ufual haunt of the buccaneers and privateers, who formerly frequented thofe feas. And there are two inftances; one of a *Mufquito Indian*, and the other of *Alexander Selkirk* a *Scotchman*, who were left by their refpective fhips, and lived alone upon this Ifland for fome years, and confequently were no ftrangers to its produce. *Selkirk*, who was the laft, after a ftay of between four and five years, was taken off the place by the Duke and Duchefs Privateers of *Briftol*, as may be feen at large in the journal of their voyage: His manner of life, during his folitude, was in moft particulars very remarkable; but there is one circumftance he relates, which was fo ftrangely verified by our own obfervation, that I cannot help reciting it. He tells us, amongft other things, as he often caught more goats than he wanted, he fometimes marked their ears and let them go. This was about thirty-two years before our arrival at the Ifland. Now it happened, that the firft goat that was killed by our people at their landing had his ears flit, whence we concluded, that he had doubt-lefs been formerly under the power of *Selkirk*. This was indeed an

animal.

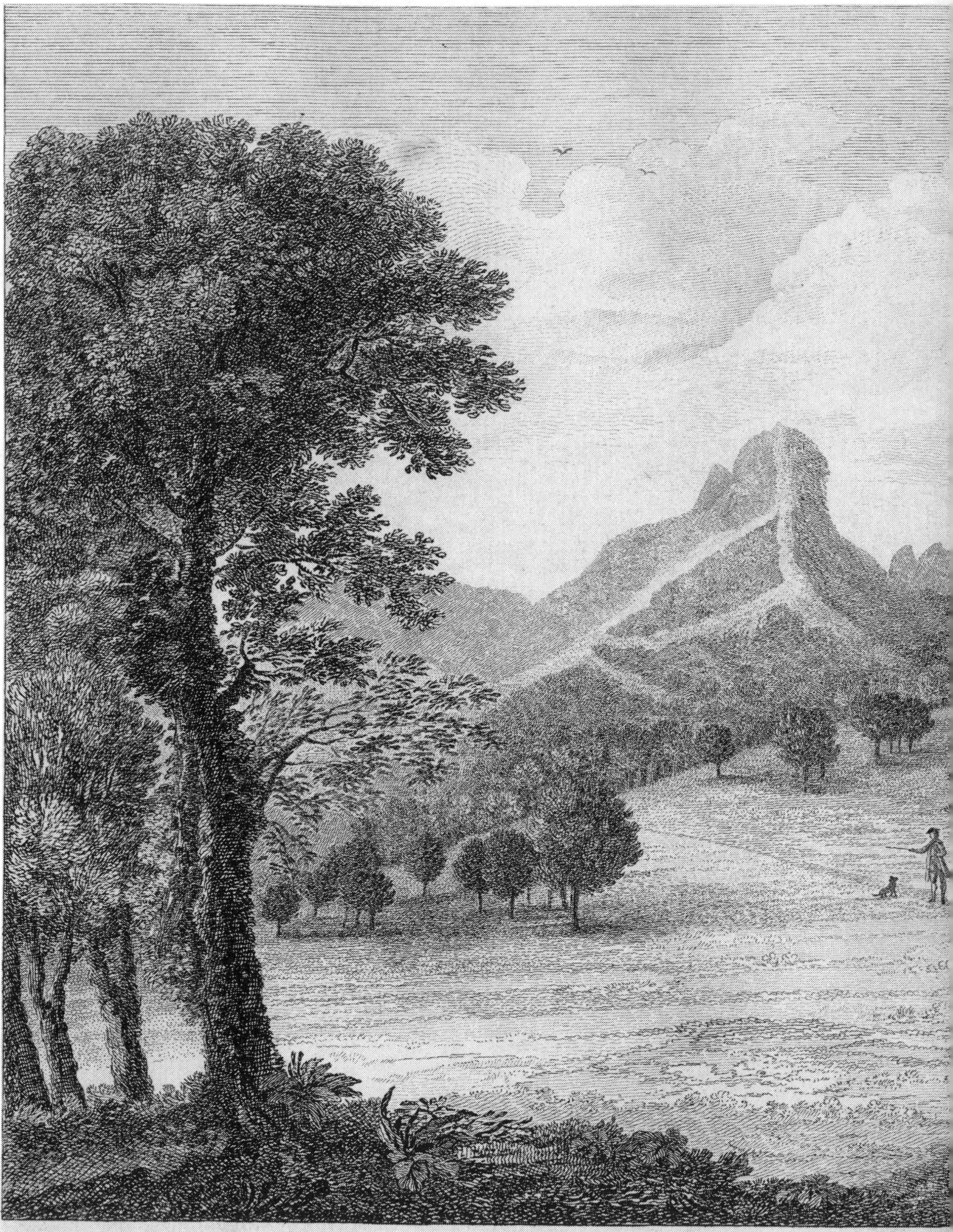

A View of the COMMODORES TEN

t the Island of JUAN FERNANDES.

J. Mason Sculp.

animal of a moſt venerable aſpect, dignified with an exceeding ma-
jeſtic beard, and with many other ſymptoms of antiquity. During
our ſtay on the Iſland, we met with others marked in the ſame
manner, all the males being diſtinguiſhed by an exuberance of
beard, and every other characteriſtick of extreme age.

But the great numbers of goats, which former writers deſcribed to
have been found upon this Iſland, are at preſent very much diminiſhed : For the *Spaniards* being informed of the advantages which
the buccaneers and privateers drew from the proviſions which goats-
fleſh here furniſhed them with, they have endeavoured to extir-
pate the breed, thereby to deprive their enemies of this relief. For
this purpoſe, they have put on ſhore great numbers of large dogs,
who have encreaſed apace, and have deſtroyed all the goats in the
acceſſible part of the country; ſo that there now remain only a few
amongſt the craggs and precipices, where the dogs cannot follow
them. Theſe are divided into ſeparate herds of twenty or thirty
each, which inhabit diſtinct faſtneſſes, and never mingle with each
other : By this means we found it extremely difficult to kill them ;
and yet we were ſo deſirous of their fleſh, which we all agreed
much reſembled veniſon, that we got knowledge, I believe, of all
their herds, and it was conceived, by comparing their numbers together, that they ſcarcely exceeded two hundred upon the whole
Iſland. I remember we had once an opportunity of obſerving a
remarkable diſpute betwixt a herd of theſe animals and a number
of dogs ; for going in our boat into the eaſtern bay, we ſaw ſome
dogs running very eagerly upon the foot, and being willing to diſcover what game they were after, we lay upon our oars ſome time
to view them, and at laſt we ſaw them take to a hill, and looking a
little further, we obſerved upon the ridge of it an herd of goats,
which ſeemed drawn up for their reception ; there was a very narrow path ſkirted on each ſide by precipices, on which the Maſter
of the herd poſted himſelf fronting the enemy, the reſt of the
goats being all behind him, where the ground was more open : As

R

this

this spot was inacceffible by any other path, excepting where this champion had placed himfelf, the dogs, though they ran up-hill with great alacrity, yet when they came within about twenty yards of him, durft not encounter him, (for he would infallibly have driven them down the precipice) but gave over the chace, and quietly laid themfelves down, panting at a great rate.

The dogs, who, as I have mentioned, are mafters of all the acceffible parts of the Ifland, are of various kinds, but fome of them very large, and are multiplied to a prodigious degree. They fometimes came down to our habitations at night, and ftole our provifion ; and once or twice they fet upon fingle perfons, but affiftance being at hand, they were driven off without doing any mifchief. As at prefent it is rare for goats to fall in their way, we conceived that they lived principally upon young feals ; and indeed fome of our people had the curiofity to kill dogs fometimes and drefs them, and they feemed to agree that they had a fifhy tafte.

Goats-flefh, as I have mentioned, being fcarce, we rarely being able to kill above one a day ; and our people growing tired of fifh, (which, as I fhall hereafter obferve, abounds at this place) they at laft condefcended to eat feals, which by degrees they came to relifh, and called it lamb. The feal, numbers of which haunt this Ifland, hath been fo often defcribed by former writers, that it is unneceffary to fay any thing particular about them in this place. But there is another amphibious creature to be met with here, called a fea-lyon, that bears fome refemblance to a feal, though it is much larger. This too we eat under the denomination of beef ; and as it is fo extraordinary an animal, I conceive, it well merits a particular annotation. They are in fize, when arrived at their full growth, from twelve to twenty feet in length, and from eight to fifteen in circumference : They are extremely fat, fo that after having cut thro' the fkin, which is about an inch in thicknefs, there is at leaft a foot of fat before you can come at either lean or bones, and we experienced more than once, that the fat of fome of the largeft afforded

us

A Sea-Lion

nd Lioness.

us a butt of oil. They are likewife very full of blood, for if they are deeply wounded in a dozen places, there will inftantly gufh out as many fountains of blood, fpouting to a confiderable diftance; and to try what quantity of blood they contained, we fhot one firft, and then cut its throat, and meafuring the blood that came from him, we found, that befides what remained in the veffels, which to be fure was confiderable, we got at leaft two hogfheads. Their fkins are covered with fhort hair of a light dun colour, but their tails, and their fins, which ferve them for feet on fhore, are almoft black; their fins or feet are divided at the ends like fingers, the web which joins them not reaching to the extremities, and each of thefe extremities is furnifhed with a nail. They have a diftant re-femblance to an overgrown feal, though in fome particulars there is a manifeft difference, efpecially in the males, who have a large fnout or trunk hanging down five or fix inches below the end of the upper jaw; this particular the females have not, and this renders the countenance of the male and female eafy to be diftinguifhed from each other, and befides, the males are of a much larger fize. The form and appearance both of the male and female are very exactly reprefented in the annexed plate, only the difproportion of their fize is not ufually fo great as is there exhibited, for the male was drawn from the life, after the largeft of thefe animals, which was found upon the Ifland: He was the mafter of the flock, and from his driving off the other males, and keeping a great number of fe-males to himfelf, he was by the feamen ludicroufly ftiled the Bafhaw. Thefe animals divide their time equally between the land and fea, continuing at fea all the fummer, and coming on fhore at the fetting in of the winter, where they refide during that whole feafon. In this interval they engender and bring forth their young, and have generally two at a birth; thefe they fuckle with their milk, they being at firft about the fize of a full-grown feal. Du-ring the time of thefe animals continuance on fhore, they feed on the grafs and verdure which grows near the bank of the frefh-wa-

ter

ter ſtreams; and, when not employed in feeding, ſleep in herds in
the moſt miry places they can find out. As they ſeem to be of a
very lethargic diſpoſition, and not eaſily awakened, each herd was
obſerved to place ſome of their males at a diſtance in the nature of
ſentinels, who never failed to alarm them, whenever our men attempted to moleſt, or even to approach them; and they were very
capable of alarming, even at a conſiderable diſtance, for the noiſe
they make is very loud and of different kinds, ſometimes grunting
like hogs, and at other times ſnorting like horſes in full vigour.
They often, eſpecially the males, have furious battles with each
other, principally about their females; and we were one day extremely ſurprized by the ſight of two animals, which at firſt appeared different from all we had ever obſerved, but, on a nearer
approach, they proved to be two ſea-lions, who had been goring
each other with their teeth, and were covered over with blood:
And the Baſhaw before-mentioned, who generally lay ſurrounded
with a ſeraglio of females, which no other male dared to approach, had not acquired that envied pre-eminence without many
bloody conteſts, of which the marks ſtill remained in the numerous ſcars which were viſible in every part of his body. We killed
many of them for food, particularly for their hearts and tongues,
which we eſteemed exceeding good eating, and preferable even to
thoſe of bullocks: And in general there was no difficulty in
killing them, for they were incapable either of eſcaping or re-
ſiſting, their motion being the moſt unweildy that can be conceived, their blubber, all the time they are moving, being agitated
in large waves under their ſkins. However, a ſailor one day being
careleſſly employed in ſkinning a young ſea-lion, the female, from
whence he had taken it, came upon him unperceived, and getting
his head in her mouth, ſhe with her teeth ſcored his ſkull in
notches in many places, and thereby wounded him ſo deſpe-
rately, that though all poſſible care was taken of him, he died in
a few days.

Theſe

These are the principal animals which we found upon the Island : For we saw but few birds, and those chiefly hawks, blackbirds, owls, and humming birds. We saw not the Pardela, which burrows in the ground, and which former writers have mentioned to be found here ; but as we often met with their holes, we supposed that the dogs had destroyed them, as they have almost done the cats, which were very numerous in *Selkirk's* time, but we saw not above one or two during our whole stay. However, the rats still keep their ground, and continue here in great numbers, and were very troublesome to us, by infesting our tents nightly.

But that which furnished us with the most delicious repasts at this Island, remains still to be described. This was the fish, with which the whole bay was most plentifully stored, and with the greatest variety : For we found here cod of a prodigious size ; and by the report of some of our crew, who had been formerly employed in the *Newfoundland* fishery, not in less plenty than is to be met with on the banks of that Island. We caught also cavallies, gropers, large breams, maids, silver fish, congers of a peculiar kind, and above all, a black fish which we most esteemed, called by some a Chimney sweeper, in shape resembling a carp. Indeed the beach is every where so full of rocks and loose stones, that there is no possibility of haling the Seyne ; but with hooks and lines we caught what numbers we pleased, so that a boat with two or three lines would return loaded with fish in about two or three hours time. The only interruption we ever met with, arose from great quantities of dog-fish and large sharks, which sometimes attended our boats and prevented our sport. Besides the fish we have already mentioned, we found here one delicacy in greater perfection, both as to size, flavour and quantity, than is perhaps to be met with in any other part of the world : This was sea cra-fish ; they generally weighed eight or nine pounds apiece, were of a most excellent taste, and lay in such abundance near the water's edge, that the

boat-

boat-hooks often ftruck into them, in putting the boat to and from the fhore.

Thefe are the moft material articles relating to the accommodations, foil, vegetables, animals, and other productions of the Ifland of *Juan Fernandes*: By which it muft appear, how properly that place was adapted for recovering us from the deplorable fituation to which our tedious and unfortunate navigation round Cape *Horn* had reduced us. And having thus given the reader fome idea of the fite and circumftances of this place, which was to be our refidence for three months, I fhall now proceed, in the next chapter, to relate all that occurred to us in that interval, refuming my narration from the 18th day of *June*, being the day in which the *Tryal* Sloop, having by a fquall been driven out to fea three days before, came again to her moorings, the day in which we finifhed the fending our fick on fhore, and about eight days after our firft anchoring at this Ifland.

CHAP.

CHAP. II.

The arrival of the *Gloucester* and the *Anne Pink* at the Island of *Juan Fernandes*, and the transactions at that place during this interval.

THE arrival of the *Tryal* Sloop at this Island, so soon after we came there ourselves, gave us great hopes of being speedily joined by the rest of the squadron; and we were for some days continually looking out; in expectation of their coming in sight. But near a fortnight being elapsed, without any of them having appeared, we began to despair of ever meeting them again; as we knew that had our ship continued so much longer at sea, we should every man of us have perished, and the vessel, occupied by dead bodies only, would have been left to the caprice of the winds and waves: And this we had great reason to fear was the fate of our consorts, as each hour added to the probability of these desponding suggestions.

But on the 21st of *June*, some of our people, from an eminence on shore, discerned a ship to leeward, with her courses even with the horizon; and they, at the same time, particularly observed, that she had no sail abroad except her courses and her main top-sail. This circumstance made them conclude that it was one of our squadron, which had probably suffered in her sails and rigging as severely as we had done: But they were prevented from forming more definite conjectures about her; for, after viewing her for a short time, the weather grew thick and hazy, and they lost sight of her. On this report, and no ship appearing for some days, we were all under the greatest concern, suspecting that her people were in the utmost distress for want of water, and so diminished and weakned by sickness, as not to be able to ply up to windward; so that we

feared,

feared that, after having been in fight of the Ifland, her whole crew would notwithftanding perifh at fea. However, on the 26th, towards noon, we difcerned a fail in the North Eaft quarter, which we conceived to be the very fame fhip that had been feen before, and our conjectures proved true ; and about one o'clock fhe approached fo near, that we could diftinguifh her to be the *Gloucefter*. As we had no doubt of her being in great diftrefs, the Commodore immediately ordered his boat to her affiftance, laden with frefh water, fifh and vegetables, which was a very feafonable relief to them ; for our apprehenfions of their calamities appeared to be but too well grounded, as perhaps there never was a crew in a more diftreffed fituation. They had already thrown over-board two thirds of their complement, and of thofe that remained alive, fcarcely any were capable of doing duty, except the officers and their fervants. They had been a confiderable time at the fmall allowance of a pint of frefh water to each man for twenty-four hours, and yet they had fo little left, that, had it not been for the fupply we fent them, they muft foon have died of thirft. The fhip plied in within three miles of the bay ; but, the winds and currents being contrary, fhe could not reach the road. However, fhe continued in the offing the next day, but had no chance of coming to an anchor, unlefs the wind and currents fhifted ; and therefore the Commodore repeated his affiftance, fending to her the *Tryal*'s boat manned with the *Centurion*'s people, and a farther fupply of water and other refrefhments. Captain *Mitchel*, the Captain of the *Gloucefter*, was under a neceffity of detaining both this boat and that fent the preceding day ; for without the help of their crews he had no longer ftrength enough to navigate the fhip. In this tantalizing fituation the *Gloucefter* continued for near a fortnight, without being able to fetch the road, though frequently attempting it, and at fome times bidding very fair for it. On the 9th of *July*, we obferved her ftretching away to the eaftward at a confiderable diftance, which we fuppofed was with a defign to get to the fouthward of the Ifland ; but as we foon loft fight of her, and fhe did

not

not appear for near a week, we were prodigiously concerned, knowing that she muft be again in extreme diftrefs for want of water. After great impatience about her, we difcovered her again on the 16th, endeavouring to come round the eaftern point of the Ifland; but the wind, ftill blowing directly from the bay, prevented her getting nearer than within four leagues of the land. On this, Captain *Mitchel* made fignals of diftrefs, and our long boat was fent to him with a ftore of water, and plenty of fifh, and other refrefhments. And the long-boat being not to be fpared, the Cockfwain had pofitive orders from the Commodore to return again immediately; but the weather proving ftormy the next day, and the boat not appearing, we much feared fhe was loft, which would have proved an irretrievable misfortune to us all: But, the 3d day after, we were relieved from this anxiety, by the joyful fight of the long-boat's fails upon the water; and we fent the Cutter immediately to her affiftance, who towed her along fide in a few hours. The crew of our long boat had taken in fix of the *Gloucefter*'s fick men to bring them on fhore, two of which had died in the boat. And now we learnt that the *Gloucefter* was in a moft dreadful condition, having fcarcely a man in health on board, except thofe they received from us; and, numbers of their fick dying daily, we found that, had it not been for the laft fupply fent by our long-boat, both the healthy and difeafed muft have all perifhed together for want of water. And thefe calamities were the more terrifying, as they appeared to be without remedy: For the *Gloucefter* had already fpent a month in her endeavours to fetch the bay, and fhe was now no farther advanced than at the firft moment fhe made the Ifland; on the contrary, the people on board her had worn out all their hopes of ever fucceeding in it, by the many experiments they had made of its difficulty. Indeed, the fame day her fituation grew more defperate than ever, for after fhe had received our laft fupply of refrefhments, we again loft fight of her; fo that we in general defpaired of her ever coming to an anchor.

S

Thus

Thus was this unhappy veſſel bandied about within a few leagues of her intended harbour, whilſt the neighbourhood of that place and of thoſe circumſtances, which could alone put an end to the calamities they laboured under, ſerved only to aggravate their diſtreſs, by torturing them with a view of the relief it was not in their power to reach. But ſhe was at laſt delivered from this dreadful ſituation, at a time when we leaſt expected it ; for after having loſt ſight of her for ſeveral days, we were pleaſingly ſurprized, on the morning of the 23d of *July*, to ſee her open the N. W point of the bay with a flowing ſail ; when we immediately diſpatched what boats we had to her aſſiſtance, and in an hour's time from our firſt perceiving her, ſhe anchored ſafe within us in the bay. And now we were more particularly convinced of the importance of the aſſiſtance and refreſhments we ſo often ſent them, and how impoſſible it would have been for a man of them to have ſurvived, had we given leſs attention to their wants ; for notwithſtanding the water, the greens, and freſh proviſions which we ſupplied them with, and the hands we ſent them to navigate the ſhip, by which the fatigue of their own people was diminiſhed, their ſick relieved, and the mortality abated ; notwithſtanding this indulgent care of the Commodore, they yet buried three fourths of their crew, and a very ſmall proportion of the remainder were capable of aſſiſting in the duty of the ſhip. On their coming to an anchor, our firſt care was to aſſiſt them in mooring, and our next to ſend the ſick on ſhore : Theſe were now reduced by deaths to leſs than fourſcore, of which we expected to loſe the greateſt part ; but whether it was, that thoſe fartheſt advanced in the diſtemper were all dead, or that the greens and freſh proviſions we had ſent on board had prepared thoſe which remained for a more ſpeedy recovery, it happened contrary to our expectations, that their ſick were in general relieved and reſtored to their ſtrength, in a much ſhorter time than our own had been when we firſt came to the Iſland, and very few of them died on ſhore.

I have

I have thus given an account of the principal events, relating to the arrival of the *Gloucester*, in one continued narration : I shall only add, that we never were joined by any other of our ships, except our Victualler, the *Anna Pink*, who came in about the middle of *August*, and whose history I shall more particularly relate hereafter. And I shall now return to the account of our own transactions on board and on shore, during the interval of the *Gloucester*'s frequent and ineffectual attempts to reach the Island.

Our next employment, after sending our sick on shore from the *Centurion*, was cleansing our ship and filling our water. The first of these measures was indispensibly necessary to our future health, as the numbers of sick, and the unavoidable negligence arising from our deplorable situation at sea, had rendered the decks most intolerably loathsome. And the filling our water was a caution that appeared not less essential to our future security, as we had reason to apprehend that accidents might oblige us to quit the Island at a very short warning ; for some Appearances, which we had discovered on shore upon our first landing, gave us grounds to believe, that there were *Spanish* cruisers in these seas, which had left the Island but a short time before our arrival, and might possibly return there again, either for a recruit of water, or in search of us ; for as we could not doubt, but that the sole business they had at sea was to intercept us, and we knew that this Island was the likeliest place, in their own opinion, to meet with us. The circumstances, which gave rise to these reflections (in part of which we were not mistaken, as shall be observed more at large hereafter) were our finding on shore several pieces of earthen jars, made use of in those seas for water and other liquids, which appeared to be fresh broken : We saw too many heaps of ashes, and near them fish-bones and pieces of fish, besides whole fish scattered here and there, which plainly appeared to have been but a short time out of the water, as they were but just beginning to decay. These appearances were certain indications that there had been ships at this place but a short time before we came there ; and as all *Spanish* Merchant-men are instructed to

S 2   avoid

avoid the Ifland, on account of its being the common rendezvous of their enemies, we concluded thofe who had touched here to be fhips of force ; and not knowing that *Pizarro* was returned to *Buenos Ayres,* and ignorant what ftrength might have been fitted out at *Callao,* we were under fome concern for our fafety, being in fo wretched and enfeebled a condition, that notwithftanding the rank of our fhip, and the fixty guns fhe carried on board, which would only have aggravated our difhonour, there was fcarcely a privateer fent to fea, that was not an over-match for us. However, our fears on this head proved imaginary, and we were not expofed to the difgrace, which might have been expected to have befallen us, had we been neceffitated (as we muft have been, had the enemy appeared) to fight our fixty-gun fhip with no more than thirty hands.

Whilft the cleaning our fhip and the filling our water went on, we fet up a large copper-oven on fhore near the fick tents, in which we baked bread every day for the fhip's company, being extremely defirous of recovering our fick as foon as poffible, and conceiving that new bread added to their greens and frefh fifh, might prove a powerful article in their relief. Indeed we had all imaginable reafon to endeavour at the augmenting our prefent ftrength, as every little accident, which to a full crew would be infignificant, was extremely alarming in our prefent helplefs fituation : Of this, we had a troublefome inftance on the 30th of *June* ; for at five in the morning, we were aftonifhed by a violent guft of wind directly off fhore, which inftantly parted our fmall bower cable about ten fathom from the ring of the anchor : The fhip at once fwung off to the beft bower, which happily ftood the violence of the jerk, and brought us up with two cables an end in eighty fathom. At this time we had not above a dozen feamen in the fhip, and we were apprehenfive, if the fquall continued, that we fhould be driven to fea in this wretched condition. However, we fent the boat on fhore, to bring off all that were capable of acting; and the wind, foon abating of its fury, gave us an opportunity of receiving the boat

back

back again with a reinforcement. With this additional ſtrength we immediately went to work, to heave in what remained of the cable, which we ſuſpected had received ſome damage from the foulneſs of the ground before it parted ; and agreeable to our conjecture, we found that ſeven fathom and a half of the outer end had been rubbed, and rendered unſerviceable. In the afternoon, we bent the cable to the ſpare anchor, and got it over the ſhip's ſide ; and the next morning, *July* 1, being favoured with the wind in gentle breezes, we warped the ſhip in again, and let go the anchor in forty-one fathom ; the eaſtermoſt point now bearing from us E. $\frac{1}{2}$ S ; the weſtermoſt N. W. by W ; and the bay as before, S. S. W ; a ſituation, in which we remained ſecure for the future. But we were much concerned for the loſs of our anchor, and ſwept frequently for it, in hopes to have recovered it ; but the buoy having ſunk at the very inſtant that the cable parted, we were never able to find it.

And now as we advanced in *July*, ſome of our men being tolerably recovered, the ſtrongeſt of them were employed in cutting down trees, and ſplitting them into billets ; while others, who were too weak for this employ, undertook to carry the billets by one at a time to the water ſide : This they performed, ſome of them with the help of crutches, and others ſupported by a ſingle ſtick. We next ſent the forge on ſhore, and employed our ſmiths, who were but juſt capable of working, in mending our chain-plates, and our other broken and decayed iron work. We began too the repairs of our rigging ; but as we had not a ſufficient quantity of junk to make ſpun-yarn, we deferred the general over-hale, in hopes of the daily arrival of the *Glouceſter*, who we knew had a great quantity of junk on board. However, that we might make as great diſpatch as poſſible in our refitting, we ſet up a large tent on the beach for the ſail-makers ; and they were immediately employed in repairing our old ſails, and making us new ones.

Theſe occupations, with our cleanſing and watering the ſhip, (which was by this time pretty well compleated) the attendance on

our

our fick, and the frequent relief fent to the *Gloucefter*, were the principal tranfactions of our infirm crew, till the arrival of the *Gloucefter* at an anchor in the bay. And then Captain *Mitchel* waiting on the Commodore, informed him, that he had been forced by the winds, in his laft abfence, as far as the fmall Ifland called *Mafa-Fuero*, lying about twenty-two leagues to the weftward of *Juan Fernandes*; and that he endeavoured to fend his boat on fhore at this place for water, of which he could obferve feveral ftreams, but the wind blew fo ftrong upon the fhore, and occafioned fuch a furf, that it was impoffible for the boat to land; though the attempt was not altogether ufelefs, as they returned with a boat-load of fifh. This Ifland had been reprefented by former Navigators as a barren rock; but Captain *Mitchel* affured the Commodore, that it was almoft every where covered with trees and verdure, and was near four miles in length; and added, that it appeared to him far from impoffible, but fome fmall bay might be found on it, which might afford fufficient fhelter for any fhip defirous of refrefhing there.

As four fhips of our fquadron were miffing, this defcription of the Ifland of *Mafa-Fuero* gave rife to a conjecture, that fome of them might poffibly have fallen in with that Ifland, and have miftaken it for the true place of our rendezvous; and this fufpicion was the more plaufible, as we had no draught of either Ifland that could be relied on. In confequence of this reafoning, Mr. *Anfon* determined to fend the *Tryal* Sloop thither, as foon as fhe could be fitted for the fea, in order to examine all its bays and creeks, that we might be fatisfied whether any of our miffing fhips were there or not. For this purpofe, fome of our beft hands were fent on board the *Tryal* the next morning, to overhale and fix her rigging; and our long boat was employed in compleating her water; and whatever ftores and neceffaries fhe wanted, were immediately fupplied, either from the *Centurion* or the *Gloucefter*. But it was the 4th of *Auguft* before the *Tryal* was in readinefs to fail, when having weighed, it foon after fell calm, and the tide fet her very near the eaftern fhore:

Captain

Captain *Saunders* hung out lights, and fired several guns to acquaint us with his danger; upon which all the boats were sent to his relief, who towed the Sloop into the bay; where she anchored until the next morning, and then weighing again, proceeded on her cruize with a fair breeze.

And now after the *Gloucester*'s arrival, we were employed in earnest in examining and repairing our rigging; but in the stripping our foremast, we were alarmed by discovering it was sprung just above the partners of the upper deck. The spring was two inches in depth, and twelve in circumference; but the Carpenters inspecting it gave it as their opinion, that fishing it with two leaves of an anchor stock, would render it as secure as ever. But our greatest difficulty in refitting was the want of cordage and canvas; for tho' we had taken to sea much greater quantities of both, than had ever been done before, yet the continued bad weather we met with, had occasioned such a consumption of these stores, that we were driven to great straits: For after working up all our junk and old shrouds, to make twice-laid cordage, we were at last obliged to unlay a cable to work into running rigging. And with all the canvas, and remnants of old sails that could be mustered, we could only make up one compleat suit.

Towards the middle of *August* our men being indifferently recovered, they were permitted to quit their sick tents, and to build separate huts for themselves, as it was imagined, that by living apart, they would be much cleanlier, and consequently likely to recover their strength the sooner; but at the same time particular orders were given, that on the firing of a gun from the ship, they should instantly repair to the water-side. Their employment on shore was now either the procuring of refreshments, the cutting of wood, or the making of oil from the blubber of the sea-lions. This oil served us for several uses, as burning in lamps, or mixing with pitch to pay the ships sides, or, when mixed with wood-ashes, to supply the use of tallow, of which we had none left, to give the ship boot-hose tops. Some of the men too were occupied in salting of cod;

for

for there being two *Newfoundland* fishermen in the *Centurion*, the Commodore made use of them in laying in a confiderable quantity of falted cod for a fea-ftore; but very little of it was made ufe of, as it was afterwards thought to be as productive of the fcurvy, as any other kind of falt provifions.

I have before-mentioned, that we had a copper-oven on fhore to bake bread for the fick; but it happened that the greateft part of the flower, for the ufe of the fquadron, was embarked on board our Victualler the *Anna Pink* : And I fhould have mentioned, that the *Tryal* Sloop, at her arrival, had informed us, that on the 9th of *May* fhe had fallen in with our Victualler, not far diftant from the Continent of *Chili*; and had kept company with her for four days, when they were parted in a hard gale of wind. This gave us fome room to hope that fhe was fafe, and that fhe might join us; but all *June* and *July* being paft without any news of her, we fuf-pected fhe was loft; and at the end of *July* the Commodore or-dered all the fhips to a fhort allowance of bread. And it was not in our bread only, that we feared a deficiency; for fince our arrival at this Ifland, we difcovered that our former Purfer had neglected to take on board large quantities of feveral kinds of provifions, which the Commodore had expreffly ordered him to receive; fo that the fuppofed lofs of our Victualler, was on all accounts a mortifying confideration. However, on *Sunday*, the 16th of *Auguft*, about noon, we efpied a fail in the northern quarter, and a gun was im-mediately fired from the *Centurion*, to call off the people from fhore; who readily obeyed the fummons, and repaired to the beach, where the boats waited to carry them on board. And now being prepared for the reception of this fhip in view, whether friend or enemy, we had various fpeculations about her; at firft, many imagined it to be the *Tryal* Sloop returned from her cruize; but as fhe drew nearer this opinion was confuted, by obferving fhe was a veffel with three mafts; and then other conjectures were eagerly canvaffed, fome judging it to be the *Severn*, others the *Pearl*, and feveral affirm-ing that it did not belong to our fquadron : But about three in the

afternoon

afternoon our difputes were ended, by an unanimous perfuafion that it was our Victualler the *Anna Pink*. This fhip, though, like the *Gloucefter*, fhe had fallen in to the northward of the Ifland, had yet the good fortune to come to an anchor in the bay, at five in the afternoon. Her arrival gave us all the fincereft joy; for each fhip's company was now reftored to their full allowance of bread, and we were now freed from the apprehenfions of our provifions falling fhort, before we could reach fome amicable port; a calamity, which in thefe feas is of all others the moft irretrievable. This was the laft fhip that joined us; and the dangers fhe encountered, and the good fortune which fhe afterwards met with, being matters worthy of a feparate narration, I fhall refer them, together with a fhort account of the other fhips of the fquadron, to the enfuing chapter.

T.

CHAP.

CHAP. III.

A short narrative of what befel the *Anna Pink* before she joined us, with an account of the loss of the *Wager*, and of the putting back of the *Severn* and *Pearl*, the two remaining ships of the squadron.

ON the first appearance of the *Anna Pink*, it seemed wonderful to us how the crew of a vessel, which came to this rendezvous two months after us, should be capable of working their ship in the manner they did, with so little appearance of debility and distress : But this difficulty was soon solved when she came to an anchor ; for we then found that they had been in harbour since the middle of *May*, which was near a month before we arrived at *Juan Fernandes* : So that their sufferings (the risque they had run of shipwreck only excepted) were greatly short of what had been undergone by the rest of the squadron. It seems, on the 16th of *May*, they fell in with the land, which was then but four leagues distant, in the latitude of 45°: 15 South. On the first fight of it they wore ship and stood to the southward, but their foretopsail splitting, and the wind being W. S. W, they drove towards the shore ; and the Captain at last, either unable to clear the land, or as others say, resolved to keep the sea no longer, steered for the coast, with a view of discovering some shelter amongst the many Islands which then appeared in sight : And about four hours after the first view of the land, the *Pink* had the good fortune to come to an anchor, to the eastward of the Island of *Inchin* ; but as they did not run sufficiently near to the East-shore of that Island, and had not hands to veer away the cable briskly, they were soon driven to the eastward, deepning their water from twenty-five fathom to thirty-five, and still continuing to drive, they, the next day, the 17th of

May,

May, let go their sheet anchor; which though it brought them up for a short time, yet, on the 18th, they drove again, till they came into sixty five fathom water, and were now within a mile of the land, and expected to be forced on shore every moment, in a place where the coast was very high and steep to, that there was not the least prospect of saving the ship or cargo; and their boats being very leaky, and there being no appearance of a landing-place, the whole crew, confisting of sixteen men and boys, gave themselves over for lost, for they apprehended, that if any of them by some extraordinary chance should get on shore, they would, in all probability, be massacred by the Savages on the coast: For these, knowing no other *Europeans* but *Spaniards*, it might be expected they would treat all strangers with the same cruelty which they had so often and so signally exerted against their *Spanish* neighbours. Under these terrifying circumstances the *Pink* drove nearer and nearer to the rocks which formed the shore; but at last, when the crew expected each instant to strike, they perceived a small opening in the land, which raised their hopes; and immediately cutting away their two anchors, they steered for it, and found it to be a small channel betwixt an Island and the Main, which led them into a most excellent harbour, which, for its security against all winds and swells, and the smoothness of its waters, may perhaps compare with any in the known world. And this place being scarcely two miles distant from the spot where they deemed their destruction inevitable, the horrors of shipwreck and of immediate death, which had so long, and so strongly possessed them, vanished almost instantaneously, and gave place to the more joyous ideas of security, repose, and refreshment.

In this harbour, discovered in this almost miraculous manner, the *Pink* came to an anchor in twenty-five fathom water, with only a hawser, and a small anchor of about three hundred weight: And here she continued for near two months, refreshing her people, who were many of them ill of the scurvy, but were soon restored to perfect health by the fresh provisions, of which they procured good

store,

ftore, and the excellent water with which the adjacent fhore abounded. But as this place may prove of the greateft importance to future Navigators, who may be forced upon this coaft by the wefterly winds, which are almoft perpetual in that part of the world, I fhall, before I enter into any farther particulars of the adventures of the *Pink*, give the beft account I could collect of this Port, its fituation, conveniencies and productions.

To facilitate the knowledge of this place to thofe who may hereafter be defirous of making ufe of it, there is annexed a plan both of the harbour itfelf, and of the large bay before it, thro' which the *Pink* drove. This plan is not perhaps in all refpects fo accurate as might be wifhed, it being compofed from the memorandums and rude fketches of the Mafter and Surgeon, who were not, I prefume, the ableft draughts-men. But as the principal parts were laid down by their eftimated diftances from each other, in which kind of eftimations it is well known the greateft part of failors are very dextrous, I fuppofe the errors are not very confiderable. Its latitude, which is indeed an important point, is not well afcertained, the *Pink* having no obfervation either the day before fhe came here, or within a day of her leaving it: But it is fuppofed that it is not very diftant from 45° 30 South, and the large extent of the bay before the harbour renders this uncertainty the lefs material. The Ifland of *Inchin* lying before the bay is fuppofed to be one of the Iflands of *Chonos*, which are mentioned in the *Spanifh* accounts, as fpreading all along that coaft; and are faid by them to be inhabited by a barbarous people, famous for their hatred of the *Spaniards*, and for their cruelties to fuch of that Nation as have fallen into their hands: And it is poffible too that the land, near which the harbour itfelf lies, may be another of thofe Iflands, and that the Continent may be confiderably farther to the eaftward. The depths of water in the different parts of the Port, and the channels by which it communicates with the bay, are fufficiently marked in the plan. But it muft be remembred, that there are two coves in it where fhips may conveniently heave down, the water being conftantly fmooth: And
there

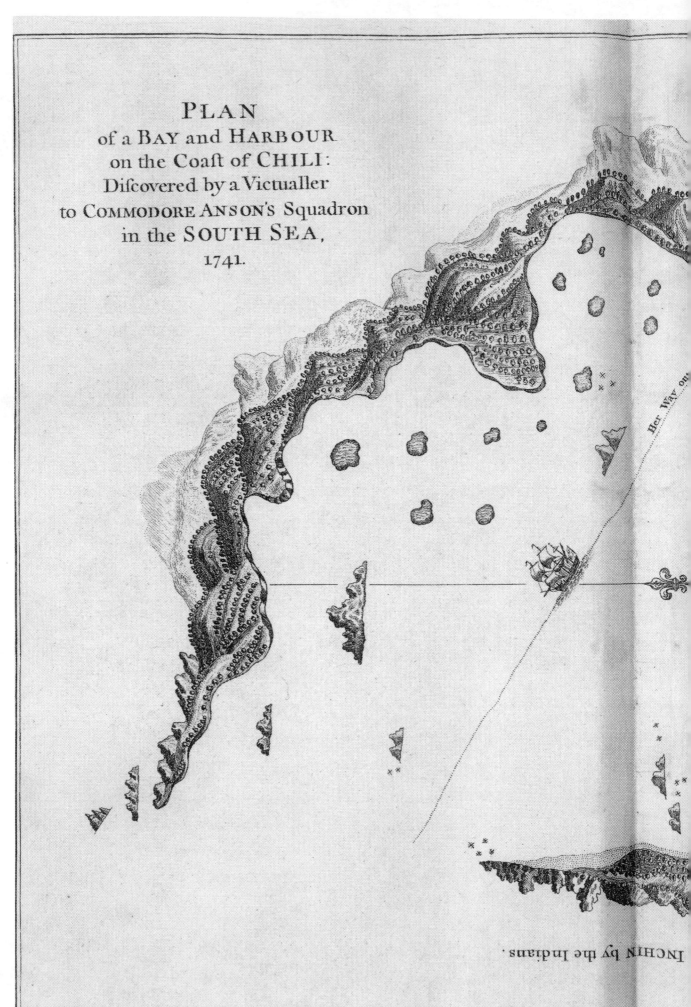

PLAN
of a BAY and HARBOUR
on the Coaft of CHILI:
Difcovered by a Victualler
to COMMODORE ANSON'S Squadron
in the SOUTH SEA,
1741.

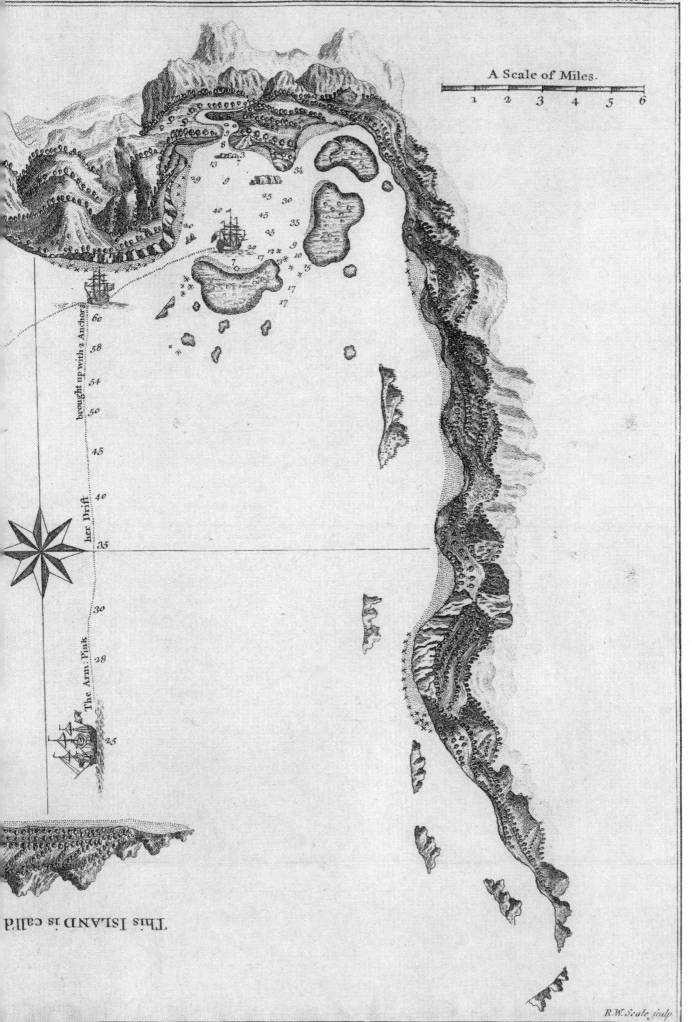

A Scale of Miles.

1 2 3 4 5 6

brought up with a Anchor

her Drift

The Arm: Pink

This ISLAND is call'd

R.W. Scale fculp.

there are several fine runs of excellent fresh water, which fall into the harbour, and some of them so luckily situated, that the casks may be filled in the long-boat with an hose: The most remarkable of these runs is the stream marked in the N. E. part of the Port. This is a fresh water river, and here the *Pink*'s people got some few mullets of an excellent flavour; and they were persuaded that, in a proper season (it being winter when they were there) it abounded with fish. The principal refreshments they met with in this port were greens, as wild celery, nettle-tops, &c. (which after so long a continuance at sea they devoured with great eagerness) shel-fish, as cockles and muscles of an extraordinary size, and extreme-ly delicious; and good store of geese, shags, and penguins. The climate, though it was the depth of winter, was not remarkably rigorous; nor the trees, and the face of the country destitute of verdure; and doubtless in the summer many other species of fresh provision, besides these here enumerated, might be found there. And notwithstanding the tales of the *Spanish* Historians, in relation to the violence and barbarity of the inhabitants, it doth not appear thar their numbers are sufficient to give the least jealousy to any ship of ordinary force, or that their disposition is by any means so mis-chievous or merciless as hath hitherto been represented: And besides all these advantages, it is so far removed from the *Spanish* frontier, and so little known to the *Spaniards* themselves, that there is reason to suppose, that with proper precautions a ship might continue here undiscovered for a long time. It is also a place of great defence; for by possessing the Island that closes up the harbour, and which is accessible in very few places, a small force might defend this Port against all the strength the *Spaniards* could muster in that part of the world; for this Island towards the harbour is steep to, and has six fathom water close to the shore, so that the *Pink* anchored within forty yards of it: Whence it is obvious how impossible it would prove, either to board or to cut out any vessel protected by a·force posted on shore within pistol-shot, and where those who were thus posted could not themselves be attacked. All these cir-

cumstances

cumftances·feem to render this place worthy of a more accurate examination; and it is to be hoped, that the important ufes which this rude account of it feems to fuggeft, may hereafter recommend it to the confideration of the Public, and to the attention of thofe who are more immediately entrufted with the conduct of our naval affairs.

After·this defcription of the place where the *Pink* lay for two months, it may be expected that I fhould relate the difcoveries made by the crew on the adjacent coaft, and the principal incidents during their ftay there : But here I muft obferve, that, being only a few in number, they did not dare to detach any of their people on diftant difcoveries; for they were perpetually terrified with the apprehenfion that they fhould be attacked either by the *Spaniards* or the *Indians*; fo that their excurfions were generally confined; to that tract of land which furrounded the Port, and where they were never out of view of the fhip. But even had they at firft known how. little foundation there was for thefe fears, yet the country in the neighbourhood was fo grown up with wood, and. traverfed with mountains, that it appeared impracticable to penetrate it : So that no. account of the inland parts could be expected from them. Indeed they were able to difprove the relations given by *Spanifh* writers, who had reprefented this coaft as inhabited by a fierce and powerful people : For they were certain that no fuch inhabitants were there to be found, at leaft during the winter feafon; fince all the time they continued there, they faw no more than one *Indian* family, which came into the harbour in a periagua, about a month after the arrival of the *Pink*, and confifted of an *Indian* near forty years old, his wife, and two children, one three years of age, and the other ftill at the breaft. They feemed to have with them all their property, which was a dog, and a cat, a fifhing-net, a hatchet, a knife, a cradle, fome bark of trees intended for the covering of a hut, a reel, fome worfted, a flint and fteel, and a few roots of a yellow hue and a very difagreeable tafte, which ferved them for bread. The Mafter of the *Pink*, as foon as he perceived them,

fent

fent his yawl, who brought them on board ; and fearing, leaft they might difcover him if they were permitted to go away, he took, as he conceived, proper precautions for fecuring them, but without any mixture of ill ufage or violence : For in the day-time they were permitted to go where they pleafed about the fhip, but at night were locked up in the fore-caftle. As they were fed in the fame manner with the reft of the crew, and were often indulged with brandy which they feemed greatly to relifh, it did not at firft appear that they were much diffatisfied with their fituation, efpecially as the Mafter took the *Indian* on fhore when he went a fhooting, (who always feemed extremely delighted when the Mafter killed his game) and as all the crew treated them with great humanity : But it was foon perceived, that though the woman continued eafy and chearful, yet the man grew penfive and reftlefs at his confinement. He feemed to be a perfon of good natural parts, and tho' not capable of converfing with the *Pink*'s people, otherwife than by figns, was yet very curious and inquifitive, and fhowed great dexterity in the manner of making himfelf underftood. In particular, feeing fo few people on board fuch a large fhip, he let them know, that he fuppofed they were once more numerous : And to reprefent to them what he imagined was become of their companions, he laid himfelf down on the deck, clofing his eyes, and ftretching himfelf out motionlefs, to imitate the appearance of a dead body. But the ftrongeft proof of his fagacity was the manner of his getting away ; for after being in cuftody on board the *Pink* eight days, the fcuttle of the fore-caftle, where he and his family were locked up every night, happened to be unnailed, and the following night being extremely dark and ftormy, he contrived to convey his wife and children through the unnailed fcuttle, and then over the fhip's fide into the yawl ; and to prevent being purfued, he cut away the long-boat and his own periagua, which were towing a-ftern, and immediately rowed afhore. All this he conducted with fo much diligence and fecrecy, that though there was a watch on the quarter-deck with loaded arms, yet he was not dif-

covered

covered by them, till the noife of his oars in the water, after he had put off from the fhip, gave them notice of his efcape ; and then it was too late either to prevent him or to purfue him ; for, their boats being all a drift, it was a confiderable time before they could contrive the means of getting on fhore themfelves to fearch for their boats. The *Indian* too by this effort, befides the recovery of his liberty, was in fome fort revenged on thofe who had confined him; both by the perplexity they were involved in from the lofs of their boats, and by the terror he threw them into at his departure ; for on the firft alarm of the watch, who cried out, *the Indians*, the whole fhip was in the utmoft confufion, believing themfelves to be boarded by a fleet of armed periagua's.

The refolution and fagacity with which the *Indian* behaved up-on this occafion, had it been exerted on a more extenfive object than the retrieving the freedom of a fingle family, might perhaps have immortalized the exploit, and have given him a rank amongft the illuftrious names of antiquity. Indeed his late Mafters did fo much juftice to his merit, as to own that it was a moft gallant enterprize, and that they were grieved they had ever been neceffita-ted, by their attention to their own fafety, to abridge the liberty of a perfon, of whofe prudence and courage they had now fuch a diftinguifhed proof. And as it was fuppofed by fome of them that he ftill continued in the woods in the neighbourhood of the port, where it was feared he might fuffer for want of provifions, they eafily prevailed upon the Mafter to leave a quantity of fuch food, as they thought would be moft agreeable to him, in a parti-cular part where they imagined he would be likely to find it : And there was reafon to conjecture, that this piece of humanity was not altogether ufelefs to him ; for, on vifiting the place fometime after, it was found that the provifion was gone, and in a manner that made them conclude it had fallen into his hands.

But however, though many of them were fatisfied that this *In-dian* ftill continued near them ; yet others would needs conclude, that he was gone to the Ifland of *Chiloe*, where they feared he

would

would alarm the *Spaniards*, and would foon return with a force fuf-ficient to furprize the *Pink* : And on this occafion the Mafter of the *Pink* was prevailed on to omit firing the evening gun; for it muft be remembered, (and there is a particular reafon hereafter for attending to this circumftance) that the Mafter, from an oftentatious imitation of the practice of men of war, had hitherto fired a gun every evening at the fetting of the watch. This he pretended was to awe the enemy, if there was any within hearing, and to convince them that the *Pink* was always on her guard; but it being now reprefented to him, that his great fecurity was his conceal-ment, and that the evening gun might poffibly difcover him, and ferve to guide the enemy to him, he was prevailed on, as has been mentioned, to omit it for the future : And his crew being now well refrefhed, and their wood and water fufficiently replenifhed, he, in a few days after the efcape of the *Indian*, put to fea, and had a for-tunate paffage to the rendezvous at the Ifland of *Juan Fernandes*, where he arrived on the 16th of *Auguft*, as hath been already men-tioned in the preceding chapter.

This veffel, the *Anna Pink*, was, as I have obferved, the laft that joined the Commodore at *Juan Fernandes*. The remaining fhips of the fquadron were the *Severn*, the *Pearl*, and the *Wager* ftore-fhip: The *Severn* and *Pearl* parted company with the fqua-dron off Cape *Noir*, and, as we afterwards learnt, put back to the *Brazils*: So that of all the fhips which came into the *South-Seas*, the *Wager*, Captain *Cheap*, was the only one that was miffing. This fhip had on board fome field-pieces mounted for land fervice, together with fome coehorn mortars, and feveral kinds of artillery, ftores and tools, intended for the operations on fhore: And there-fore, as the enterprize on *Baldivia* had been refolved on for the firft undertaking of the fquadron, Captain *Cheap* was extremely folici-tous that thefe materials, which were in his cuftody, might be ready before *Baldivia*; that if the fquadron fhould poffibly rendezvous there, (as he knew not the condition they were then reduced to) no delay nor difappointment might be imputed to him.

U   But

But whilft the *Wager*, with thefe views, was making the beft of her way to her firft rendezvous off the Ifland of *Socoro*, whence (as there was little probability of meeting any of the fquadron there) fhe propofed to fteer directly for *Baldivia*, fhe made the land on the 14th of *May*, about the latitude of 47° South; and, the Captain exerting himfelf on this occafion, in order to get clear of it, he had the misfortune to fall down the after-ladder, and thereby diflocated his fhoulder, which rendered him incapable of acting. This accident, together with the crazy condition of the fhip, which was little better than a wreck, prevented her from getting off to fea, and entangled her more and more with the land, fo that the next morning, at day-break, fhe ftruck on a funken rock, and foon after bilged, and grounded between two fmall Iflands, at about a mufquet-fhot from the fhore.

In this fituation the fhip continued entire a long time, fo that all the crew had it in their power to get fafe on fhore; but a general confufion taking place, numbers of them, inftead of confulting their fafety, or reflecting on their calamitous condition, fell to pillaging the fhip, arming themfelves with the firft weapons that came to hand, and threatning to murder all who fhould oppofe them. This frenzy was greatly heightned by the liquors they found on board, with which they got fo extremely drunk, that fome of them tumbling down between decks were drowned, as the water flowed in, being incapable of getting up and retreating to other places where the water had not yet entered : And the Captain, having done his utmoft to get the whole crew on fhore, was at laft obliged to leave thefe mutineers behind him, and to follow his officers, and fuch as he had been able to prevail on ; but he did not fail to fend back the boats, to perfuade thofe who remained, to have fome regard to their prefervation ; tho' all his efforts were for fome time without fuccefs. However, the weather next day proving ftormy, and there being great danger of the fhip's parting, they began to be alarmed with the fears of perifhing, and were defirous of getting to land ; but it feems their madnefs had not yet left

them,

them, for the boat not appearing to fetch them off fo foon as they expected, they at laft pointed a four pounder, which was on the quarter-deck, againft the hut, where they knew the Captain refided on fhore, and fired two fhot which paffed but juft over it.

From this fpecimen of the behaviour of part of the crew, it will not be difficult to frame fome conjecture of the diforder and anarchy which took place, when they at laft got all on fhore. For the men conceived, that by the lofs of the fhip, the authority of the officers was at an end; and, they being now on a defolate coaft, where fcarcely any other provifions could be got, except what fhould be faved out of the wreck, this was another infurmountable fource of difcord: For as the working upon the wreck, and the fecuring the provifions, fo that they might be preferved for future exigencies as much as poffible, and the taking care that what was neceffary for immediate fubfiftance might be fparingly and equally diftributed, were matters not to be brought about but by difcipline and fubordination; the mutinous difpofition of the people, ftimulated by the impulfes of immediate hunger, rendered every regulation made for this purpofe ineffectual: So that there were continual concealments, frauds and thefts, which animated each man againft his fellow, and produced infinite feuds and contefts. And hence there was conftantly kept on foot a perverfe and malevolent turn of temper, which rendered them utterly ungovernable.

But befides thefe heart-burnings occafioned by petulance and hunger, there was another important point, which fet the greateft part of the people at variance with the Captain. This was their differing with him in opinion, on the meafures to be purfued in the prefent exigency: For the Captain was determined, if poffible, to fit up the boats in the beft manner he could, and to proceed with them to the northward. For having with him above an hundred men in health, and having gotten fome fire-arms and ammunition from the wreck, he did not doubt but they could mafter any *Spanifh* veffel they fhould meet with in thofe feas: And he thought he could not fail of meeting with one in the neighbourhood of *Chiloe*

or

or *Baldivia*, in which, when he had taken her, he intended to proceed to the rendezvous at *Juan Fernandes*; and he farther infifted, that fhould they meet with no prize by the way, yet the boats alone would eafily carry them there. But this was a fcheme that, however prudent, was no ways relifhed by the generality of his people; for, being quite jaded with the diftreffes and dangers they had already run through, they could not think of profecuting an enterprize farther, which had hitherto proved fo difaftrous : And therefore the common refolution was to lengthen the long-boat, and with that and the reft of the boats to fteer to the fouthward, to pafs through the Streights of *Magellan*, and to range along the Eaft fide of *South America*, till they fhould arrive at *Brazil*, where they doubted not to be well received, and to procure a paffage to *Great-Britain*. This project was at firft fight infinitely more hazardous and tedious than what was propofed by the Captain; but as it had the air of returning home, and flattered them with the hopes of bringing them once more to their native country, this circumftance alone rendered them inattentive to all its inconveniencies, and made them adhere to it with infurmountable obftinacy; fo that the Captain himfelf, though he never changed his opinion, was yet obliged to give way to the torrent, and in appearance to acquiefce in this refolution, whilft he endeavoured under-hand to give it all the obftruction he could; particularly in the lengthning of the long-boat, which he contrived fhould be of fuch a fize, that though it might ferve to carry them to *Juan Fernandes*, would yet, he hoped, appear incapable of fo long a navigation, as that to the coaft of *Brazil*.

But the Captain, by his fteady oppofition at firft to this favourite project, had much embittered the people againft him; to which likewife the following unhappy accident greatly contributed. There was a Midfhipman whofe name was *Cozens*, who had appeared the foremoft in all the refractory proceedings of the crew. He had involved himfelf in brawls with moft of the officers who had adhered to the Captain's authority, and had even treated the Captain

himfelf

himself with great abuse and infolence. As his turbulence and brutality grew every day more and more intolerable, it was not in the leaft doubted, but there were fome violent meafures in agitation, in which *Cozens* was engaged as the ringleader: For which reafon the Captain, and thofe about him, conftantly kept themfelves on their guard. But at laft the Purfer, having, by the Captain's order, ftopped the allowance of a fellow who would not work, *Cozens*, though the man did not complain to him, intermedled in the affair with great eagernefs; and groflly infulting the Purfer, who was then delivering out provifions juft by the Captain's tent, and was himfelf fufficiently violent, the Purfer, enraged by his fcurrility, and perhaps piqued by former quarrels, cried out *a mutiny*, adding, *that the dog had piftols*, and then himfelf fired a piftol at *Cozens*, which however mift him: But the Captain, on this outcry and the report of the piftol, rufhed out of his tent; and, not doubting but it had been fired by *Cozens* as the commencement of a mutiny, he immediately fhot him in the head without farther deliberation, and though he did not kill him on the fpot, yet the wound proved mortal, and he died about fourteen days after.

This incident, however difpleafing to the people, did yet, for a confiderable time, awe them to their duty, and rendered them more fubmiffive to the Captain's authority; but at laft, when towards the middle of *October* the long-boat was nearly compleated, and they were preparing to put to fea, the additional provocation he gave them by covertly traverfing their project of proceeding through the Streights of *Magellan*, and their fears that he might at length engage a party fufficient to overturn this favourite meafure, made them refolve to make ufe of the death of *Cozens* as a reafon for depriving him of his command, under pretence of carrying him a prifoner to *England*, to be tried for murder; and he was accordingly confined under a guard. But they never intended to carry him with them, as they too well knew what they had to apprehend on their return to *England*, if their Commander fhould be prefent to confront them:

And

And therefore, when they were juft ready to put to fea, they fet him at liberty, leaving him and the few who chofe to take their fortunes with him, no other embarkation but the yawl, to which the barge was afterwards added, by the people on board her being prevailed on to return back.

When the fhip was wreckt, there remained alive on board the *Wager* near an hundred and thirty perfons; of thefe above thirty died during their ftay upon the place, and near eighty went off in the long-boat, and the Cutter to the fouthward: So that there remained with the Captain, after their departure, no more than nineteen perfons, which however was as many as the barge and the yawl, the only embarkations left them, could well carry off. It was the 13th of *October*, five months after the fhipwreck, that the long-boat, converted into a fchooner, weighed, and ftood to the fouthward, giving the Captain, who, with Lieutenant *Hamilton* of the land-forces and the furgeon, was then on the beach, three cheers at their departure. It was the 29th of *January* following before they arrived at *Rio Grande*, on the coaft of *Brazil*: And having, by various accidents, left about twenty of their people on fhore at the different places they touched at, and a greater number having perifhed by hunger during the courfe of their navigation, there were no more than thirty of them left, when they arrived in that Port. Indeed, the undertaking of itfelf was a moft extraordinary one; for, not to mention the length of the run, the veffel was fcarcely able to contain the number that firft put to fea in her; and their ftock of provifions (being only what they had faved out of the fhip) was extremely flender, and the Cutter, the only boat they had with them, foon broke away from the ftern, and was ftaved to pieces; fo that when their provifion and their water failed them, they had frequently no means of getting on fhore to fearch for a frefh fupply.

When the long-boat and Cutter were gone, the Captain, and thofe who were left with him, propofed to pafs to the north-

ward

ward in the barge and yawl : But the weather was fo bad, and the difficulty of fubfifting fo great, that it was two months after the departure of the long-boat before he was able to put to fea. It feems, the place, where the *Wager* was caft away, was not a part of the Continent, as was firft imagined, but an Ifland at fome diftance from the Main, which afforded no other forts of provifion but fhelfifh, and a few herbs; and as the greateft part of what they had gotten from the fhip was carried off in the long-boat, the Captain and his people were often in great neceffity, efpecially as they chofe to pre-ferve, what little fea-provifions remained, for their ftore when they fhould go to the northward. During their refidence at this Ifland, which was by the feamen denominated *Wager's Ifland*, they had now and then a ftraggling canoe or two of *Indians*, which came and bartered their fifh and other provifions with our people. This was indeed fome little fuccour, and at another feafon might perhaps have been greater; for as there were feveral *Indian* huts on the fhore, it was fuppofed that in fome years, during the height of fummer, many of thefe favages might refort thither to fifh : And from what has been related in the account of the *Anna Pink*, it fhould feem to be the general practice of thofe *Indians* to frequent this coaft in the fummer time for the benefit of fifhing, and to retire in the winter into a better climate, more to the northward.

And on this mention of the *Anna Pink*, I cannot but obferve, how much it is to be lamented, that the *Wager's* people had no knowledge of her being fo near them on the coaft; for as fhe was not above thirty leagues diftant from them, and came into their neighbourhood about the fame time the *Wager* was loft, and was a fine roomy fhip, fhe could eafily have taken them all on board, and have carried them to *Juan Fernandes*. Indeed, I fufpect fhe was ftill nearer to them than what is here eftimated; for feveral of the *Wager's* people, at different times, heard the report of a cannon, which I conceive could be no other than the evening gun fired from

the

the *Anna Pink*, efpecially as what was heard at *Wager's Ifland* was about the fame time of the day. But to return to Captain *Cheap*.

Upon the 14th of *December*, the Captain and his people embarked in the barge and the yawl, in order to proceed to the northward, taking on board with them all the provifions they could amafs from the wreck of the fhip; but they had fcarcely been an hour at fea, when the wind began to blow hard, and the fea ran fo high, that they were obliged to throw the greateft part of their provifions over-board, to avoid immediate deftruction. This was a terrible misfortune, in a part of the world where food is fo difficult to be got: However, they ftill perfifted in their defign, putting on fhore as often as they could to feek fubfiftance. But about a fort-night after, another dreadful accident befel them, for the yawl funk at an anchor, and one of the men in her was drowned; and as the barge was incapable of carrying the whole company, they were now reduced to the hard neceffity of leaving four marines behind them on that defolate fhore. But they ftill kept on their courfe to the northward, ftruggling with their difafters, and greatly delayed by the perverfenefs of the winds, and the frequent interruptions which their fearch after food occafioned: Till at laft, about the end of *January*, having made three unfuccefsful attempts to double a head-land, which they fuppofed to be what the *Spaniards* called Cape *Tres Montes*, it was unanimoufly refolved to give over this expedition, the difficulties of which appeared infuperable, and to return again to *Wager Ifland*, where they got back about the middle of *February*, quite difheartned and dejected with their reiterated difappointments, and almoft perifhing with hunger and fatigue.

However, on their return they had the good luck to meet with feveral pieces of beef, which had been wafhed out of the fhip, and were fwimming in the fea. This was a moft feafonable relief to them after the hardfhips they had endured: And to compleat their good fortune, there came, in a fhort time, two canoes of *Indians*,

amongft

amongſt which was a native of *Chiloe*, who ſpoke a little *Spaniſh*; and the ſurgeon, who was with Captain *Cheap*, underſtanding that language, he made a bargain with the *Indian*, that if he would carry the Captain and his people to *Chiloe* in the barge, he ſhould have her, and all that belonged to her for his pains. Accordingly, on the 6th of *March*, the eleven perſons to which the company was now reduced, embarked in the barge on this new expedition; but after having proceeded for a few days, the Captain and four of his principal officers being on ſhore, the ſix, who together with an *Indian* remained in the barge, put off with her to ſea, and did not return.

By this means there were left on ſhore Captain *Cheap*, Mr. *Hamilton* Lieutenant of marines, the Honourable Mr. *Byron* and Mr. *Campbel*, Midſhipmen, and Mr. *Elliot* the ſurgeon. One would have thought that their diſtreſſes had long before this time been incapable of augmentation; but they found, on reflection, that their preſent ſituation was much more diſmaying than any thing they had yet gone through, being left on a deſolate coaſt without any proviſion, or the means of procuring any; for their arms, ammunition, and every conveniency they were maſters of, except the tattered habits they had on, were all carried away in the barge.

But when they had ſufficiently revolved in their own minds the various circumſtances of this unexpected calamity, and were perſuaded that they had no relief to hope for, they perceived a canoe at a diſtance, which proved to be that of the *Indian*, who had undertaken to carry them to *Chiloe*, he and his family being then on board it. He made no difficulty of coming to them; for it ſeems he had left Captain *Cheap* and his people a little before to go a fiſhing, and had in the mean time committed them to the care of the other *Indian*, whom the ſailors had carried to ſea in the barge. But when he came on ſhore, and found the barge gone and his companion miſſing, he was extremely concerned, and could with difficulty be perſuaded that the other *Indian* was not murthered; but, being

X   at laſt

at laſt ſatisfied with the account that was given him, he ſtill undertook to carry them to the *Spaniſh* ſettlements, and (as the *Indians*
are well ſkilled in fiſhing and fowling) to procure them proviſions
by the way.

About the middle of *March*, Captain *Cheap* and the four who
were left with him ſet out for *Chiloe*, the *Indian* having procured
a number of canoes, and gotten many of his neighbours together
for that purpoſe. Soon after they embarked, Mr. *Elliot* the ſurgeon died, ſo that there now remained only four of the whole company At laſt, after a very complicated paſſage by land and water, Captain *Cheap*, Mr. *Byron*, and Mr. *Campbel*, arrived in the
beginning of *June* at the Iſland of *Chiloe*, where they were received by the *Spaniards* with great humanity; but, on account of
ſome quarrel among the *Indians*, Mr. *Hamilton* did not get thither
till two months after. Thus, above a twelvemonth after the loſs
of the *Wager*, ended this fatiguing peregrination, which by a variety of misfortunes had diminiſhed the company from twenty to
no more than four, and thoſe too brought ſo low, that had their
diſtreſſes continued but a few days longer, in all probability none of
them would have ſurvived. For the Captain himſelf was with
difficulty recovered ; and the reſt were ſo reduced by the ſeverity
of the weather, their labour, and their want of all kinds of neceſſaries, that it was wonderful how they ſupported themſelves ſo
long. After ſome ſtay at *Chiloe*, the Captain and the three who
were with him were ſent to *Valparaiſo*, and thence to St. *Jago*,
the Capitol of *Chili*, where they continued above a year : But on
the advice of a cartel being ſettled betwixt *Great-Britain* and
Spain, Captain *Cheap*, Mr. *Byron*, and Mr. *Hamilton*, were permitted to return to *Europe* on board a *French* ſhip. The
other Midſhipman, Mr. *Campbel*, having changed his religion,
whilſt at St. *Jago*, choſe to go back to *Buenos Ayres* with *Pizarro* and his officers, with whom he went afterwards to *Spain* on
board the *Aſia* ; and there having failed in his endeavours to

procure

procure a commiffion from the Court of *Spain*, he returned to *England*, and attempted to get reinftated in the *Britifh* Navy; and has fince publifhed a narration of his adventures, in which he complains of the injuftice that had been done him, and ftrongly difavows his ever being in the *Spanifh* fervice : But as the change of his religion, and his offering himfelf to the Court of *Spain*, (though not accepted) are matters which, he is confcious, are capable of being inconteftably proved; on thefe two heads, he has been entirely filent. And now, after this account of the accidents which befel the *Anna Pink*, and the cataftrophe of the *Wager*, I fhall again refume the thread of our own ftory.

CHAP.

CHAP. IV.

Conclusion of our proceedings at *Juan Fernandes*, from the arrival of the *Anna Pink*, to our final departure from thence.

ABOUT a week after the arrival of our Victualler, the *Tryal* Sloop, that had been sent to the Island of *Mafa-Fuero*, returned to an anchor at *Juan Fernandes*, after having been round that Island, without meeting any part of our squadron. As, upon this occasion, the Island of *Mafa-Fuero* was more particularly examined, than I dare say it had ever been before, or perhaps ever will be again; and as the knowledge of it may, in certain circumstances, be of great consequence hereafter, I think it incumbent on me to insert the accounts given of this place, by the officers of the *Tryal* Sloop.

The *Spaniards* have generally mentioned two Islands, under the name of *Juan Fernandes*, stiling them the greater and the less: The greater being that Island where we anchored, and the less being the Island we are now describing, which, because it is more distant from the Continent, they have distinguished by the name of *Mafa-Fuero*. The *Tryal* Sloop found that it bore from the greater *Juan Fernandes* W. by S, and was about twenty-two leagues distant. It is much larger than has been generally reported; for former writers have represented it as a barren rock, destitute of wood and water, and altogether inaccessible; whereas our people found it was covered with trees, and that there were several fine falls of water pouring down its sides into the sea : They found too, that there was a place where a ship might come to an anchor on the North side of it, though indeed the anchorage is inconvenient; for the bank extends but a little way, is steep to, and has very deep

water

A view of the north east side of MASA-

RO *lying in the latitude of 33° 5' S°*

J.S. Muller Sculp.

A view of the west side of N

Plate XXII.

A-FUERO *distant 4 miles.*

water upon it, so that you must come to an anchor very near the shore, and there lie exposed to all the winds but a southerly one: And besides the inconvenience of the anchorage, there is also a reef of rocks running off the eastern point of the Island, about two miles in length; but there is little danger to be feared from them, because they are always to be seen by the seas breaking over them. This place has at present one advantage beyond the Island of *Juan Fernandes*; for it abounds with goats, who, not being accustomed to be disturbed, were no ways shy or apprehensive of danger, till they had been frequently fired at. These animals reside here in great tranquillity, the *Spaniards* having not thought the Island considerable enough to be frequented by their enemies, and therefore they have not been solicitous in destroying the provisions upon it; so that no dogs have been hitherto set on shore there. And besides the goats, our people found there vast numbers of seals and sealions: And upon the whole, they seemed to imagine, that though it was not the most eligible place for a ship to refresh at, yet in case of necessity it might afford some sort of shelter, and prove of considerable use, especially to a single ship, who might apprehend meeting with a superior force at *Fernandes*. The appearance of its N. E. side, and also of its West side, may be seen in the two annexed plates. This may suffice in relation to the Island of *Masa-Fuero*.

The latter part of the month of *August* was spent in unloading the provisions from the *Anna Pink*; and here we had the mortification to find that great quantities of our provisions, as bread, rice, groats, &c. were decayed, and unfit for use. This was owing to the water the *Pink* had made by her working and straining in bad weather; for hereby several of her casks had rotted, and her bags were soaked through. And now, as we had no farther occasion for her service, the Commodore, pursuant to his orders from the board of Admiralty, sent notice to Mr. *Gerard* her Master, that he discharged the *Anna Pink* from the service of attending the squadron; and gave him, at the same time, a certificate, specifying
how

how long fhe had been employed. In confequence of this difmif-
fion, her Mafter was at liberty, either to return directly to *England*,
or to make the beft of his way to any Port, where he thought he
could take in fuch a cargoe, as would anfwer the intereft of his
Owners. But the Mafter, being fenfible of the bad condition of the
fhip and of her unfitnefs for any fuch voyage, wrote the next day
an anfwer to the Commodore's meffage, acquainting Mr. *Anfon*,
that from the great quantity of water the *Pink* had made in her
paffage round Cape *Horn*, and fince that, in the tempeftuous weather he had met with on the coaft of *Chili*, he had reafon to apprehend that her bottom was very much decayed; and that befides,
her upper works were rotten abaft; that fhe was extremely leaky;
that her fore-beam was broke; and that, in his opinion, it was
impoffible to proceed to fea with her before fhe had been thoroughly refitted: He therefore requefted the Commodore, that the
Carpenters of the fquadron might be directed to furvey her, that
their judgment of her condition might be known. In compliance
with this defire, Mr. *Anfon* immediately ordered the Carpenters to
take a careful and ftrict furvey of the *Anna Pink*, and to give him
a faithful report under their hands of the condition in which they
found her, directing them at the fame time to proceed herein with
fuch circumfpection, that, if they fhould be hereafter called upon,
they might be able to make oath of the veracity of their proceedings. Purfuant to thefe orders, the Carpenters immediately fet
about the examination, and the next day made their report; which
was, that the *Pink* had no lefs than fourteen knees and twelve
beams broken and decayed; that one breaft-hook was broken, and
another rotten; that her water-ways were open and decayed; that
two ftandards were broken, as alfo feveral clamps, befides others
which were rotten; that all her iron-work was greatly decayed;
that her fpirkiting and timbers were very rotten; and that, having
ripped off part of her fheathing, they found her wales and outfide
planks extremely defective, and her bows and decks very leaky;
and in confequence of thefe defects and decays they certified, that

in

in their opinion she could not depart from the Island without great hazard, unless she was first of all thoroughly refitted.

The thorough refitting of the *Anna Pink*, proposed by the Carpenters, was, in our present situation, impossible to be complied with, as all the plank and iron in the squadron was insufficient for that purpose. And now the Master finding his own sentiments confirmed by the opinion of all the Carpenters, he offered a petition to the Commodore in behalf of his Owners, desiring that, since it appeared he was incapable of leaving the Island, Mr. *Anson* would please to purchase the hull and furniture of the *Pink* for the use of the squadron. Hereupon the Commodore ordered an inventory to be taken of every particular belonging to the *Pink*, with its just value: And as by this inventory it appeared, that there were many stores which would be useful in refitting the other ships, and which were at present very scarce in the squadron, by reason of the great quantities that had been already expended, he agreed with Mr. *Gerard* to purchase the whole together for 300 *l.* The *Pink* being thus broken up, Mr. *Gerard*, with the hands belonging to the *Pink*, were sent on board the *Gloucester*; as that ship had buried the greatest number of men in proportion to her complement. But afterwards, one or two of them were received on board the *Centurion* on their own petition, they being extremely averse to failing in the same ship with their old Master, on account of some particular ill usage they conceived they had suffered from him.

This transaction brought us down to the beginning of *September*, and our people by this time were so far recovered of the scurvy, that there was little danger of burying any more at present; and therefore I shall now sum up the total of our loss since our departure from *England*, the better to convey some idea of our past sufferings, and of our present strength. We had buried on board the *Centurion*, since our leaving St. *Helens*, two hundred and ninety-two, and had now remaining on board two hundred and fourteen. This will doubtless appear a most extraordinary mortality: But yet on board the *Gloucester* it had been much greater; for out of a much

smaller

smaller crew than ours they had buried the same number, and had only eighty-two remaining alive. It might be expected that on board the *Tryal*, the slaughter would have been the most terrible, as her decks were almost constantly knee-deep in water; but it happened otherwise, for she escaped more favourably than the rest, since she only buried forty-two, and had now thirty-nine remaining alive. The havock of this disease had fallen still severer on the invalids and marines than on the sailors; for on board the *Centurion*, out of fifty invalids and seventy-nine marines, there remained only four invalids, including officers, and eleven marines; and on board the *Gloucester* every invalid perished; and out of forty-eight marines, only two escaped. From this account it appears, that the three ships together departed from *England* with nine hundred and sixty-one men on board, of whom six hundred and twenty-six were dead before this time; so that the whole of our remaining crews, which were now to be distributed amongst three ships, amounted to no more than three hundred and thirty-five men and boys; a number, greatly insufficient for the manning the *Centurion* alone, and barely capable of navigating all the three, with the utmost exertion of their strength and vigour. This prodigious reduction of our men was still the more terrifying, as we were hitherto uncertain of the fate of *Pizarro*'s squadron, and had reason to suppose, that some part of it at least had got round into these seas: Indeed, we were satisfied from our own experience, that they must have suffered greatly in their passage; but then every port in the *South-Seas* was open to them, and the whole power of *Chili* and *Peru* would doubtless be united in refreshing and refitting them, and recruiting the numbers they had lost. Besides, we had some obscure knowledge of a force to be fitted out from *Callao*; and, however contemptible the ships and sailors of this part of the world may have been generally esteemed, it was scarcely possible for any thing, bearing the name of a ship of force, to be feebler or less considerable than ourselves. And had there been nothing to be apprehended from the naval power of the *Spaniards* in this part of the world, yet our enfeebled

feebled condition would neverthelefs give us the greateft uneafinefs, as we were incapable of attempting any of their confiderable places; for the rifquing of twenty men, weak as we then were, was rifquing the fafety of the whole: So that we conceived we fhould be neceffitated to content ourfelves with what few prizes we could pick up at fea, before we were difcovered; after which, we fhould in all probability be obliged to depart with precipitation, and ef-teem ourfelves fortunate to regain our native country, leaving our enemies to triumph on the inconfiderable mifchief they had re-ceived from a fquadron, whofe equipment had filled them with fuch dreadful apprehenfions. This was a fubject, on which we had reafon to imagine the *Spanifh* oftentation would remarkably exert itfelf; though the caufes of our difappointment and their fecurity were neither to be fought for in their valour nor our mifconduct.

Such were the defponding reflections which at that time arofe on the review and comparifon of our remaining ftrength with our original numbers: Indeed our fears were far from being groundlefs, or difproportioned to our feeble and almoft defperate fituation. It is true, the final event proved more honourable than we had fore-boded; but the intermediate calamities did likewife greatly furpafs our moft gloomy apprehenfions, and could they have been pre-dicted to us at this Ifland of *Juan Fernandes*, they would doubtlefs have appeared infurmountable. But to return from this digreffion.

In the beginning of *September*, as has been already mentioned, our men were tolerably well recovered; and now, the time of navi-gation in this climate drawing near, we exerted ourfelves in getting our fhips in readinefs for the fea. We converted the fore-maft of the Victualler into a main-maft for the *Tryal* Sloop; and ftill flat-tering ourfelves with the poffibility of the arrival of fome other fhips of our fquadron, we intended to leave the main-maft of the Victualler, to make a mizen-maft for the *Wager*. Thus all hands being employed in forwarding our departure, we, on the 8th, about eleven in the morning, efpied a fail to the N. E, which continued to approach us, till her courfes appeared even with the

Y   horizon.

horizon. In this interval we all had hopes fhe might prove one of our own fquadron; but at length finding fhe fteered away to the eaftward, without haling in for the Ifland, we concluded fhe muft be a *Spaniard*. And now great difputes were fet on foot about the poffibility of her having difcovered our tents on fhore, fome of us ftrongly infifting, that fhe had doubtlefs been near enough to have perceived fomething that had given her a jealoufy of an enemy, which had occafioned her ftanding to the eaftward without haling in; but leaving thefe contefts to be fettled afterwards, it was refolved to purfue her, and, the *Centurion* being in the greateft forwardnefs, we immediately got all our hands on board, fet up our rigging, bent our fails, and by five in the afternoon got under fail. We had at this time very little wind, fo that all the boats were employed to tow us out of the bay; and even what wind there was lafted only long enough to give us an offing of two or three leagues, when it flatted to a calm. The night coming on we loft fight of the chace, and were extremely impatient for the return of day-light, in hopes to find that fhe had been becalmed as well as we; though I muft confefs, that her greater diftance from the land was a reafonable ground for fufpecting the contrary, as we indeed found in the morning to our great mortification; for though the weather continued perfectly clear, we had no fight of the fhip from the maft-head. But as we were now fatisfied that it was an enemy, and the firft we had feen in thefe feas, we refolved not to give over the fearch lightly; and, a fmall breeze fpringing up from the W. N. W, we got up our top-gallant mafts and yards, fet all the fails, and fteered to the S. E, in hopes of retrieving our chace, which we imagined to be bound to *Valparaifo*. We continued on this courfe all that day and the next, and then not getting fight of our chace we gave over the purfuit, conceiving that by that time fhe muft, in all probability, have reached her Port. And now we prepared to return to *Juan Fernandes*, and haled up to the S. W. with that view, having but very little wind till the 12th, when, at three in the morning, there fprung up a frefh gale from

the

the W. S. W, and we tacked and ftood to the N. W: And at day-break we were agreeably furprized with the fight of a fail on our weather-bow, between four and five leagues diftant. On this we crouded all the fail we could, and ftood after her, and foon perceived it not to be the fame fhip we originally gave chace to. She at firft bore down upon us, fhowing *Spanifh* colours, and making a fignal as to her confort; but obferving that we did not anfwer her fignal, fhe inftantly loofed clofe to the wind, and ftood to the fouthward. Our people were now all in fpirits, and put the fhip about with great alacrity; and as the chace appeared to be a large fhip, and had miftaken us for her confort, we conceived that fhe was a man of war, and probably one of *Pizarro's* fquadron: This induced the Commodore to order all the officers cabins to be knocked down and thrown over-board, with feveral cafks of wa-ter and provifions which ftood between the guns; fo that we had foon a clear fhip, ready for an engagement. About nine o'clock we had thick hazy weather and a fhower of rain, during which we loft fight of the chace; and we were apprehenfive, if the weather fhould continue, that by going upon the other tack, or by fome other artifice, fhe might efcape us; but it clearing up in lefs than an hour, we found that we had both weathered and fore-reached upon her confiderably, and now we were near enough to difcover that fhe was only a Merchantman, without fo much as a fingle tire of guns. About half an hour after twelve, being then within a reafonable diftance of her, we fired four fhot amongft her rigging; on which, they lowered their top-fails, and bore down to us, but in very great confufion, their top-gallant fails and ftay-fails all flutter-ing in the wind: This was owing to their having let run their fheets and halyards juft as we fired at them; after which, not a man amongft them had courage enough to venture aloft (for there the fhot had paffed but juft before) to take them in. As foon as the veffel came within hail of us, the Commodore ordered them to bring to under his lee-quarter, and then hoifted out the boat, and fent Mr. *Saumarez*, his firft Lieutenant, to take poffeffion of the

prize,

prize, with directions to send all the prisoners on board the *Centurion*, but first the officers and passengers. When Mr. *Saumarez* came on board them, they received him at the side with the strongest tokens of the most abject submission; for they were all of them (especially the passengers, who were twenty-five in number) extremely terrified, and under the greatest apprehensions of meeting with very severe and cruel usage; but the Lieutenant endeavoured, with great courtesy, to dissipate their fright, assuring them, that their fears were altogether groundless, and that they would find a generous enemy in the Commodore, who was not less remarkable for his lenity and humanity, than for his resolution and courage. The prisoners, who were first sent on board the *Centurion*, informed us, that our prize was called *Nuestra Senora del Monte Carmelo*, and was commanded by Don *Manuel Zamorra*. Her cargoe consisted chiefly of sugar, and great quantities of blue cloth made in the province of *Quito*, somewhat resembling our *English* coarse broad-cloths, but inferiour to them. They had besides several bales of a coarser sort of cloth, of different colours, somewhat like *Colchester* bays, called by them *Pannia da Tierra*, with a few bales of cotton and tobacco; which, though strong, was not ill flavoured. These were the principal goods on board her; but we found besides, what was to us much more valuable than the rest of the cargoe: This was some trunks of wrought plate, and twenty-three serons of dollars, each weighing upwards of 200 *l.* averdupois. The ship's burthen was about four hundred and fifty tuns; she had fifty-three sailors on board, both whites and blacks; she came from *Callao*, and had been twenty-seven days at sea, before she fell into our hands. She was bound to the port of *Valparaiso* in the kingdom of *Chili*, and proposed to have returned from thence loaded with corn and *Chili* wine, some gold, dried beef, and small cordage, which at *Callao* they convert into larger rope. Our prize had been built upwards of thirty years; yet as they lie in harbour all the winter months, and the climate is favourable, they esteemed it no very great age. Her rigging was very indifferent, as were likewise her

sails,

fails, which were made of Cotton. She had only three four pounders, which were altogether unferviceable, their carriages being
fcarcely able to fupport them: And there were no fmall arms on
board, except a few piftols belonging to the paffengers. The prifoners
informed us, that they left *Callao* in company with two other fhips,
whom they had parted with fome days before, and that at firft
they conceived us to be one of their company ; and by the defcription we gave them of the fhip we had chafed from *Juan Fernandes*,
they affured us, fhe was of their number, but that the coming in
fight of that Ifland was directly repugnant to the Merchant's inftructions, who had expreffly forbid it, as knowing that if any
Englifh fquadron was in thofe feas, the Ifland of *Fernandes* was
moft probably the place of their rendezvous.

And now, after this fhort account of the fhip and her cargoe, it
is neceffary that I fhould relate the important intelligence which we
met with on board her, partly from the information of the prifoners, and partly from the letters and papers which fell into
our hands. We here firft learnt with certainty the force and deftination of that fquadron, which cruifed off the *Maderas* at our arrival there, and afterwards chafed the *Pearl* in our paffage to port
St. *Julian*. This we now knew was a fquadron compofed of five
large *Spanifh* fhips, commanded by Admiral *Pizarro*, and purpofely
fitted out to traverfe our defigns, as hath been already more amply related in the 3d chapter of the 1ft book. And we had, at the fame time,
the fatisfaction to find, that *Pizarro*, after his utmoft endeavours to
gain his paffage into thefe feas, had been forced back again into the
river of *Plate*, with the lofs of two of his largeft fhips : And befides
this difappointment of *Pizarro*, which, confidering our great debility, was no unacceptable intelligence, we farther learnt, that an embargo had been laid upon all fhipping in thefe feas, by the Viceroy
of *Peru*, in the month of *May* preceding, on a fuppofition that
about that time we might arrive upon the coaft. But on the account fent over-land by *Pizarro* of his own diftreffes, part of which
they knew we muft have encountered, as we were at fea during the

fame

fame time, and on their having no news of us in eight months after we were known to fet fail from St. *Catherine's*, they were fully perfuaded that we were either fhip-wreck'd, or had perifhed at fea, or at leaft had been obliged to put back again; for it was conceived impoffible for any fhips to continue at fea during fo long an interval: And therefore, on the application of the Merchants, and the firm perfuafion of our having mifcarried, the embargo had been lately taken off.

This laft article made us flatter ourfelves, that, as the enemy was ftill a ftranger to our having got round Cape *Horn*, and the navigation of thcfe feas was reftored, we might meet with fome confiderable captures, and might thereby indemnify ourfelves for the incapacity we were now under of attempting any of their confiderable fettlements on fhore. And thus much we were certain of, from the information of our prifoners, that, whatever our fuccefs might be as to the prizes we might light on, we had nothing to fear, weak as we were, from the *Spanifh* force in this part of the world; though we difcovered that we had been in moft imminent peril from the enemy, when we leaft apprehended it, and when our other diftreffes were at the greateft height; for we learnt, from the letters on board, that *Pizarro*, in the exprefs he difpatched to the Viceroy of *Peru*, after his return to the river of *Plate*, had intimated to him, that it was poffible fome part at leaft of the *Englifh* fquadron might get round; but that, as he was certain from his own experience, that if they did arrive in thofe feas it muft be in a very weak and defencelefs condition, he advifed the Viceroy, in order to be fecure at all events, to fit out what fhips of force he had, and fend them to the fouthward, where, in all probability, they would intercept us fingly, and before we had an opportunity of touching any where for refrefhment; in which cafe, he doubted not but we fhould prove an eafy conqueft. The Viceroy of *Peru* approved of this advice, and immediately fitted out four fhips of force from *Callao*; one of fifty guns, two of forty guns, and one of twenty-four guns: Three of them were ftationed off the Port of *Conception*, and one of them

at

at the Iſland of *Fernandes* ; and in theſe ſtations they continued cruiſing for us till the 6th of *June*, when not ſeeing any thing of us, and conceiving it to be impoſſible that we could have kept the ſeas ſo long, they quitted their cruiſe and returned to *Callao*, fully ſatisfied that we had either periſhed, or at leaſt had been driven back. As the time of their quitting their ſtation was but a few days before our arrival at the Iſland of *Fernandes*, it is evident, that had we made that Iſland on our firſt ſearch for it, without haling in for the main to ſecure our eaſting, (a circumſtance, which at that time we conſidered as very unfortunate to us, on account of the numbers which we loſt by our longer continuance at ſea) had we, I ſay, made the Iſland on the 28th of *May*, when we firſt expected to ſee it, and were in reality very near it, we had doubt-leſs fallen in with ſome part of the *Spaniſh* ſquadron ; and in the diſtreſſed condition we were then in, the meeting with a healthy well provided enemy, was an incident that could not but have been perplexing, and might perhaps have proved fatal, not only to us, but to the *Tryal*, the *Glouceſter*, and the *Anna Pink*, who ſepa-rately joined us, and who were each of them leſs capable than we were of making any conſiderable reſiſtance. I ſhall only add, that theſe *Spaniſh* ſhips ſent out to intercept us, had been greatly ſhat-tered by a ſtorm during their cruiſe ; and that, after their arrival at *Callao*, they had been laid up. And our priſoners aſſured us, that whenever intelligence was received at *Lima*, of our being in theſe ſeas, it would be at leaſt two months before this armament could be again fitted out.

The whole of this intelligence was as favourable, as we in our reduced circumſtances could wiſh for. And now we were fully ſa-tisfied as to the broken jars, aſhes, and fiſh-bones, which we had obſerved at our firſt landing at *Juan Fernandes*, theſe things being doubtleſs the relicts of the cruiſers ſtationed off that Port. Having thus ſatisfied ourſelves in the material articles, and having gotten on board the *Centurion* moſt of the priſoners, and all the ſilver, we, at eight in the ſame evening, made ſail to the northward, in company

with

with our prize, and at fix the next morning difcovered the Ifland of *Fernandes*, where, the next day, both we and our prize came to an anchor.

And here I cannot omit one remarkable incident which occurred, when the prize and her crew came into the bay, where the reft of the fquadron lay. The *Spaniards* in the *Carmelo* had been fufficiently informed of the diftreffes we had gone through, and were greatly furprized that we had ever furmounted them : But when they faw the *Tryal* Sloop at anchor, they were ftill more aftonifhed, that after all our fatigues we had the induftry (befides refitting our other fhips) to compleat fuch a veffel in fo fhort a time, they taking it for granted that fhe had been built upon the fpot. And it was with great difficulty they were prevailed on to believe, that fhe came from *England* with the reft of the fquadron ; they at firft infifting, that it was impoffible fuch a bawble as that could pafs round Cape *Horn*, when the beft fhips of *Spain* were obliged to put back.

By the time we arrived at *Juan Fernandes*, the letters found on board our prize were more minutely examined : And, it appearing from them, and from the accounts of our prifoners, that feveral other Merchantmen were bound from *Callao* to *Valparaifo*, Mr. *Anfon* difpatched the *Tryal* Sloop the very next morning, to cruife off the laft-mentioned Port, reinforcing him with ten hands from on board his own fhip. Mr. *Anfon* likewife refolved, on the intelligence recited above, to feparate the fhips under his command, and employ them in diftinct cruifes, as he thought that by this means we fhould not only encreafe our chance for prizes, but that we fhould likewife run a lefs rifque of alarming the coaft, and of being difcovered. And now the fpirits of our people being greatly raifed, and their defpondency diffipated by this earneft of fuccefs, they forgot all their paft diftreffes, and refumed their wonted alacrity, and laboured indefatigably in compleating our water, receiving our lumber, and in preparing to take our farewel of the Ifland : But as thefe occupations took us up four or five days with all our induftry, the Commodore, in that interval, directed that the guns belonging to

the

the *Anna Pink*, being four fix pounders, four four pounders, and two fwivels, fhould be mounted on board the *Carmelo* our prize : And having fent on board the *Gloucefter* fix paffengers, and twenty-three feamen to affift in navigating the fhip, he directed Captain *Mitchel* to leave the Ifland as foon as poffible, the fervice requiring the utmoft difpatch, ordering him to proceed to the latitude of five degrees South, and there to cruife off the highland of *Paita*, at fuch a diftance from fhore, as fhould prevent his being difcovered. On this ftation he was to continue till he fhould be joined by the Commodore, which would be whenever it fhould be known that the Viceroy had fitted out the fhips at *Callao*, or on Mr. *Anfon*'s receiving any other intelligence, that fhould make it neceffary to unite our ftrength. Thefe orders being delivered to the Captain of the *Gloucefter*, and all our bufinefs compleated, we, on the *Saturday* following, being the 19th of *September*, weighed our anchor, in company with our prize, and got out of the bay, taking our laft leave of the Ifland of *Juan Fernandes*, and fteering to the eaftward, with an intention of joining the *Tryal* Sloop in her ftation off *Valparaifo*.

Z   C H A P.

CHAP. V.

Our cruife from the time of our leaving *Juan Fernandes*, to the taking the town of *Paita*.

ALTHOUGH the *Centurion*, with her prize the *Carmelo*, weighed from the bay of *Juan Fernandes* on the 19th of *September*, leaving the *Gloucefter* at anchor behind her; yet, by the irregularity and fluctuation of the winds in the offing, it was the 22d of the fame month in the evening, before we loft fight of the Ifland: After which, we continued our courfe to the eaftward, in order to reach our ftation, and to join the *Tryal* off *Valparaifo*. The next night, the weather proved fqually, and we fplit our maintop-fail, which we handed for the prefent, but got it repaired, and fet it again the next morning. And now, on the 24th, a little before fun-fet, we faw two fail to the eaftward; on which, our prize ftood directly from us, to avoid giving any fufpicion of our being cruifers; whilft we, in the mean time, made ourfelves ready for an engagement, and fteered towards the two fhips we had difcovered with all our canvas. We foon perceived that one of thefe, which had the appearance of being a very ftout fhip, made directly for us, whilft the other kept at a very great diftance. By feven o'clock we were within piftol-fhot of the neareft, and had a broad-fide ready to pour into her, the Gunners having their matches in their hands, and only waiting for orders to fire; but as we knew it was now impoffible for her to efcape us, Mr. *Anfon*, before he permitted them to fire, ordered the Mafter to hail the fhip in *Spanifh*; on which the commanding officer on board her, who proved to be Mr. *Hughs*, Lieutenant of the *Tryal*, anfwered us in *Englifh*, and informed us, that fhe was a prize taken by the *Tryal* a few days before, and that the other fail at a diftance was the *Tryal* herfelf difabled in her

masts.

mafts. We were foon after joined by the *Tryal*; and Captain *Saunders* her Commander came on board the *Centurion*. He informed the Commodore, that he had taken this fhip the 18th inftant; that fhe was a prime failor, and had coft him thirty-fix hours chace, before he could come up with her; that for fome time he gained fo little upon her, that he began to defpair of taking her; and the *Spaniards* though alarmed at firft with feeing nothing but a cloud of fail in purfuit of them, the *Tryal*'s hull being fo low in the water that no part of it appeared, yet knowing the goodnefs of their fhip, and finding how little the *Tryal* neared them, they at length laid afide their fears, and, recommending themfelves to the blefled Virgin for protection, began to think themfelves fecure. And indeed their fuccefs was very near doing honour to their *Ave Marias*; for, altering their courfe in the night, and fhutting up their windows to prevent any of their lights from being feen, they had fome chance of efcaping; but a fmall crevice in one of the fhutters rendered all their invocations ineffectual; for through this crevice the people on board the *Tryal* perceived a light, which they chafed, till they arrived within gun-fhot; and then Captain *Saunders* alarmed them unexpectedly with a broadfide, when they flattered themfelves they were got out of his reach: However, for fome time after they ftill kept the fame fail abroad, and it was not obferved that this firft falute had made any impreffion on them; but, juft as the *Tryal* was preparing to repeat her broadfide, the *Spaniards* crept from their holes, lowered their fails, and fubmitted without any oppofition. She was one of the largeft Merchantmen employed in thofe feas, being about fix hundred tuns burthen, and was called the *Arranzazu*. She was bound from *Callao* to *Valparaifo*, and had much the fame cargoe with the *Carmelo* we had taken before, except that her filver amounted only to about 5000 *l.* fterling.

But to balance this fuccefs, we had the misfortune to find that the *Tryal* had fprung her main-maft, and that her maintop-maft had come by the board; and as we were all of us ftanding to the eaftward the next morning, with a frefh gale at South, fhe had the ad-

ditional

ditional ill-luck to ſpring her fore-maſt : So that now ſhe had not a maſt left, on which ſhe could carry ſail. Theſe unhappy incidents were ſtill aggravated by the impoſſibility we were juſt then under of aſſiſting her ; for the wind blew ſo hard, and raiſed ſuch a hollow ſea, that we could not venture to hoiſt out our boat, and conſequently could have no communication with her ; ſo that we were obliged to lie to for the greateſt part of forty-eight hours to attend her, as we could have no thought of leaving her to herſelf in her preſent unhappy ſituation : And as an accumulation to our misfortunes, we were all the while driving to the leeward of our ſtation, at the very time when, by our intelligence, we had reaſon to expect ſeveral of the enemy's ſhips would appear upon the coaſt, who would now gain the port of *Valparaiſo* without obſtruction. And I am verily perſuaded, that the embaraſſment we received from the diſmaſting of the *Tryal*, and our abſence from our intended ſtation occaſioned thereby, deprived us of ſome very conſiderable captures.

The weather proving ſomewhat more moderate on the 27th, we ſent our boat for the Captain of the *Tryal*, who, when he came on board us, produced an inſtrument, ſigned by himſelf and all his officers, repreſenting that the Sloop, beſides being diſmaſted, was ſo very leaky in her hull, that even in moderate weather it was neceſſary to keep the pumps conſtantly at work, and that they were then ſcarcely ſufficient to keep her free ; ſo that in the late gale, though they had all been engaged at the pumps by turns, yet the water had encreaſed upon them ; and, upon the whole, they apprehended her to be at preſent ſo very defective, that if they met with much bad weather, they muſt all inevitably periſh ; and therefore they petitioned the Commodore to take ſome meaſures for their future ſafety. But the refitting of the *Tryal*, and the repairing of her defects, was an undertaking that in the preſent conjuncture greatly exceeded his power ; for we had no maſts to ſpare her, we had no ſtores to compleat her rigging, nor had we any port where ſhe might be hove down, and her bottom examined : Beſides, had a port and proper requiſites for this purpoſe been in our

poſſeſſion,

poffeffion ; yet it would have been extream imprudence, in fo critical a conjuncture, to have loitered away fo much time, as would have been neceffary for thefe operations. The Commodore therefore had no choice left him, but that of taking out her people, and deftroying her : But, at the fame time, as he conceived it neceffary for his Majefty's fervice to keep up the appearance of our force, he appointed the *Tryal*'s prize (which had been often employed by the Viceroy of *Peru* as a man of war) to be a frigate in his Majefty's fervice, manning her with the *Tryal*'s crew, and giving new commiffions to the Captain and all the inferior officers accordingly. This new frigate, when in the *Spanifh* fervice, had mounted thirty-two guns ; but fhe was now to have only twenty, which were the twelve that were on board the *Tryal*, and eight that had belonged to the *Anna Pink*. When this affair was thus far regulated, Mr. *Anfon* gave orders to Captain *Saunders* to put it in execution, directing him to take out of the Sloop the arms, ftores, ammunition, and every thing that could be of any ufe to the other fhips, and then to fcuttle her and fink her. And after Captain *Saunders* had feen her deftroyed, he was to proceed with his new frigate (to be called the *Tryal*'s Prize) and to cruife off the highland of *Valparaifo*, keeping it from him N. N. W, at the diftance of twelve or fourteen leagues : For as all fhips bound from *Valparaifo* to the northward fteer that courfe, Mr. *Anfon* propofed by this means to ftop any intelligence, that might be difpatched to *Callao*, of two of their fhips being miffing, which might give them apprehenfions of the *Englifh* fquadron being in their neighbourhood. The *Tryal*'s Prize was to continue on this ftation twenty-four days, and, if not joined by the Commodore at the expiration of that term, fhe was then to proceed down the coaft to *Pifco* or *Nafca*, where fhe would be certain to meet with Mr. *Anfon*. The Commodore likewife ordered Lieutenant *Saumarez*, who commanded the *Centurion*'s prize, to keep company with Captain *Saunders*, both to affift him in unloading the Sloop, and alfo that by fpreading in their cruife, there might be lefs danger of any of the enemy's my's ships slipping by unobserved. These orders being dispatched, the *Centurion* parted from them at eleven in the evening, on the 27th of *September*, directing her course to the southward, with a view of cruising for some days to the windward of *Valparaiso*.

And now by this disposition of our ships we flattered ourselves, that we had taken all the advantages of the enemy that we possibly could with our small force, since our disposition was doubtless the most prudent that could be projected. For, as we might suppose the *Gloucester* by this time to be drawing near her station off the highland of *Paita*, we were enabled, by our separate stations, to intercept all vessels employed either betwixt *Peru* and *Chili* to the southward, or betwixt *Panama* and *Peru* to the northward: Since the principal trade from *Peru* to *Chili* being carried on to the port of *Valparaiso*, the *Centurion* cruising to the windward of *Valparaiso*, would, in all probability, meet with them, as it is the constant practice of those ships to fall in with the coast, to the windward of that port: And the *Gloucester* would, in like manner, be in the way of the trade bound from *Panama* or the northward, to any part of *Peru*; since the highland off which she was stationed is constantly made by all ships in that voyage. And whilst the *Centurion* and *Gloucester* were thus situated for interrupting the enemy's trade, the *Tryal*'s Prize and *Centurion*'s Prize were as conveniently stationed for preventing all intelligence, by intercepting all ships bound from *Valparaiso* to the northward; for it was on board these vessels that it was to be feared some account of us might possibly be sent to *Peru*.

But the most prudent dispositions carry with them only a probability of success, and can never ensure its certainty: Since those chances, which it was reasonable to overlook in deliberations, are sometimes of most powerful influence in execution. Thus in the present case, the distress of the *Tryal*, and the quitting our station to assist her (events which no degree of prudence could either foresee or obviate) gave an opportunity to all the ships, bound to *Valparaiso*, to reach that port without molestation, during this unlucky interval.

interval. So that though, after leaving Captain *Saunders*, we were very expeditious in regaining our ftation, where we got the 29th at noon, yet in plying on and off till the 6th of *October*, we had not the good fortune to difcover a fail of any fort: And then having loft all hopes of making any advantage by a longer ftay, we made fail to the leeward of the port, in order to join our prizes; but when we arrived on the ftation appointed for them, we did not meet with them, though we continued there four or five days. We fuppofed that fome chace had occafioned their leaving their ftation, and therefore we proceeded down the coaft to the highland of *Nafca*, where Captain *Saunders* was directed to join us. Here we arrived on the 21ft, and were in great expectation of meeting with fome of the enemy's fhips on the coaft, as both the accounts of former voyages, and the information of our prifoners affured us, that all fhips bound to *Callao* conftantly make this land, to prevent the danger of running to the leeward of the port. But notwithftanding the advantages of this ftation, we faw no fail till the 2d of *November*, when two fhips appeared in fight together; we immediately gave them chace, but foon perceived that they were the *Tryal*'s and *Centurion*'s prizes: As they had the wind of us, we brought to and waited their coming up; when Captain *Saunders* came on board us, and acquainted the Commodore, that he had cleared the *Tryal* purfuant to his orders, and having fcuttled her, he remained by her till fhe funk, but that it was the 4th of *October* before this was effected; for there ran fo large and hollow a fea, that the Sloop, having neither mafts nor fails to fteady her, rolled and pitched fo violently, that it was impoffible for a boat to lay a long-fide of her, for the greateft part of the time: And during this attendance on the Sloop, they were all driven fo far to the North-weft, that they were afterwards obliged to ftretch a long way to the weftward to regain the ground they had loft; which was the reafon that we had not met with them on their ftation as we expected. We found they had not been more fortunate in their cruife than we were, for they had feen no veffel fince they fepa-

rated

rated from us. The little fuccefs we all had, and our certainty, that had any fhips been ftirring in thefe feas for fome time paft we muft have met with them, made us believe, that the enemy at *Valparaifo*, on the miffing of the two fhips we had taken, had fuf-pected us to be in the neighbourhood, and had confequently laid an embargo on all the trade in the fouthern parts. We likewife ap-prehended, that they might by this time be fitting out the men of war at *Callao*; for we knew that it was no uncommon thing for an exprefs from *Valparaifo* to reach *Lima* in twenty-nine or thirty days, and it was now more than fifty fince we had taken our firft prize. Thefe apprehenfions of an embargo along the coaft, and of the equip-ment of the *Spanifh* fquadron at *Callao*, determined the Commo-dore to haften down to the leeward of *Callao*, and to join Captain *Mitchel* (who was ftationed off *Paita*) as foon as poffible, that our ftrength being united, we might be prepared to give the fhips from *Callao* a warm reception, if they dared to put to fea. With this view we bore away the fame afternoon, taking particular care to keep at fuch a diftance from the fhore, that there might be no danger of our being difcovered from thence; for we knew that all the country fhips were commanded, under the fevereft penalty, not to fail by the port of *Callao* without ftopping; and as this order was conftantly complied with, we fhould undoubtedly be known for enemies, if we were feen to act contrary to it. In this new navigation, not being certain whether we might not meet the *Spa-nifh* fquadron in our route, the Commodore took on board the *Centurion* part of his crew, with which he had formerly manned the *Carmelo*. And now ftanding to the northward, we, before night came on, had a view of the fmall Ifland called St. *Gallan*, which bore from us N. N. E. ¼ E, about feven leagues diftant. This Ifland lies in the latitude of about fourteen degrees South, and about five miles to the northward of a highland, called *Morro veijo*, or the old man's head. I mention this Ifland, and the highland near it, more particularly, becaufe between them is the moft eligible fta-tion on that coaft for cruifing upon the enemy; as all fhips bound

to

to *Callao*, whether from the northward or the southward, run well in with the land in this part. By the 5th of *November*, at three in the afternoon, we were advanced within view of the high land of *Barranca*, lying in the latitude of 10° : 36′ South, bearing from us N. E. by E, diftant eight or nine leagues; and an hour and an half afterwards we had the fatisfaction we had fo long wifhed for, of feeing a fail. She firft appeared to leeward, and we all immediately gave her chace; but the *Centurion* fo much outfailed the two prizes, that we foon ran them out of fight, and gained confiderably on the chace : However, night coming on before we came up with her, we, about feven o'clock, loft fight of her, and were in fome perplexity what courfe to fteer; but at laft Mr. *Anfon* refolved, as we were then before the wind, to keep all his fails fet, and not to change his courfe : For though we had no doubt but the chace would alter her courfe in the night; yet, as it was uncertain what tack fhe would go upon, it was thought more prudent to keep on our courfe, as we muft by this means unavoidably near her, than to change it on conjecture; when, if we fhould miftake, we muft infallibly lofe her. Thus then we continued the chace about an hour and an half in the dark, fome one or other on board us conftantly imagining they difcerned her fails right a-head of us; but at laft Mr. *Brett*, then our fecond Lieutenant, did really difcover her about four points on the larboard-bow, fteering off to the feaward : We immediately clapped the helm a weather, and ftood for her; and in lefs than an hour came up with her, and having fired fourteen fhot at her, fhe ftruck. Our third Lieutenant, Mr. *Dennis*, was fent in the boat with fixteen men, to take poffeffion of the prize, and to return the prifoners to our fhip. This fhip was named the *Santa Terefa de Jefus*, built at *Guaiaquil*, of about three hundred tuns burthen, and was commanded by *Bartolome Urrunaga*, a *Bifcayer* : She was bound from *Guaiaquil* to *Callao*; her loading confifted of timber, cocao, coco nuts, tobacco, hides, *Pito* thread (which is very ftrong, and is made of a fpecies of grafs) *Quito* cloth, wax, &c. The fpecies on board her was in-

A a   confiderable,

confiderable, being principally fmall filver money, and not amounting to more than 170 *l.* fterling It is true, her cargoe was of
great value, could we have difpofed of it; but, the *Spaniards* having ftrict orders never to ranfom their fhips, all the goods that we
took in thefe feas, except what little we had occafion for ourfelves,
were of no advantage to us. Indeed, though we could make no
profit thereby ourfelves, it was fome fatisfaction to us to confider,
that it was fo much really loft to the enemy, and that the defpoil-
ing them was no contemptible branch of that fervice, in which we
were now employed by our country.

Befides our prize's crew, which amounted to forty-five hands,
there were on board her ten paffengers, confifting of four men and
three women, who were natives of the country, born of *Spanifh*
parents, and three black female flaves that attended them. The
women were a mother and her two daughters, the eldeft about
twenty-one, and the youngeft about fourteen. It is not to be wondered at, that women of thefe years fhould be exceffively alarmed at
the falling into the hands of an enemy, whom, from the former
outrages of the Buccaneers, and by the artful infinuations of their
Priefts, they had been taught to confider as the moft terrible and
brutal of all mankind. Thefe apprehenfions too were in the prefent inftance exaggerated by the fingular beauty of the youngeft of
the women, and the riotous difpofition which they might well expect to find in a fet of failors, that had not feen a woman for near
a twelvemonth. Full of thefe terrors, the women all hid themfelves when our officer went on board, and when they were found
out, it was with great difficulty that he could perfuade them to approach the light: However, he foon fatisfied them, by the humanity of his conduct and his affurances of their future fecurity and
honourable treatment, that they had nothing to fear. And the
Commodore being informed of the matter fent directions that
they fhould be continued on board their own fhip, with the ufe of
the fame apartments, and with all the other conveniencies they had
enjoyed before, giving ftrict orders that they fhould receive no kind

of

of inquietude or moleſtation whatever : And that they might be the more certain of having theſe orders complied with, or of complaining if they were not, the Commodore permitted the Pilot, who in *Spaniſh* ſhips is generally the ſecond perſon on board, to ſtay with them, as their guardian and protector. He was particularly choſen for this purpoſe by Mr. *Anſon*, as he ſeemed to be extremely intereſted in all that concerned the women, and had at firſt declared that he was married to the youngeſt of them ; though it afterwards appeared, both from the information of the reſt of the priſoners, and other circumſtances, that he had aſſerted this with a view, the better to ſecure them from the inſults they expected on their firſt falling into our hands. By this compaſſionate and indulgent behaviour of the Commodore, the conſternation of our female priſoners entirely ſubſided, and they continued eaſy and chearful during the whole time they were with us, as I ſhall have occaſion to mention more particularly hereafter.

I have before obſerved, that at the beginning of this chace the *Centurion* ran her two conſorts out of ſight, for which reaſon we lay by all the night, after we had taken the prize, for Captain *Saunders* and Lieutenant *Saumarez* to join us, firing guns, and making falſe fires every half hour, to prevent their paſſing us unobſerved ; but they were ſo far a-ſtern, that they neither heard nor ſaw any of our ſignals, and were not able to come up with us till broad day-light. When they had joined us we proceeded together to the northward, being now four ſail in company. We here found the ſea, for many miles round us, of a beautiful red colour : This, upon examination, we imputed to an immenſe quantity of ſpawn ſpread upon its ſurface ; and taking up ſome of the water in a wine-glaſs, it ſoon changed from a dirty aſpect to a clear chryſtal, with only ſome red globules of a ſlimy nature floating on the top. And now having a ſupply of timber on board our new prize, the Commodore ordered our boats to be repaired, and a ſwivel gunſtock to be fixed in the bow both of the barge and pinnace, in order

to

to encreafe their force, in cafe we fhould be obliged to have recourfe
to them for boarding fhips, or for any attempts on fhore.

As we ftood from hence to the northward, nothing remarkable
occurred for two or three days, though we fpread our fhips in fuch
a manner, that it was not probable any veffel of the enemy could
efcape us. In our run along this coaft we generally obferved, that
there was a current which fet us to the northward, at the rate of
ten or twelve miles each day. And now being in about eight degrees of South latitude, we began to be attended with vaft numbers of flying fifh and bonitos, which were the firft we faw after
our departure from the coaft of *Brazil.* But it is remarkable, that
on the Eaft fide of South *America* they extended to a much higher
latitude than they do on the Weft fide ; for we did not lofe them
on the coaft of *Brazil,* till we approached the fouthern tropic. The
reafon for this diverfity is doubtlefs the different degrees of heat
obtaining in the fame latitude on different fides of that Continent.
And on this occafion, I muft beg leave to make a fhort digreffion
on the heat and cold of different climates, and on the varieties which
occur in the fame place in different parts of the year, and in different places lying in the fame degree of latitude.

The Ancients, as appears in many places, conceived that of the
five zones, into which they divided the furface of the globe, two
only were habitable, fuppofing that all between the tropics was too
hot, and all within the polar circle too cold to be fupported by
mankind. The falfhood of this reafoning has been long evinced ;
but the particular comparifons of the heat and cold of thefe various
climates, has as yet been very imperfectly confidered. However,
enough is known fafely to determine this pofition, that all places
between the tropics are far from being the hotteft on the globe, as
many of thofe within the polar circles are far from enduring that
extreme degree of cold, to which their fituation fhould feem to fubject them : That is to fay, in other words, that the temperature of
a place depends much more upon other circumftances, than upon
its diftance from the pole, or its proximity to the equinoctial.

This

This propofition relates to the general temperature of places, taking the whole year round; and in this fenfe it cannot be denied, but that the city of *London*, for inftance, enjoys much warmer fea-fons than the bottom of *Hudfon*'s bay, which is nearly in the fame latitude with it ; for there the feverity of the winter is fo great, that it will fcarcely permit the hardieft of our garden plants to live. And if the comparifon be made between the coaft of *Brazil* and the weftern fhore of South *America*, as, for example, betwixt *Bahia* and *Lima*, the difference will be ftill more remarkable ; for though the coaft of *Brazil* is extremely fultry, yet the coaft of the *South-Seas* in the fame latitude is perhaps as temperate and tolerable as any part of the globe ; fince in ranging along it, we did not once meet with fo warm weather, as is frequent in a fummer's day in *England*: And this was the more remarkable, as there never fell any rains to refrefh and cool the air.

The caufes of this temperature in the *South-Seas* are not difficult to be affigned, and fhall be hereafter mentioned. I am now only folicitous to eftablifh the truth of this affertion, that the latitude of a place alone is no rule whereby to judge of the degree of heat and cold which obtains there. Perhaps this pofition might be more briefly confirmed, by obferving, that on the tops of the *Andes*, though under the equinoctial, the fnow never melts the whole year round ; a criterion of cold, ftronger than what is known to take place in many parts far removed within the polar circle.

I have hitherto confidered the temperature of the air all the year through, and the grofs eftimations of heat and cold which every one makes from his own fenfation. If this matter be examined by means of Thermometers, which in refpect to the abfolute degree of heat and cold are doubtlefs the moft unerring evidences ; if this be done, the refult will be indeed moft wonderful : For it will appear that the heat in very high latitudes, as at *Peterfburgh* for inftance, is at particular times much greater than any that has been hitherto obferved between the tropics ; and that even at *London* in the year 1746, there was the part of one day confiderably hotter than what

was

was at any time felt by a ſhip of Mr. *Anſon*'s ſquadron, in running
from hence to Cape *Horn* and back again, and paſſing twice under
the ſun; for in the ſummer of that year, the thermometer in *London* (being one of thoſe graduated according to the method of *Faren-
heit*) ſtood once at 78°; and the greateſt height at which a thermometer of the ſame kind ſtood in the foregoing ſhip, I find to be
76°: This was at St. *Catherine*'s, in the latter end of *December*,
when the ſun was within about three degrees of the vertex. And
as to *Peterſburgh*, I find, by the acts of the academy eſtabliſhed
there, that in the year 1734, on the 20th and 25th of *July*, the
thermometer roſe to 98° in the ſhade, that is, it was twenty-two diviſions higher than it was found to be at St. *Catherine's*; which is
a degree of heat that, were it not authoriſed by the regularity and
circumſpection with which the obſervations ſeem to have been
made, would appear altogether incredible.

If it ſhould be aſked, how it comes to paſs then, that the heat
in many places between the tropics is eſteemed ſo violent and inſuf-
ferable, when it appears by theſe inſtances, that it is ſometimes rivalled or exceeded in very high latitudes not far from the polar circle? I ſhould anſwer, that the eſtimation of heat in any particular
place, ought not to be founded upon that degree of heat which
may now and then obtain there, but is rather to be deduced from
the medium obſerved in a whole ſeaſon, or perhaps in a whole year:
And in this light it will eaſily appear, how much more intenſe the
ſame degree of heat may prove, by being long continued without
remarkable variation. For inſtance, in comparing together St. *Catherine's* and *Peterſburgh*, we will ſuppoſe the ſummer heat at St.
Catherine's to be 76°, and the winter heat to be twenty diviſions
ſhort of it: I do not make uſe of this laſt conjecture upon ſufficient obſervation; but I am apt to ſuſpect, that the allowance is full
large. Upon this ſuppoſition then, the medium heat all the year
round will be 66°, and this perhaps by night as well as day, with
no great variation: Now thoſe who have attended to thermometers will readily own, that a continuation of this degree of heat

for

for a length of time would by the generality of mankind be ſtiled violent and ſuffocating. But now at *Peterſburgh*, though a few times in the year the heat, by the thermometer, may be conſiderably greater than at St. *Catherine's*, yet, as at other times the cold is immenſely ſharper, the medium for a year, or even for one ſeaſon only, would be far ſhort of 66°. For I find, that the variation of the thermometer at *Peterſburgh* is at leaſt five times greater, from its higheſt to its loweſt point, than what I have ſuppoſed to take place at St. *Catherine's*.

But beſides this eſtimation of the heat of a place, by taking the medium for a conſiderable time together, there is another circumſtance which will ſtill augment the apparent heat of the warmer climates, and diminiſh that of the colder, though I do not remember to have ſeen it remarked in any author. To explain myſelf more diſtinctly upon this head, I muſt obſerve, that the meaſure of abſolute heat, marked by the thermometer, is not the certain criterion of the ſenſation of heat, with which human bodies are affected : For as the preſence and perpetual ſucceſſion of freſh air is neceſſary to our reſpiration, ſo there is a ſpecies of tainted or ſtagnated air, which is often produced by the continuance of great heats, which never fails to excite in us an idea of ſultrineſs and ſuffocating warmth, much beyond what the mere heat of the air alone, ſuppoſing it pure and agitated, would occaſion. Hence it follows, that the mere inſpection of the thermometer will never determine the heat which the human body feels from this cauſe ; and hence it follows too, that the heat in moſt places between the tropics muſt be much more troubleſome and uneaſy, than the ſame degree of abſolute heat in a high latitude : For the equability and duration of the tropical heat contribute to impregnate the air with a multitude of ſteams and vapours from the ſoil and water, and theſe being, many of them, of an impure and noxious kind, and being not eaſily removed, by reaſon of the regularity of the winds in thoſe parts, which only ſhift the exhalations from place to place, without diſperſing them, the atmoſphere is by this means rendered leſs

proper

proper for refpiration, and mankind are confequently affected with what they ftile a moft intenfe and ftifling heat: Whereas in the higher latitudes thefe vapours are probably raifed in fmaller quantities, and the irregularity and violence of the winds frequently difperfe them; fo that, the air being in general pure and lefs ftagnant, the fame degree of abfolute heat is not attended with that uneafy and fuffocating fenfation. This may fuffice in general with refpect to the prefent fpeculation; but I cannot help wifhing, as it is a fubject in which mankind, efpecially travellers of all forts, are very much interefted, that it were more thoroughly and accurately examined, and that all fhips bound to the warmer climates would furnifh themfelves with thermometers of a known fabric, and would obferve them daily, and regifter their obfervations; for confidering the turn to philofophical fubjects, which has obtained in *Europe* for the laft fourfcore years, it is incredible how very rarely any thing of this kind hath been attended to. For my own part, I do not recollect that I have ever feen any obfervations of the heat and cold, either in the *Eaft* or *Weft-Indies*, which were made by mariners or officers of veffels, except thofe made by Mr. *Anfon*'s order, on board the *Centurion*, and by Captain *Leg* on board the *Severn*, which was another fhip of our fquadron.

This digreffion I have been in fome meafure drawn into, by the confideration of the fine weather we met with on the coaft of *Peru*, even under the equinoctial itfelf, but the particularities of this weather I have not yet defcribed: I fhall now therefore add, that in this climate every circumftance concurred, that could render the open air and the day-light defirable. For in other countries the fcorching heat of the fun in fummer renders the greater part of the day unapt either for labour or amufement; and the frequent rains are not lefs troublefome in the more temperate parts of the year. But in this happy climate the fun rarely appears: Not that the heavens have at any time a dark and gloomy look; but there is conftantly a chearful grey fky, juft fufficient to fcreen the fun, and to mitigate the violence of its perpendicular rays, without obfcuring

the

the air, or tinging the day-light with an unpleafant or melancholy hue. By this means all parts of the day are proper for labour or exercife abroad, nor is there wanting that refrefhment and pleafing refrigeration of the air, which is fometimes produced in other climates by rains; for here the fame effect is brought about, by the frefh breezes from the cooler regions to the fouthward. It is reafonable to fuppofe, that this fortunate complexion of the heavens is principally owing to the neighbourhood of thofe vaft hills, called the *Andes*, which running nearly parallel to the fhore, and at a fmall diftance from it, and extending themfelves immenfely higher than any other mountains upon the globe, form upon their fides and declivities a prodigious tract of country, where, according to the different approaches to the fummit, all kinds of climates may at all feafons of the year be found. Thefe mountains, by intercepting great part of the eaftern winds which generally blow over the Continent of South *America*, and by cooling that part of the air which forces its way over their tops, and by keeping befides a prodigious extent of the atmofphere perpetually cool, by its contiguity to the fnows with which they are covered; thefe hills, I fay, by thus extending the influence of their frozen crefts to the neighbouring coafts and feas of *Peru*, are doubtlefs the caufe of the temperature and equability which conftantly prevail there. For when we were advanced beyond the equinoctial, where thefe mountains left us, and had nothing to fcreen us to the eaftward, but the high lands on the Ifthmus of *Panama*, which are but mole-hills to the *Andes*, we then foon found that in a fhort run we had totally changed our climate, paffing in two or three days from the temperate air of *Peru*, to the fultry burning atmofphere of the *Weft-Indies*. But it is time to return to our narration.

On the 10th of *November* we were three leagues South of the fouthermoft Ifland of *Lobos*, lying in the latitude of 6° : 27′ South : There are two Iflands of this name; this called *Lobos de la Mar*; and another, which lies to the northward of it, very much refembling it in fhape and appearance, and often miftaken for it, called

B b *Lobos*

Lobos de tierra. We were now drawing near to the station appointed to the *Gloucester*, for which reason, fearing to miss her, we made an easy sail all night. The next morning, at day-break, we saw a ship in shore, and to windward, plying up the coast: She had passed by us with the favour of the night, and we soon perceiving her not to be the *Gloucester*, got our tacks on board, and gave her chace; but it proving very little wind, so that neither of us could make much way, the Commodore ordered the barge, his pinnace, and the *Tryal's* pinnace to be manned and armed, and to pursue the chace and board her. Lieutenant *Brett*, who commanded the barge, came up with her first, about nine o clock, and running along side of her, he fired a volley of small shot between the masts, just over the heads of the people on board, and then instantly entered with the greatest part of his men; but the enemy made no resistance, being sufficiently frightened by the dazzling of the cutlasses, and the volley they had just received. Lieutenant *Brett* ordered the sails to be trimmed, and bore down to the Commodore, taking up in his way the two pinnaces. When he was arrived within about four miles of us he put off in the barge, bringing with him a number of the prisoners, who had given him some material intelligence, which he was desirous the Commodore should be acquainted with as soon as possible. On his arrival we learnt, that the prize was called *Nuestra Senora del Carmin*, of about two hundred and seventy tuns burthen; she was commanded by *Marcos Morena*, a native of *Venice*, and had on board forty-three mariners: She was deep laden with steel, iron, wax, pepper, cedar, plank, snuff, rosarios, *European* bale goods, powder blue, cinnamon, *Romish* indulgences, and other species of merchandize: And though this cargoe, in our present circumstances, was but of little value to us, yet with respect to the *Spaniards*, it was the most considerable capture that fell into our hands in this part of the world; for it amounted to upwards of 400,000 dollars prime cost at *Panama*. This ship was bound to *Callao*, and had stopped at *Paita* in her passage, to take in a recruit of water and provisions, and had not

left

left that place above twenty-four hours, before she fell into our hands.

I have mentioned that Mr. *Brett* had received some important intelligence from the prisoners, which he endeavoured to acquaint the Commodore with immediately. The first person he received it from (though upon further examination it was confirmed by the other prisoners) was one *John Williams* an *Irishman*, whom he found on board the *Spanish* vessel. *Williams* was a Papist, who worked his passage from *Cadiz*, and had travelled over all the kingdom of *Mexico* as a Pedlar : He pretended, that by this business he had got 4 or 5000 dollars; but that he was embarrassed by the Priests, who knew he had money, and was at last stript of all he had. He was indeed at present all in rags, being but just got out of *Paita* goal, where he had been confined for some misdemeanor ; he expressed great joy upon seeing his countrymen, and immediately informed them, that, a few days before, a vessel came into *Paita*, where the Master of her informed the Governor, that he had been chased in the offing by a very large ship, which from her size, and the colour of her sails, he was persuaded must be one of the *English* squadron : This we then conjectured to have been the *Gloucester*, as we afterwards found it was. The Governor, upon examining the Master, was fully satisfied of his relation, and immediately sent away an express to *Lima* to acquaint the Viceroy therewith : And the Royal Officer residing at *Paita*, being apprehensive of a visit from the *English*, was busily employed in removing the King's treasure and his own to *Piura*, a town within land, about fourteen leagues distant. We further learnt from our prisoners, that there was a very considerable sum of money belonging to some Merchants at *Lima*, that was now lodged at the Custom house at *Paita* ; and that this was intended to be shipped on board a vessel, which was then in the port of *Paita*, and was preparing to sail with the utmost expedition, being bound for the bay of *Sonsonnate*, on the coast of *Mexico*, in order to purchase a part of the cargoe of the *Manila* ship. This vessel at *Paita* was esteemed a prime sailor, and had

B b 2

just

juſt received a new coat of tallow on her bottom ; and, in the opinion of the priſoners, ſhe might be able to ſail the ſucceeding morning. The character they gave us of this veſſel, on which the money was to be ſhipped, left us little reaſon to believe that our ſhip,
which had been in the water near two years, could have any chance
of coming up with her, if we once ſuffered her to eſcape out of the
Port. And therefore, as we were now diſcovered, and the coaſt
would be ſoon alarmed, and as our cruiſing in theſe parts any longer
would anſwer no purpoſe, the Commodore reſolved to ſurprize the
place, having firſt minutely informed himſelf of its ſtrength and
condition, and being fully ſatisfied, that there was little danger of
loſing many of our men in the attempt. This ſurprize of *Paita*,
beſides the treaſure it promiſed us, and its being the only enterprize
it was in our power to undertake, had theſe other advantages attending it, that we ſhould in all probability ſupply ourſelves with great
quantities of live proviſion, of which we were at this time in want:
And we ſhould likewiſe have an opportunity of ſetting our priſoners
on ſhore, who were now very numerous, and made a greater conſumption of our food than our ſtock that remained was capable of
furniſhing long. In all theſe lights the attempt was a moſt eligible
one, and what our neceſſities, our ſituation, and every prudential
conſideration, prompted us to. How it ſucceeded, and how far it
anſwered our expectations, ſhall be the ſubject of the following
chapter.

CHAP.

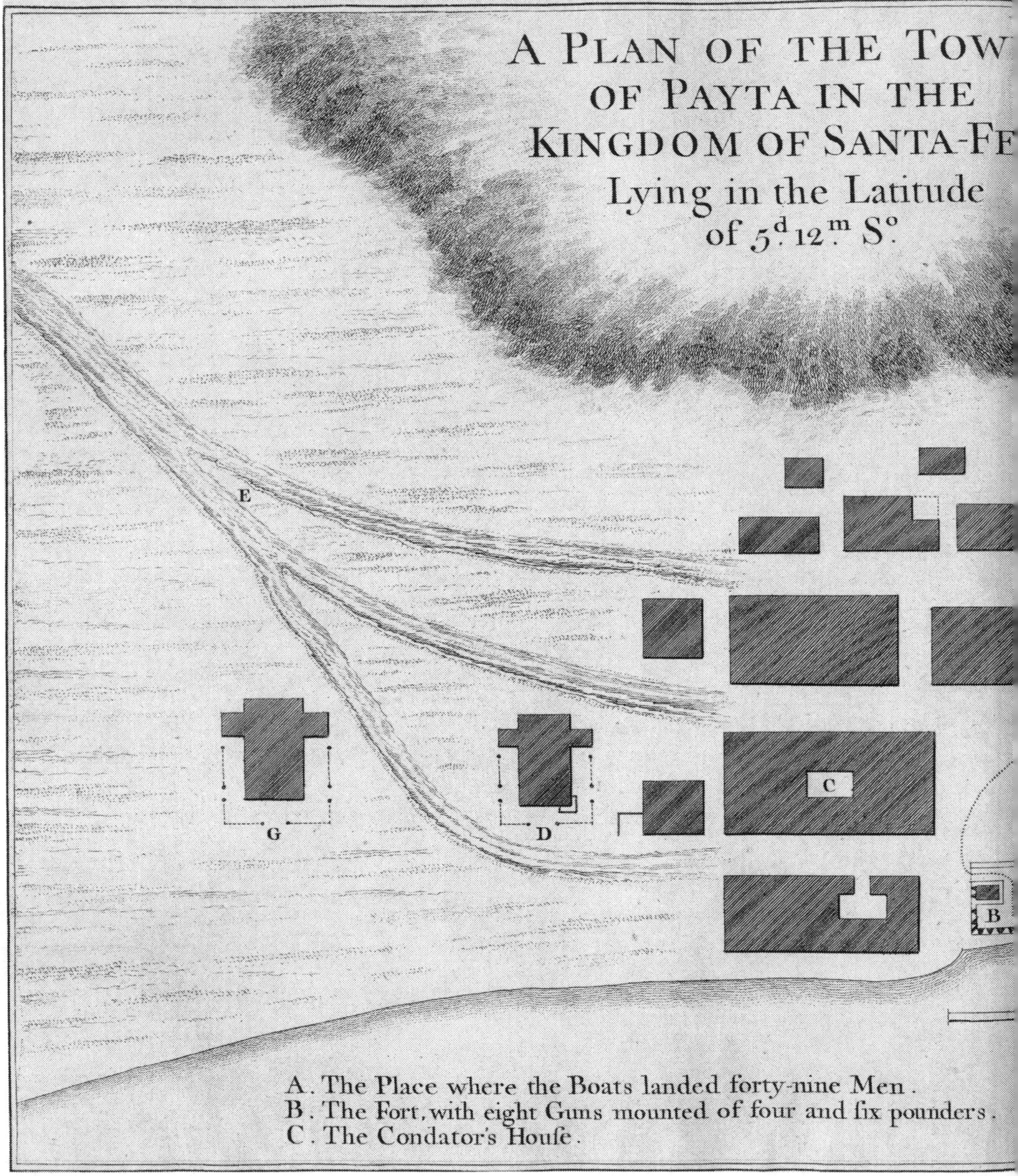

A PLAN OF THE TOW
OF PAYTA IN THE
KINGDOM OF SANTA-FE
Lying in the Latitude
of $5^d 12^m S^o$.

E

G

D

C

B

A. The Place where the Boats landed forty-nine Men.
B. The Fort, with eight Guns mounted of four and six pounders.
C. The Condator's House.

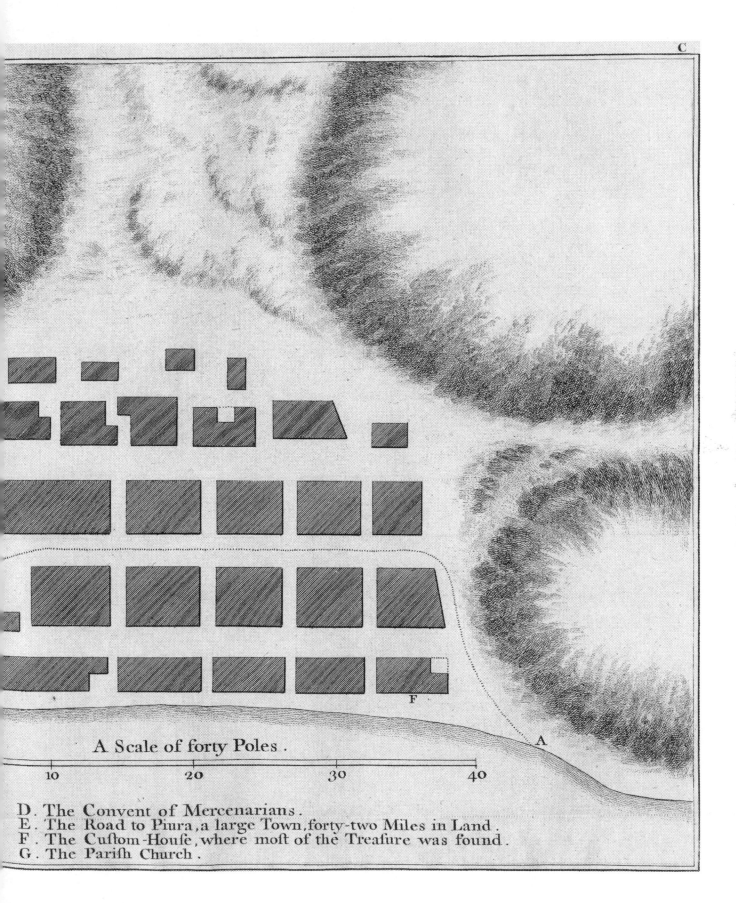

A Scale of forty Poles.

10   20   30   40

D. The Convent of Mercenarians.
E. The Road to Piura, a large Town, forty-two Miles in Land.
F. The Custom-House, where most of the Treasure was found.
G. The Parish Church.

CHAP. VI.

The taking of *Paita*, and our proceedings till we left the coaſt of *Peru*.

THE town of *Paita* is ſituated in the latitude of 5° : 12 South, in a moſt barren ſoil, compoſed only of ſand and ſlate: The extent of it (as may be ſeen in the annexed plan of it) is but ſmall, containing in all leſs than two hundred families. The houſes are only ground-floors; the walls built of ſplit cane and mud, and the roofs thatched with leaves: Theſe edifices, though extremely ſlight, are abundantly ſufficient for a climate, where rain is conſidered as a prodigy, and is not ſeen in many years: So that it is ſaid, that a ſmall quantity of rain falling in this country in the year 1728, it ruined a great number of buildings, which mouldered away, and as it were melted before it. The inhabitants of *Paita* are principally *Indians* and black ſlaves, or at leaſt a mixed breed, the whites being very few. The port of *Paita*, though in reality little more than a bay, is eſteemed the beſt on that part of the coaſt; and is indeed a very ſecure and commodious anchorage. It is greatly frequented by all veſſels coming from the North; ſince it is here only that the ſhips from *Acapulco, Sonſonnate, Realeijo* and *Panama*, can touch and refreſh in their paſſage to *Callao* : And the length of theſe voyages (the wind for the greateſt part of the year being full againſt them) renders it impoſſible to perform them without calling upon the coaſt for a recruit of freſh water. It is true, *Paita* is ſituated on ſo parched a ſpot, that it does not itſelf furniſh a drop of freſh water, or any kind of greens or proviſions, except fiſh and a few goats: But there is an *Indian* town called *Colan*, about two or three leagues diſtant to the northward, from whence water, maize, greens, fowls, &c. are brought to *Paita* on

balſas

balfas or floats, for the conveniency of the ships that touch here; and cattle are sometimes brought from *Piura*, a town which lies about fourteen leagues up in the country. The water brought from *Colan* is whitish, and of a disagreeable appearance, but is said to be very wholsome: For it is pretended by the inhabitants, that it runs through large woods of sarsaparilla, and that it is sensibly impregnated therewith. This port of *Paita*, besides furnishing the northern trade bound to *Callao*, with water and necessaries, is the usual place where passengers from *Acapulco* or *Panama*, bound to *Lima*, disembark; for, as it is two hundred leagues from hence to *Callao*, the port of *Lima*, and as the wind is generally contrary, the passage by sea is very tedious and fatiguing, but by land there is a tolerable good road parallel to the coast, with many stations and villages for the accommodation of travellers.

It appears by the plan, that the town of *Paita* is itself an open place; so that its sole protection and defence is the fort marked (B) in the plan. It was of consequence to us to be well informed of the fabrick and strength of this fort; and by the examination of our prisoners we found, that there were eight pieces of cannon mounted in it, but that it had neither ditch nor outwork, being only surrounded by a plain brick wall; and that the garrison consisted of only one weak company, but the town itself might possibly arm three hundred men more.

Mr. *Anson* having informed himself of the strength of the place, resolved (as hath been said in the preceding chapter) to attempt it that very night. We were then about twelve leagues distant from the shore, far enough to prevent our being discovered; yet not so far, but that by making all the sail we could, we might arrive in the bay with our ships in the night. However, the Commodore prudently considered, that this would be an improper method of proceeding, as our ships, being such large bodies, might be easily discovered at a distance even in the night, and might thereby alarm the inhabitants, and give them an opportunity of removing their valuable effects. He therefore, as the strength of the place did not

require

require our whole force, refolved to attempt it with our boats only, ordering the eighteen oared barge, and our own and the *Tryal's* pinnaces on that fervice ; and having picked out fifty-eight men to man them, well provided with arms and ammunition, he gave the command of the expedition to Lieutenant *Brett*, and gave him his neceffary orders. And the better to prevent the difappointment and confufion which might arife from the darknefs of the night, and the ignorance of the ftreets and paffages of the place, two of the *Spanifh* Pilots were ordered to attend the Lieutenant, and to conduct him to the moft convenient landing-place, and were afterwards to be his guides on fhore ; and that we might have the greater fecurity for their faithful behaviour on this occafion, the Commodore took care to affure all our prifoners, that, if the Pilots acted properly, they fhould all of them be releafed, and fet on fhore at this place ; but in cafe of any mifconduct or treachery, he threatened them that the Pilots fhould be inftantly fhot, and that he would carry all the reft of the *Spaniards*, who were on board him, prifoners to *England*. So that the prifoners themfelves were interefted in our fuccefs, and therefore we had no reafon to fufpect our Conductors either of negligence or perfidy.

And on this occafion I cannot but remark a fingular circumftance of one of the Pilots employed by us in this bufinefs. It feems (as we afterwards learnt) he had been taken by Captain *Clipperton* above twenty years before, and had been forced to lead *Clipperton* and his people to the furprize of *Truxillo*, a town within land to the fouthward of *Paita*, where however he contrived to alarm his countrymen, and to fave them, though the place was taken. Now that the only two attempts on fhore, which were made at fo long an interval from each other, fhould be guided by the fame perfon, and he too a prifoner both times, and forced upon the employ contrary to his inclination, is an incident fo very extraordinary, that I could not help taking notice of it. But to return to the matter in hand.

During

During our preparations, the ſhips themſelves ſtood towards the Port with all the ſail they could make, being ſecure that we were yet at too great a diſtance to be ſeen. But about ten o'clock at night, the ſhips being then within five leagues of the place, Lieutenant *Brett*, with the boats under his command, put off, and arrived at the mouth of the bay without being diſcovered; but no ſooner had he entered it, than ſome of the people, on board a veſſel riding at anchor there, perceived him, who inſtantly put off in their boat, rowing towards the fort, ſhouting and crying, *the Engliſh, the Engliſh dogs,* &c. by which the whole town was ſuddenly alarmed, and our people ſoon obſerved ſeveral lights hurrying backwards and forwards in the fort, and other marks of the inhabitants being in great motion. Lieutenant *Brett*, on this, encouraged his men to pull briſkly up to the ſhore, that they might give the enemy as little time as poſſible to prepare for their defence. However, before our boats could reach the ſhore, the people in the fort had got ready ſome of their cannon, and pointed them towards the landing-place; and though in the darkneſs of the night it might be well ſuppoſed that chance had a greater ſhare than ſkill in their direction, yet the firſt ſhot paſſed extremely near one of the boats, whiſtling juſt over the heads of the crew. This made our people redouble their efforts; ſo that they had reached the ſhore, and were in part diſembarked by the time the ſecond gun fired. As ſoon as our men landed, they were conducted by one of the *Spaniſh* Pilots to the entrance of a narrow ſtreet, not above fifty yards diſtant from the beach, where they were covered from the fire of the fort; and being formed in the beſt manner the ſhortneſs of the time would allow, they immediately marched for the parade, which was a large ſquare at the end of this ſtreet, the fort being one ſide of the ſquare, and the Governor's houſe another, as may be ſeen more diſtinctly in the plan, where likewiſe the road they took from their landing to the fort is marked out by a prickt line. In this march (though performed with tolerable regularity) the ſhouts and clamours of threeſcore ſailors, who had been confined ſo long on

ſhipboard,

shipboard, and were now for the first time on shore in an enemy's country, joyous as they always are, when they land, and animated besides in the present case with the hopes of an immense pillage ; the huzza's, I say, of this spirited detachment, joined with the noise of their drums, and favoured by the night, had augmented their numbers, in the opinion of the enemy, to at least three hundred ; by which persuasion the inhabitants were so greatly intimidated, that they were much more solicitous about the means of their flight than of their resistance : So that though upon entering the parade, our people received a volley from the Merchants who owned the treasure then in the town, and who, with a few others, had ranged themselves in a gallery that ran round the Governor's house, yet that post was immediately abandoned upon the first fire made by our people, who were thereby left in quiet possession of the parade.

On this success Lieutenant *Brett* divided his men into two parties, ordering one of them to surround the Governor's house, and if possible to secure the Governor, whilst he himself with the other marched to the fort, with an intent to force it. But, contrary to his expectation, he entered it without opposition ; for the enemy, on his approach, abandoned it, and made their escape over the walls. By this means the whole place was mastered in less than a quarter of an hour's time from the first landing, with no other loss than that of one man killed on the spot, and two wounded ; one of which was the *Spanish* Pilot of the *Teresa*, who received a slight bruise by a ball which grazed on his wrist : Indeed another of the company, the Honourable Mr. *Kepple*, son to the Earl of *Albemarle*, had a very narrow escape ; for having on a jocky cap, one side of the peak was shaved off close to his temple by a ball, which however did him no other injury.

And now Lieutenant *Brett*, after this success, placed a guard at the fort, and another at the Governor's house, and appointed centinels at all the avenues of the town, both to prevent any surprize from the enemy, and to secure the effects in the place from being

C c

embezzled.

embezzled. And this being done, his next care was to feize on the Cuftom-houfe where the treafure lay, and to examine if any of the inhabitants remained in the town, that he might know what farther precautions it was neceffary to take; but he foon found that the numbers left behind were no ways formidable: For the greateft part of them (being in bed when the place was furprized) had run away with fo much precipitation, that they had not given themfelves time to put on their cloaths. And in this precipitate rout the Governor was not the laft to fecure himfelf, for he fled betimes half naked, leaving his wife, a young Lady of about feventeen years of age, to whom he had been married but three or four days, behind him, though fhe too was afterwards carried off in her fhift by a couple of centinels, juft as the detachment, ordered to inveft the houfe, arrived before it. This efcape of the Governor was an unpleafing circumftance, as Mr. *Anfon* had particularly recommended it to Lieutenant *Brett* to fecure his perfon, if poffible, in hopes that by that means we might be able to treat for the ranfom of the place: But it feems his alertnefs rendered it impoffible to feize him. The few inhabitants who remained were confined in one of the churches under a guard, except fome ftout Negroes which were found in the place; thefe, inftead of being fhut up, were employed the remaining part of the night to affift in carrying the treafure from the Cuftom-houfe and other places to the fort: However, there was care taken that they fhould be always attended by a file of mufqueteers.

The tranfporting the treafure from the Cuftom-houfe to the fort, was the principal occupation of Mr. *Brett*'s people, after he had got poffeffion of the place. But the failors, while they were thus employed, could not be prevented from entring the houfes which lay near them, in fearch of private pillage. And the firft things which occurred to them, being the cloaths which the *Spaniards* in their flight had left behind them, and which, according to the cuftom of the country, were moft of them either embroidered or laced, our people eagerly feized thefe glittering habits, and put

them

them on over their own dirty trowfers and jackets, not forgetting, at the fame time, the tye or bag-wig and laced hat, which were generally found with the cloaths; and when this practice was once begun, there was no preventing the whole detachment from imitating it : And thofe, who came lateft into the fafhion, not finding mens cloaths fufficient to equip themfelves, they were obliged to take up with womens gowns and petticoats, which (provided there was finery enough) they made no fcruple of putting on, and blending with their own greafy drefs. So that when a party of them thus ridiculoufly metamorphifed firft appeared before Mr. *Brett*, he was extreamly furprized at their appearance, and could not immediately be fatisfied they were his own people.

Thefe were the tranfactions of our detachment on fhore at *Paita* the firft night. And now to return to what was done on board the *Centurion* in that interval. I muft obferve, that after the boats were gone off, we lay by till one o'clock in the morning, and then fuppofing our detachment to be near landing, we made an eafy fail for the bay. About feven in the morning we began to open the bay, and foon after we had a view of the town; and though we had no reafon to doubt of the fuccefs of the enterprize, yet it was with great joy that we firft difcovered an infallible fignal of the certainty of our hopes; this was by means of our perfpectives, for through them we faw an *Englifh* flag hoifted on the flag-ftaff of the fort, which to us was an inconteftable proof that our people had got poffeffion of the town. We plied into the bay with as much expedition as the wind, which then blew off fhore, would permit us: And at eleven, the *Tryal*'s boat came on board us, loaden with dollars and church-plate; and the officer who commanded her informed us of the preceding night's tranfactions, fuch as we have already related them. About two in the afternoon we came to an anchor in ten fathom and a half, at a mile and a half diftance from the town, and were confequently near enough to have a more immediate intercourfe with thofe on fhore. And now we found that Mr. *Brett* had hitherto gone on in collecting and removing the

treafure

treafure without interruption ; but that the enemy had rendezvoufed from all parts of the country on a hill, at the back of the town, where they made no inconfiderable appearance : For amongft the reft of their force, there were two hundred horfe feemingly very well armed, and mounted, and, as we conceived, properly trained and regimented, being furnifhed with trumpets, drums and ftandards. Thefe troops paraded about the hill with great oftentation, founding their military mufick, and practifing every art to intimidate us, (as our numbers on fhore were by this time not unknown to them) in hopes that we might be induced by our fears to abandon the place before the pillage was compleated. But we were not fo ignorant as to believe, that this body of horfe, which feemed to be what the enemy principally depended on, would dare to venture in ftreets and amongft houfes, even had their numbers been three times as great ; and therefore, notwithftanding their menaces, we went on, as long as the day-light lafted, calmly, in fending off the treafure, and in employing the boats to carry on board the refrefhments, fuch as hogs, fowls, &c. which we found here in great abundance. But at night, to prevent any furprize, the Commodore fent on fhore a reinforcement, who pofted themfelves in all the ftreets, leading to the parade ; and for their greater fecurity, they traverfed the ftreets with barricadoes fix feet high : And the enemy continuing quiet all night, we, at day-break, returned again to our labour of loading the boats, and fending them off.

By this time we were convinced of what confequence it would have been to us, had fortune feconded the prudent views of the Commodore, by permitting us to have fecured the Governor. For we found in the place many ftore-houfes full of valuable effects, which were ufelefs to us at prefent, and fuch as we could not find room for on board. But had the Governor been in our power, he would, in all probability, have treated for a ranfom, which would have been extremely advantageous both to him and us : Whereas, he being now at liberty, and having collected all the force of the country for many leagues round, and having even got a body of mi-

litia

litia from *Piura*, which was fourteen leagues diftant, he was fo elated with his numbers, and fo fond of his new military command, that he feemed not to trouble himfelf about the fate of his Government. So that though Mr. *Anfon* fent feveral meffages to him by the inhabitants, who were in our power, defiring him to enter into a treaty for the ranfom of the town and goods, giving him, at the fame time, an intimation that he fhould be far from infifting on a rigorous equivalent, but perhaps might be fatisfied with fome live cattle, and a few neceffaries for the ufe of the fquadron, and affuring him too, that if he would not condefcend at leaft to treat, he would fet fire to the town, and all the warehoufes: Yet the Governor was fo imprudent and arrogant, that he defpifed all thefe reiterated applications, and did not deign even to return the leaft anfwer to them.

On the fecond day of our being in poffeffion of the place, feveral Negro flaves deferted from the enemy on the hill, and coming into the town, voluntarily entered into our fervice: One of thefe was well known to a Gentleman on board, who remembered him formerly at *Panama*. And the *Spaniards* without the town being in extreme want of water, many of their flaves crept into the place by ftealth, and carried away feveral jars of water to their mafters on the hill; and though fome of them were feized by our men in the attempt, yet the thirft amongft the enemy was fo preffing, that they continued this practice till we left the place. And now, on this fecond day we were affured, both by the deferters and by thefe prifoners we took, that the *Spaniards* on the hill, who were by this time encreafed to a formidable number, had refolved to ftorm the town and fort the fucceeding night; and that one *Gordon*, a *Scotch* Papift, and Captain of a fhip in thofe feas, was to have the command of this enterprize. But we, notwithftanding, continued fending off our boats, and profecuted our work without the leaft hurry or precipitation till the evening; and then a reinforcement was again fent on fhore by the Commodore, and Lieutenant *Brett* doubled his guards at each of the barricadoes; and our pofts being

connected

connected by the means of centinels placed within call of each other, and the whole being visited by frequent rounds, attended with a drum, these marks of our vigilance which the enemy could not be ignorant of, as they could doubtless hear the drum, if not the calls of the centinels; these marks, I say, of our vigilance, and of our readiness to receive them, cooled their resolution, and made them forget the vaunts of the preceding day; so that we passed this second night with as little molestation as we had done the first.

We had finished sending the treasure on board the *Centurion* the evening before ; so that the third morning, being the 15th of *November*, the boats were employed in carrying off the most valuable part of the effects that remained in the town. And the Commodore i tending to sail this day, he, about ten o'clock, pursuant to his promise, sent all his prisoners, amounting to eighty-eight, on shore, giving orders to Lieutenant *Brett* to secure them in one of the churches under a strict guard, till he was ready to embark his men. Mr. *Brett* was at the same time ordered to set the whole town on fire, except the two churches (which by good fortune stood at some distance from the other houses) and then he was to abandon the place, and to come on board. These orders were punctually complied with; for Mr. *Brett* immediately set his men to work, to distribute pitch, tar, and other combustibles (of which great quantities were found here) into houses situated in different streets of the town, so that, the place being fired in many quarters at the same time, the destruction might be more violent and sudden, and the enemy, after our departure, might not be able to extinguish it. These preparations being made, he, in the next place, ordered the cannon, which he found in the fort, to be nailed up ; and then setting fire to those houses which were most to windward, he collected his men, and marched towards the beach, where the boats waited to carry them off. And the part of the beach where he intended to embark being an open place without the town, near where the churches are marked in the foregoing plan, the *Spaniards* on the hill perceiving he was retreating, resolved to try if they

could

could not precipitate his departure, and thereby lay fome foundation for their future boafting. And for this purpofe a fmall fquadron of their horfe, confifting of about fixty, picked out, as I fuppofe, for this fervice, marched down the hill with much feeming refolution ; fo that, had we not been prepoffeffed with a jufter opinion of their prowefs, we might have fufpected, that now we were on the open beach with no advantage of fituation, they would certainly have charged us : But we prefumed (and we were not miftaken) that this was mere oftentation. For, notwithftanding the pomp and parade they advanced with, Mr. *Brett* had no fooner ordered his men to halt and face about, but the enemy ftopped their career, and never dared to advance a ftep further.

When our people were arrived at their boats, and were ready to go on board, they were for fome time delayed, by miffing one of their number ; but being unable, by their mutual enquiries amongft each other, to inform themfelves where he was left, or by what accident he was detained, they, after a confiderable delay, refolved to get into their boats, and to put off without him. And the laft man was actually embarked, and the boats juft putting off, when they heard him calling to them to take him in. The town was by this time fo thoroughly on fire, and the fmoke covered the beach fo effectually, that they could fcarcely fee him, though they heard his voice. The Lieutenant inftantly ordered one of the boats to his relief, who found him up to the chin in water, for he had waded as far as he durft, being extremely frightned with the apprehenfions of falling into the hands of an enemy, enraged, as they doubtlefs were, with the pillage and deftruction of their town. On enquiring into the caufe of his ftaying behind, it was found that he had taken that morning too large a dofe of brandy, which had thrown him into fo found a fleep, that he did not awake till the fire came near enough to fcorch him. He was ftrangely amazed on firft opening his-eyes, to fee the place all in a blaze on one fide, and feveral *Spaniards* and *Indians* not far from him on the other. The greatnefs and fuddennefs of his fright inftantly reduced him to a ftate

of

of fobriety, and gave him fufficient prefence of mind to pufh thro the thickeft of the fmoke as the likelieft means to efcape the ene-my ; and making the beft of his way to the beach, he ran as far into the water as he durft, (for he could not fwim) before he ventured to look back

And here I cannot but obferve, to the honour of our people, that though there were great quantities of wine and fpirituous li-quors found in the place, yet this man was the only one who was known to have fo far neglected his duty, as to get drunk. Indeed their whole behaviour, while they were on fhore was much more regular than could well have been expected from failors, who had been fo long confined to a fhip : And though part of this prudent demeanor muft doubtlefs be imputed to the diligence of their Officers, and to the excellent difcipline to which they had been long inured on board the Commodore, yet it was doubtlefs no fmall reputation to the men, that they fhould in general refrain from in-dulging themfelves in thofe intoxicating liquors, which they found ready to their hands in almoft every warehoufe.

And having mentioned this fingle inftance of drunkennefs, I cannot pafs by another overfight, which was likewife the only one of its kind, and which was attended with very particular circum-ftances. There was an *Englifl:man*, who had formerly wrought as a fhip-carpenter in the yard at *Portfmouth*, but leaving his country, had afterwards entered into the *Spanifh* fervice, and was employed by them at the port of *Guaiaquil*; and it being well known to his friends in *England* that he was then in that part of the world, they put letters on board the *Centurion*, directed to him. This man be-ing then by accident amongft the *Spaniards*, who were retired to the hill at *Paita*, he was defirous (as it fhould feem) of acquiring fome reputation amongft his new Mafters. With this view he came down unarmed to a centinel of ours, who was placed at fome diftance from the fort towards the enemy, and pretended to be de-firous of furrendring himfelf, and of entering into our fervice. Our centinel had a cock'd piftol, but being deceived by the other's

fair

Plate XXIV.

S. Müller Sculp.

of PAYTA on the Coast of

SOUTH SEA.

REFERENCES.

e e. Two of the Vice Roys Gallies of 36 Oars each.
f. {The Fort with eight Guns mounted, and capable of mounting thirteen.
g. The place where the Boats landed 49 Men.
h. The Convent of Mercenarians.
i. The Parish Church.

fair fpeeches, he was fo imprudent as to let him approach much nearer than he ought; fo that the Shipwright, watching his opportunity, rufhed on the centinel, and feizing his piftol, wrenched it out of his hand, and inftantly ran away with it up the hill. By this time, two of our people, who feeing the fellow advance had fufpected his intention, were making towards him, and were thereby prepared to purfue him; but he got to the top of the hill before they could reach him, and then turning about fired the piftol; at which inftant his purfuers fired at him, and though he was at a great diftance, and the creft of the hill hid him as foon as they had fired, fo that they took it for granted they had miffed him, yet we afterwards learnt that he was fhot through the body, and had fallen down dead the very next ftep he took after he was out of fight. The centinel too, who had been thus groffly impofed upon, did not efcape unpunifhed; for he was ordered to be feverely whipt for being thus fhamefully furprized upon his poft, and for having given an example of careleffnefs, which, if followed in other inftances, might prove fatal to us all. But to return:

By the time our people had taken their comrade out of the water, and were making the beft of their way for the fquadron, the flames had taken poffeffion of every part of the town, and had got fuch hold, both by means of combuftibles that had been diftributed for that purpofe, and by the flightnefs of the materials of which the houfes were compofed, and their aptitude to take fire, that it was fufficiently apparent, no efforts of the enemy (though they flocked down in great numbers) could poffibly put a ftop to it, or prevent the entire deftruction of the place, and all the merchandize contained therein. A whole town on fire at once, efpecially a place that burnt with fuch facility and violence, being a very fingular fpectacle, Mr. *Brett* had the curiofity to delineate its appearance, together with that of the fhips in the harbour, which may be feen in the annexed plate.

D d

Our

Our detachment under Lieutenant *Brett* having fafely joined the fquadron, the Commodore prepared to leave the place the fame evening. He found, when he firft came into the bay, fix veffels of the enemy at anchor; one of which was the fhip, which, according to our intelligence, was to have failed with the treafure to the coaft of *Mexico*, and which, as we were perfuaded fhe was a good failor, we refolved to take with us: The others were two Snows, a Bark, and two Row-gallies of thirty-fix oars a-piece: Thefe laft, as we were afterwards informed, with many others of the fame kind built at different ports, were intended to prevent our landing in the neighbourhood of *Callao*: For the *Spaniards*, on the firft intelligence of our fquadron and its force, expected that we would attempt the city of *Lima*. The Commodore, having no occafion for thefe other veffels, had ordered the mafts of all five of them to be cut away on his firft arrival; and now, at his leaving the place, they were towed out of the harbour, and fcuttled and funk; and the command of the remaining fhip, called the *Soli-dad*, being given to Mr. *Hughs* the Lieutenant of the *Tryal*, who had with him a crew of ten men to navigate her, the fquadron, towards midnight, weighed anchor, and failed out of the bay, being now augmented to fix fail, that is, the *Centurion* and the *Tryal* Prize, together with the *Carmelo*, the *Terefa*, the *Carmin*, and our laft acquired veffel the *Solidad*.

And now, before I entirely quit the account of our tranfactions at this place, it may not perhaps be improper to give a fuccinct relation of the booty we made here, and of the lofs the *Spaniards* fuftained. I have before obferved, that there were great quantities of valuable effects in the town; but as the greateft part of them were what we could neither difpofe of nor carry away, the total amount of this merchandize can only be rudely gueffed at. But the *Spaniards*, in the reprefentations they made to the Court of *Madrid*, (as we were afterwards affured) eftimated their whole lofs at a million and a half of dollars: And when it is confidered, that no fmall part of the goods we burnt there were of the richeft and moft

expenfive

expenſive ſpecies, as broad-cloaths, ſilks, cambrics, velvets, &c. I cannot but think their valuation ſufficiently moderate. As to our parts, our acquiſition, though inconſiderable in compariſon of what we deſtroyed, was yet in itſelf far from deſpicable ; for the wrought plate dollars and other coin which fell into our hands amounted to upwards of 30,000 l. ſterling, beſides ſeveral rings, bracelets, and jewels, whoſe intrinſick value we could not then determine ; and over and above all this, the plunder, which became the property of the immediate captors, was very great ; ſo that upon the whole it was by much the moſt important booty we made upon that coaſt.

There remains, before I take leave of this place, another particularity to be mentioned, which, on account of the great honour which our national character in thoſe parts has thence received, and the reputation which our Commodore in particular has thereby acquired, merits a diſtinct and circumſtantial diſcuſſion. It has been already related, that all the priſoners taken by us in our preceding prizes were put on ſhore, and diſcharged at this place ; amongſt which, there were ſome perſons of conſiderable diſtinction, particularly a youth of about ſeventeen years of age, ſon of the Vice-Preſident of the Council of Chili. As the barbarity of the Buccaneers, and the artful uſe the Eccleſiaſticks had made of it, had filled the natives of thoſe countries with the moſt terrible ideas of the Engliſh cruelty, we always found our priſoners, at their firſt coming on board us, to be extremely dejected, and under great horror and anxiety. In particular this youth, whom I laſt mentioned, having never been from home before, lamented his captivity in the moſt moving manner, regretting, in very plaintive terms, his parents, his brothers, his ſiſters, and his native country ; of all which he was fully perſuaded he had taken his laſt farewel, believing that he was now devoted, for the remaining part of his life, to an abject and cruel ſervitude ; nor was he ſingular in his fears, for his companions on board, and indeed all the Spaniards that came into our power, had the ſame deſponding opinion of their ſituation. Mr. Anſon conſtantly exerted his utmoſt endeavours to efface theſe

inhuman

inhuman impreffions they had received of us; always taking care, that as many of the principal people among them as there was room for fhould dine at his table by turns; and giving the ftricteft orders too, that they fhould at all times, and in every circumftance, be treated with the utmoft decency and humanity. But notwithftanding this precaution, it was generally obferved, that for the firft day or two they did not quit their fears, but fufpected the gentlenefs of their ufage to be only preparatory to fome unthought of calamity. However, being confirmed by time, they grew perfectly eafy in their fituation and remarkably chearful, fo that it was often difputable, whether or no they confidered their being detained by us as a misfortune. For the youth I have above-mentioned, who was near two months on board us, had at laft fo far conquered his melancholy furmifes, and had taken fuch an affection to Mr. *Anfon*, and feemed fo much pleafed with the manner of life, totally different from all he had ever feen before, that it is doubtful to me whether, if his own opinion had been taken, he would not have preferred a voyage to *England* in the *Centurion*, to the being fet on fhore at *Paita*, where he was at liberty to return to his country and his friends.

This conduct of the Commodore to his prifoners, which was continued without interruption or deviation, gave them all the higheft idea of his humanity and benevolence, and induced them likewife (as mankind are fond of forming general opinions) to entertain very favourable thoughts of the whole *Englifh* Nation. But whatever they might be difpofed to think of Mr. *Anfon* before the taking of the *Terefa*, their veneration for him was prodigioufly increafed by his conduct towards thofe women, whom (as I have already mentioned) he took in that veffel: For the leaving them in the poffeffion of their apartments, the ftrict orders given to prevent all his people on board from approaching them, and the permitting the Pilot to ftay with them as their guardian, were meafures that feemed fo different from what might be expected from an enemy and an heretick, that the *Spaniards* on board, though they

had

had themselves experienced his beneficence, were surprized at this new instance of it, and the more so, as all this was done without his ever having seen the women, though the two daughters were both esteemed handsome, and the youngest was celebrated for her uncommon beauty. The women themselves too were so sensible of the obligations they owed him, for the care and attention with which he had protected them, that they absolutely refused to go on shore at *Paita*, till they had been permitted to wait on him on board the *Centurion*, to return him thanks in person. Indeed, all the prisoners left us with the strongest assurances of their grateful remembrance of his uncommon treatment. A Jesuit in particular, whom the Commodore had taken, and who was an Ecclesiastick of some distinction, could not help expressing himself with great thankfulness for the civilities he and his countrymen had found on board, declaring, that he should consider it as his duty to do Mr. *Anson* justice at all times; adding, that his usage of the men prisoners was such as could never be forgot, and such as he could never fail to acknowledge and recite upon all occasions : But that his behaviour to the women was so extraordinary, and so extremely honourable, that he doubted all the regard due to his own ecclesiastical character, would be scarcely sufficient to render it credible. And indeed we were afterwards informed, that both he and the rest of our prisoners had not been silent on this head, but had, both at *Lima* and at other places, given the greatest encomiums to our Commodore; the Jesuit in particular, as we were told, having, on his account, interpreted in a lax and hypothetical sense that article of his Church, which asserts the impossibility of hereticks being saved.

And let it not be imagined, that the impressions which the *Spaniards* hence received to our advantage, is a matter of small import; for, not to mention several of our countrymen who have already felt the good effects of these prepossessions, the *Spaniards* are a Nation, whose good opinion of us is doubtless of more consequence than that of all the world besides : Not only as the commerce we have formerly carried on with them, and perhaps may again hereafter,

after, is fo extremely valuable ; but alfo as the tranfacting it does fo immediately depend on the honour and good faith of thofe who are entrufted with its management. But however, had no national conveniencies attended it, the Commodore's equity and good temper would not lefs have deterred him from all tyranny and cruelty to thofe, whom the fortune of war had put into his hands. I fhall only add, that by his conftant attachment to thefe humane and prudent maxims, he has acquired a diftinguifhed reputation amongft the *Creolian Spaniards*, which is not confined merely to the coaft of the *South-Seas*, but is extended through all the *Spanifh* fettlements in *America* ; fo that his name is frequently to be met with in the mouths of moft of the *Spanifh* inhabitants of that prodigious Empire.

CHAP.

C H A P. VII.

From our departure from *Paita*, to our arrival at *Quibo*.

WHEN we got under fail from the road of *Paita*, (which, as I have already obferved, was about midnight, on the 16th of *November*) we flood to the weftward, and in the morning the Commodore gave orders, that the whole fquadron fhould fpread themfelves, in order to look out for the *Gloucefter*. For we now drew near to the ftation where Captain *Mitchel* had been directed to cruife, and hourly expected to get fight of him ; but the whole day paffed without feeing him.

And now a jealoufy, which had taken its rife at *Paita*, between thofe who had been ordered on fhore for the attack, and thofe who had continued on board, grew to fuch a height, that the Commodore, being made acquainted with it, thought it neceffary to interpofe his authority to appeafe it. The ground of this animofity was the plunder gotten at *Paita*, which thofe who had acted on fhore had appropriated to themfelves, and confidered it as a reward for the rifques they had run, and the refolution they had fhown in that fervice. But thofe, who had remained on board, confidered this as a very partial and unjuft procedure, urging, that had it been left to their choice, they fhould have preferred the acting on fhore to the continuing on board ; that their duty, while their comrades were on fhore, was extremely fatiguing ; for befides the labour of the day, they were conftantly under arms all night to fecure the prifoners, whofe numbers exceeded their own, and of whom it was then neceffary to be extremely watchful, to prevent any attempts they might have formed in that critical conjuncture : That upon the whole it could not be denied, but that the prefence of a fufficient

force

force on board was as neceſſary to the ſucceſs of the enterprize, as
the action of the others on ſhore, and therefore thoſe who had
continued on board inſiſted, that they could not be deprived of
their ſhare of the plunder, without manifeſt injuſtice. Theſe were
the conteſts amongſt our men, which were carried on with great
heat on both ſides: And though the plunder in queſtion was a very
trifle, in compariſon of the treaſure taken in the place, (in which
there was no doubt but thoſe on board had an equal right)
yet as the obſtinacy of ſailors is not always regulated by the importance of the matter in diſpute, the Commodore thought it neceſſary to put a ſtop to this ferment betimes. And accordingly, the
morning after our leaving of *Paita*, he ordered all hands upon the
quarter-deck; where, addreſſing himſelf to thoſe who had been
detached on ſhore, he commended their behaviour, and thanked
them for their ſervices on that occaſion: But then repreſenting to
them the reaſons urged, by thoſe who had continued on board, for
an equal diſtribution of the plunder, he told them, that he thought
theſe reaſons very concluſive, and that the expectations of their
comrades were juſtly founded; and therefore he ordered, that not
only the men, but all the officers likewiſe, who had been employed in taking the place, ſhould produce the whole of their plunder
immediately upon the quarter-deck; and that it ſhould be impar-
tially divided amongſt the whole crew, in proportion to each man's
rank and commiſſion: And to prevent thoſe who had been in poſſeſſion of the plunder from murmuring at this diminution of their
ſhare, the Commodore added, that as an encouragement to others
who might be hereafter employed on like ſerviçes, he would give
his entire ſhare to be diſtributed amongſt thoſe who had been detached for the attack of the place. Thus this troubleſome affair,
which if permitted to have gone on, might perhaps have been attended with miſchievous conſequences, was by the Commodore's
prudence ſoon appeaſed, to the general ſatisfaction of the ſhip's
company: Not but there were ſome few, whoſe ſelfiſh diſpoſitions
were uninfluenced by the juſtice of this procedure, and who were

incapable

incapable of difcerning the force of equity, however glaring, when it tended to deprive them of any part of what they had once got into their hands.

This important bufinefs employed the beft part of the day, after we came from *Paita*. And now, at night, having no fight of the *Gloucefter*, the Commodore ordered the fquadron to bring to, that we might not pafs her in the dark. The next morning we again looked out for her, and at ten we faw a fail, to which we gave chace; and at two in the afternoon we came near enough to her to difcover her to be the *Gloucefter*, with a fmall veffel in tow. About an hour after, we were joined by them; and then we learnt that Captain *Mitchel*, in the whole time of his cruife, had only taken two prizes; one of them being a fmall Snow, whofe cargoe confifted chiefly of wine, brandy, and olives in jars, with about 7000 *l.* in fpecie; and the other a large boat or launch, which the *Gloucefter*'s barge came up with near the fhore. The prifoners on board this veffel alledged, that they were very poor, and that their loading confifted only of cotton; though the circumftances in which the barge furprized them, feemed to infinuate that they were more opulent than they pretended to be; for the *Gloucefter*'s people found them at dinner upon pidgeon-pye, ferved up in filver difhes. However, the Officer who commanded the barge having opened feveral of the jars on board, to fatisfy his curiofity, and finding nothing in them but cotton, he was inclined to believe the account the prifoners gave him: But the cargoe being taken into the *Gloucefter*, and there examined more ftrictly, they were agreeably furprized to find, that the whole was a very extraordinary piece of falfe package; and that there was concealed amongft the cotton, in every jar, a confiderable quantity of double doubloons and dollars, to the amount in the whole of near 12,000 *l.* This treafure was going to *Paita*, and belonged to the fame Merchants who were the proprietors of the greateft part of the money we had taken there; fo that had this boat efcaped the *Gloucefter*, it is probable her cargoe would have fallen into our hands. Befides thefe two prizes which

E e

we

we have mentioned, the *Gloucester*'s people told us, that they had been in fight of two or three other ships of the enemy which had escaped them ; and one of them we had reason to believe, from some of our intelligence, was of an immense value.

Being now joined by the *Gloucester* and her prize, it was resolved that we should stand to the northward, and make the best of our way either to Cape St. *Lucas* on *California*, or to Cape *Corientes* on the coast of *Mexico*. Indeed the Commodore, when at *Juan Fernandes*, had determined with himself to touch in the neighbourhood of *Panama*, and to endeavour to get some correspondence over land with the fleet under the command of Admiral *Vernon*. For when we departed from *England*, we left a large force at *Portsmouth*, which was intended to be sent to the *West-Indies*, there to be employed in an expedition against some of the *Spanish* settlements. And Mr. *Anson* taking it for granted, that this enterprize had succeeded, and that *Porto Bello* perhaps might be then garrisoned by *British* troops, he hoped, that on his arrival at the *Isthmus*, he should easily procure an intercourse with our countrymen on the other side, either by the *Indians*, who were greatly disposed in our favour, or even by the *Spaniards* themselves, some of whom, for proper rewards, might be induced to carry on this intelligence, which, after it was once begun, might be continued with very little difficulty ; so that Mr. *Anson* flattered himself, that he might by this means have received a reinforcement of men from the other side, and that by settling a prudent plan of operations with our Commanders in the *West-Indies*, he might have taken even *Panama* itself ; which would have given to the *British* Nation the possession of that *Isthmus*, whereby we should have been in effect masters of all the treasures of *Peru*; and should have had in our hands an equivalent for any demands, however extraordinary, which we might have been induced to have made on either of the branches of the House of *Bourbon*.

Such were the projects which the Commodore revolved in his thoughts at the Island of *Juan Fernandes*, notwithstanding the

feeble

feeble condition to which he was then reduced. And indeed, had the fuccefs of our force in the *Weft-Indies* been anfwerable to the general expectation, it cannot be denied but thefe views would have been the moft prudent that could have been thought of. But in examining the papers which were found on board the *Carmelo*, the firft prize we took, we learnt (though I then omitted to mention it) that our attempt againft *Carthagena* had failed, and that there was no probability that our fleet, in that part of the world, would engage in any new enterprize, that would at all facilitate this plan. And therefore Mr. *Anfon* gave over all hopes of being reinforced a-crofs the *Ifthmus*, and confequently had no inducement at prefent to proceed to *Panama*, as he was incapable of attacking the place; and there was great reafon to believe, that by this time there was a general embargo on all the coaft.

The only feafible meafure then which was left us, was to get as foon as poffible to the fouthern parts of *California*, or to the adjacent coaft of *Mexico*, there to cruife for the *Manila* Galeon, which we knew was now at fea, bound to the port of *Acapulco*. And we doubted not to get on that ftation, time enough to intercept her; for this fhip does not actually arrive at *Acapulco* till towards the middle of *January*, and we were now but in the middle of *November*, and did not conceive that our paffage thither would coft us above a month or five weeks; fo that we imagined, we had near twice as much time as was neceffary for our purpofe. Indeed there was a bufinefs which we forefaw would occafion fome delay, but we flattered ourfelves that it would be difpatched in four or five days, and therefore could not interrupt our project. This was the recruiting of our water; for the number of prifoners we had entertained on board, fince our leaving the Ifland of *Fernandes*, had fo far exhaufted our ftock, that it was impoffible to think of venturing upon this paffage to the coaft of *Mexico*, till we had procured a frefh fupply; efpecially as at *Paita*, where we had fome hopes of getting a quantity, we did not find enough for our confumption during the time we ftayed there. It was for fome time

a mat-

a matter of deliberation, where we should take in this necessary article; but by confulting the accounts of former Navigators, and
examining our prifoners, we at laft refolved for the Ifland of *Quibo*,
fituated at the mouth of the bay of *Panama* : Nor was it but on
good grounds that the Commodore conceived this to be the propereft place for watering the fquadron. Indeed, there was a fmall
Ifland called *Cocos*, which was lefs out of our way than *Quibo*,
where fome of the Buccaneers have pretended they found water;
but none of our prifoners knew any thing of it, and it was thought
too hazardous to rifque the fafety of the fquadron, and expofe ourfelves to the hazard of not meeting with water when we came there,
on the mere authority of thefe legendary writers, of whofe mifrepre-
fentations and falfities we had almoft daily experience. Befides, by
going to *Quibo* we were not without hopes that fome of the enemies
fhips bound to or from *Panama* might fall into our hands, particularly
fuch of them as were put to fea, before they had any intelligence of
our fquadron.

Having determined therefore to go to *Quibo*, we directed our
courfe to the northward, being eight fail in company, and confequently having the appearance of a very formidable fleet; and on
the 19th, at day-break, we difcovered Cape *Blanco*, bearing
S. S. E. ½ E, feven miles diftant. This Cape lies in the latitude of
4° : 15' South, and is always made by fhips bound either to windward or to leeward; fo that off this Cape is a moft excellent ftation
to cruife upon the enemy. By this time we found that our laft
prize, the *Solidad* was far from anfwering the character given her
of a good failor; and fhe and the *Santa Terefa* delaying us confiderably, the Commodore ordered them both to be cleared of every
thing that might prove ufeful to the reft of the fhips, and then to
be burnt; and having given proper inftructions, and a rendezvous
to the *Gloucefter* and the other prizes, we proceeded in our courfe
for *Quibo*; and, on the 22d in the morning, faw the Ifland of
Plata, bearing Eaft, diftant four leagues. Here one of our prizes
was ordered to ftand clofe in with it, both to difcover if there
were

were any fhips between that Ifland and the Continent, and likewife to look out for a ftream of frefh water, which was reported to be there, and which would have faved us the trouble of going to *Quibo*; but fhe returned without having feen any fhip, or finding any water. At three in the afternoon point *Manta* bore S. E. by E. feven miles diftant; and there being a town of the fame name in the neighbourhood, Captain *Mitchel* took this opportunity of fending away feveral of his prifoners from the *Gloucefter* in the *Spanifh* launch. The boats were now daily employed in diftributing provifions on board the *Tryal* and other prizes, to compleat their ftock for fix months: And that the *Centurion* might be the better prepared to give the *Manila* fhip (one of which we were told was of an immenfe fize) a warm reception, the Carpenters were ordered to fix eight ftocks in the main and fore-tops, which were properly fitted for the mounting of fwivel guns.

On the 25th we had a fight of the Ifland of *Gallo*, bearing E. S. E. ¼ E, four leagues diftant; and from hence we croffed the bay of *Panama* with a N. W. courfe, hoping that this would have carried us in a direct line to the Ifland of *Quibo*. But we afterwards found that we ought to have ftood more to the weftward; for the winds in a fhort time began to incline to that quarter, and made it difficult for us to gain the Ifland. And now, after paffing the equinoctial, (which we did on the 22d) and leaving the neighbourhood of the *Cordilleras*, and ftanding more and more towards the *Ifthmus*, where the communication of the atmofphere to the eaftward and the weftward was no longer interrupted, we found in very few days an extraordinary alteration in the climate. For inftead of that uniform temperature, where neither the excefs of heat or cold was to be complained of, we had now for feveral days together clofe and fultry weather, refembling what we had before met with on the coaft of *Brazil*, and in other parts between the tropics on the eaftern fide of *America*. We had befides frequent calms and heavy rains; which we at firft afcribed to the neighbourhood of the line, where this kind of weather is generally found to

prevail

prevail at all seasons of the year; but observing that it attended us to the latitude of seven degrees North, we were at length induced to believe, that the stormy season, or, as the *Spaniards* call it, the Vandevals, was not yet over; though many writers, particularly Captain *Shelvocke*, positively affert, that this season begins in *June*, and is ended in *November*; and our prisoners all affirmed the same thing. But perhaps its end may not be always constant, and it might last this year longer than usual.

On the 27th, Captain *Mitchel* having finished the clearing of his largest prize, she was scuttled, and set on fire; but we still consisted of five ships, and were fortunate enough to find them all good sailors; so that we never occasioned any delay to each other. Being now in a rainy climate, which we had been long disused to, we found it necessary to caulk the decks and sides of the *Centurion*, to prevent the rain-water from running into her.

On the 3d of *December* we had a view of the Island of *Quibo*; the East end of which then bore from us N. N. W, four leagues distant, and the Island of *Quicara* W.N.W, at about the same distance. Here we struck ground with sixty-five fathom of line, and found the bottom to consist of grey sand, with black specks. There is hereafter inserted (being contained in the same plate with the view of the hill of *Petaplan*) a view of these two Islands, where (*a*) represents the S. E. end of *Quibo*, bearing N. by W. four leagues distant: And (*b*) the Island of *Quicara*, which bears from the point (*a*) W.S.W. ½ S, and is distant from it four leagues, the point (*a*) being itself in the latitude of 7° : 20 North. When we had thus got sight of the land, we found the wind to hang westerly; and therefore, night coming on, we thought it adviseable to stand off till morning, as there are said to be some shoals in the entrance of the channel. At six the next morning point *Mariato* bore N. E. ½ N, three or four leagues distant. In weathering this point all the squadron, except the *Centurion*, were very near it; and the *Gloucester* being the leewardmost ship, was forced to tack and stand to the southward, so that we lost sight of her. At nine, the Island *Sebaco*

baco bore N. W. by N, four leagues diftant ; but the wind ftill proving unfavourable, we were obliged to ply on and off for the fucceeding twenty-four hours, and were frequently taken aback. However, at eleven the next morning the wind happily fettled in the S. S. W, and we bore away for the S. S. E. end of the Ifland, and about three in the afternoon entered the *Canal Bueno*, paffing round a fhoal which ftretches off about two miles from the South point of the Ifland. This *Canal Bueno*, or *Good Channel*, is at leaft fix miles in breadth ; and as we had the wind large, we kept in a good depth of water, generally from twenty-eight to thirty-three fathom, and came not within a mile and a half diftance of the breakers ; though, in all probability, if it had been neceffary, we might have ventured much nearer, without incurring the leaft dan- ger. At feven in the evening we came to an anchor in thirty-three fathom muddy ground ; the South point of the Ifland bearing S. E. by S, a remarkable high part of the Ifland W. by N, and the Ifland *Sebaco* E. by N. Being thus arrived at this Ifland of *Quibo*, the account of the place, and of our tranfactions there, fhall be referred to the enfuing chapter.

C H A P.

CHAP. VIII.

Our proceedings at *Quibo*, with an account of the place.

THE next morning, after our coming to an anchor, an officer was difpatched on fhore to difcover the watering place, who having found it, returned before noon; and then we fent the long boat for a load of water, and at the fame time we weighed and ftood farther in with our fhips. At two we came again to an anchor in twenty-two fathom, with a bottom of rough gravel intermixed with broken fhells, the watering place now bearing from us N. W. ½ N, only three quarters of a mile diftant. A plan of the road where we lay and of the Eaft-end Ifland is annexed, where the foundings are laid down, fuch as we found them, the latitude of the S. E. point of the Ifland being, as hath been already mentioned, 7°: 20′ North.

This Ifland of *Quibo* is extremely convenient for wooding and watering; for the trees grow clofe to the high-water mark, and a large rapid ftream of frefh water runs over the fandy beach into the fea: So that we were little more than two days in laying in all the wood and water we wanted. The whole Ifland is of a very moderate height, excepting one part. It confifts of a continued wood fpread over the whole furface of the country, which preferves its verdure all the year round. Amongft the other wood, we found there abundance of caffia, and a few lime-trees. It appeared fingular to us, that confidering the climate and the fhelter, we fhould fee no other birds there than parrots, parroquets, and mackaws; indeed of thefe laft there were prodigious flights. Next to thefe birds, the animals we found there in moft plenty were monkeys and guanos, and thefe we frequently killed for food; for tho'

there

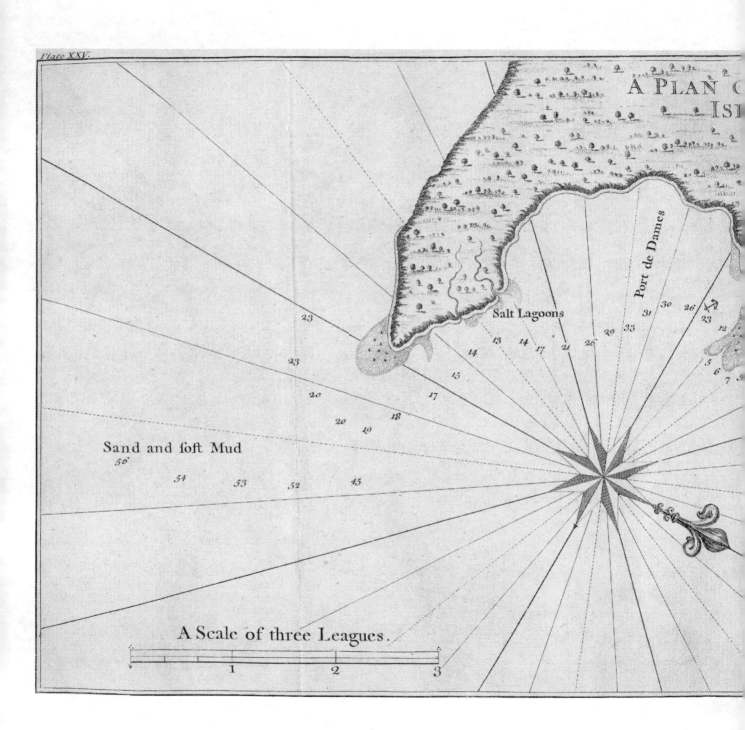

Plate XXV.

A PLAN O
ISI

Port de Dames

Salt Lagoons

Sand and soft Mud

A Scale of three Leagues.

1   2   3

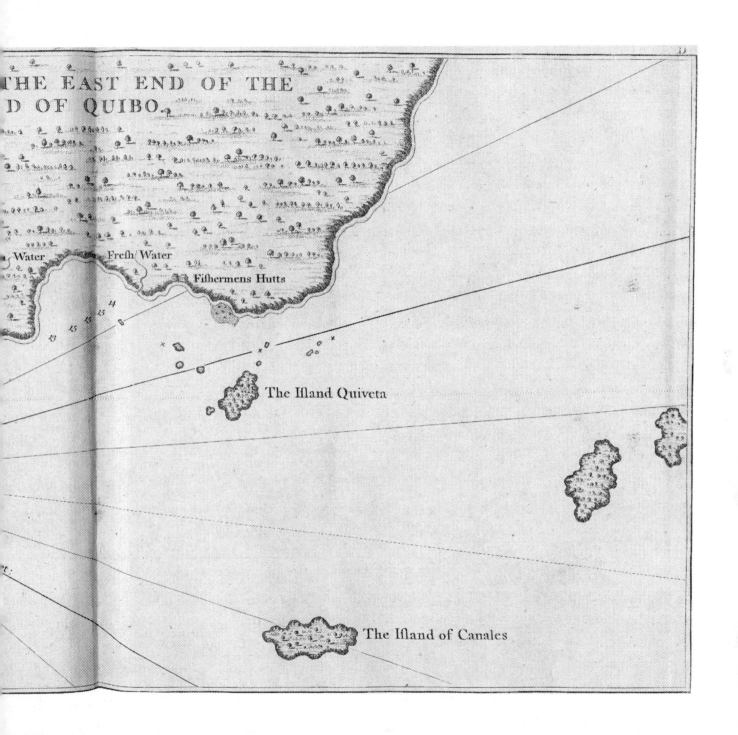

THE EAST END OF THE

D OF QUIBO.

Water   Fresh Water

Fishermens Hutts

14

13   15 15

13

The Island Quiveta

The Island of Canales

though there were many herds of deer upon the place, yet the difficulty of penetrating the woods prevented our coming near them, so that though we saw them often, we killed only two during our stay. Our prisoners assured us, that this Island abounded with tygers ; and we did once discover the print of a tyger's paw upon the beach, but the tygers themselves we never saw. The *Spaniards* too informed us, that there was often found in the woods a most mischievous serpent, called the flying snake, which they said darted itself from the boughs of trees on either man or beast that came within its reach ; and whose sting, they believed, to be inevitable death. Besides these mischievous land-animals, the sea hereabouts is infested with great numbers of alligators of an extraordinary size ; and we often observed a large kind of flat-fish, jumping a considerable height out of the water, which we supposed to be the fish that is said frequently to destroy the pearl-divers, by clasping them in its fins as they rise from the bottom ; and we were told that the divers, for their security, are now always armed with a sharp knife, which, when they are entangled, they stick into the belly of the fish, and thereby disengage themselves from its embraces.

Whilst the ship continued here at anchor, the Commodore, attended by some of his officers, went in a boat to examine a bay which lay to the northward ; and they afterwards ranged all along the eastern side of the Island. And in the places where they put on shore in the course of this expedition, they generally found the soil to be extremely rich, and met with great plenty of excellent water. In particular, near the N. E. point of the Island, they discovered a natural cascade, which surpassed, as they conceived, every thing of this kind, which human art or industry hath hitherto produced. It was a river of transparent water, about forty yards wide, which ran down a declivity of near a hundred and fifty yards in length. The channel it ran in was very irregular ; for it was entirely formed of rock, both its sides and bottom being made up of large detached blocks ; and by these the course of the water was frequently inter-

F f

rupted :

rupted : For in fome places it ran floping with a rapid but uniform motion, while in other parts it tumbled over the ledges of rocks with a perpendicular defcent. All the neighbourhood of this ftream was a fine wood ; and even the huge maffes of rock which overhung the water, and which, by their various projections, formed the inequalities of the channel, were covered with lofty foreft trees. Whilft the Commodore, and thofe who were with him, were attentively viewing this place, and were remarking the different blendings of the water, the rocks and the wood, there came in fight (as it were with an intent ftill to heighten and animate the profpect) a prodigious flight of mackaws, which hovering over this fpot, and often wheeling and playing on the wing about it, afforded a moft brilliant appearance, by the glittering of the fun on their variegated plumage ; fo that fome of the fpectators cannot refrain from a kind of tranfport, when they recount the complicated beauties which occurred in this extraordinary water-fall.

In this expedition, which the boat made along the eaftern fide of the Ifland, though they met with no inhabitants, yet they faw many huts upon the fhore, and great heaps of fhells of fine mother of pearl fcattered up and down in different places : Thefe were the remains left by the pearl-fifhers from *Panama*, who often frequent this place in the fummer feafon ; for the pearl oyfters, which are to be met with every where in the bay of *Panama*, are fo plenty at *Quibo*, that by advancing a very little way into the fea, you might ftoop down and reach them from the bottom. They are ufually very large, and out of curiofity we opened fome of them with a view of tafting them, but we found them extremely tough and unpalatable. And having mentioned thefe oyfters and the pearl-fifhery, I muft beg leave to recite a few particulars relating thereto.

The oyfters moft productive of pearls are thofe found in confiderable depths ; for though what are taken up by wading near fhore are of the fame fpecies, yet the pearls found in them are very rare and very fmall. It is faid too, that the pearl partakes in fome degree

gree of the quality of the bottom on which the oyfter is found; fo that if the bottom be muddy, the pearl is dark and ill-coloured.

The taking up oyfters from great depths for the fake of the pearls they contain, is a work performed by Negro flaves, of which the inhabitants of *Panama* and the neighbouring coaft formerly kept great numbers, which were carefully trained to this bufinefs. And thefe are faid not to be efteemed compleat divers, till they have by degrees been able to protract their ftay under water fo long, that the blood gufhes out from their nofe, mouth and ears. And it is the tradition of the country, that when this accident has once be-fallen them, they dive for the future with much greater facility than before; and they have no apprehenfion either that any inconveni-ence can attend it, the bleeding generally ftopping of itfelf, or that there is any probability of their being ever fubject to it a fecond time. But to return from this digreffion.

Though the pearl oyfter, as hath been faid, was incapable of being eaten, yet the fea at this place furnifhed us with another dain-ty, in the greateft plenty and perfection: This was the turtle, of which we took here what quantity we pleafed. There are gene-rally reckoned four fpecies of turtle; that is, the trunk turtle, the loggerhead, the hawkfbill, and the green turtle. The two firft are rank and unwholefome; the hawkfbill (which furnifhes the tor-tois-fhell) is but indifferent food, though better than the other two; but the green turtle is generally efteemed, by the greateft part of thofe who are acquainted with its tafte, to be the moft delicious of all eatables; and that it is a moft wholfome food, we are amply con-vinced by our own experience: For we fed on this laft fpecies, or the green turtle, for near four months, and confequently had it been in any degree noxious, its ill effects could not poffibly have efcaped us. At this Ifland we took what quantity we pleafed with great facility; for as they are an amphibious animal, and get on fhore to lay their eggs, which they generally depofit in a large hole in the fand, juft above the high-water mark, covering them up, and leaving them to be hatched by the heat of the fun, we ufually

difperfed

difperfed feveral of our men along the beach, whofe bufinefs it was
to turn them on their backs when they came to land; and the turtle being thereby prevented from getting away, we carried them off
at our leifure: By this means we not only fecured a fufficient
ftock for the time we ftayed on the Ifland, but we took a number
of them with us to fea, which proved of great fervice both in
lengthning out our ftore of provifion, and in heartning the whole
crew with an almoft conftant fupply of frefh and palatable food;
for the turtle being large, they generally weighing about 200 *lb.*
weight each, thofe we took with us lafted us near a month, and by
that time we met with a frefh recruit on the coaft of *Mexico,*
where we often faw them in the heat of the day floating in great
numbers on the furface of the water faft afleep; when we difcovered them, we ufually fent out our boat with a man in the bow, who
was a dextrous diver, and when the boat came within a few yards
of the turtle, the diver plunged into the water, and took care to
rife clofe upon it; and feizing the fhell near the tail, and preffing
down the hinder parts, the turtle, when awakened, began to
ftrike with its claws, which motion fupported both it and the diver,
till the boat came up and took them in. By this management we
never wanted turtle for the fucceeding four months in which we continued at fea; and though, when at *Quibo,* we had already been
three months on board, without otherwife putting our foot on fhore,
than in the few days we ftayed at this Ifland of *Quibo,* (except thofe
employed in the attack of *Paita*) yet in the whole feven months,
from our leaving *Juan Fernandes* to our anchoring in the harbour
of *Chequetan,* we buried no more in the whole fquadron than two
men; a moft inconteftable proof, that the turtle, on which we fed
for the laft four months of this term, was at leaft innocent, if not
fomething more.

Confidering the fcarcity of provifions on fome part of the coaft
of thefe feas, it appears wonderful, that a fpecies of food fo very
palatable and falubrious as turtle, and fo much abounding in thofe
parts, fhould be profcribed by the *Spaniards* as unwholefome, and
little

little lefs than poifonous. Perhaps the ftrange appearance of this animal may have been the foundation of this ridiculous and fuperftitious averfion, which is ftrongly rooted in all the inhabitants of that coaft, and of which we had many inftances in the courfe of this navigation. I have already obferved, that we put our *Spanifh* prifoners on fhore at *Paita*, and that the *Gloucefter* fent theirs to *Manta*; but as we had taken in our prizes fome *Indian* and Negro flaves, we did not fet thefe on fhore with their mafters, but continued them on board, as our crews were thin, to affift in navigating our fhips. Thefe poor people being poffeffed with the prejudices of the country they came from, were aftonifhed at our feeding on turtle, and feemed fully perfuaded that it would foon deftroy us; but finding that none of us died, nor even fuffered in our health by a continuation of this diet, they at laft got fo far the better of their averfion, as to be perfuaded to tafte it, to which the abfence of all other kinds of frefh provifions might not a little contribute. However, it was with great reluctance, and very fparingly, that they firft began to eat of it: But the relifh improving upon them by degrees, they at laft grew extremely fond of it, and preferred it to every other kind of food, and often felicitated each other on the happy experience they had acquired, and the delicious and plentiful repafts it would be always in their power to procure, when they fhould again return back to their country. Thofe who are acquainted with the manner of life of thefe unhappy wretches, need not be told, that next to large draughts of fpirituous liquors, plenty of tolerable food is the greateft joy they know, and confequently the difcovering a method which would always fupply them with what quantity they pleafed, of a food more luxurious to the palate than any their haughty Lords and Mafters could indulge in, was doubtlefs a circumftance, which they confidered as the moft fortunate that could befal them.

After this digreffion, which the prodigious quantity of turtle on this Ifland of *Quibo*, and the ftore of it we thence took to fea, in fome meafure led me into, I fhall now return to our own proceedings.

In

In three days time we had compleated our bufinefs at this place, and were extremely impatient to put to fea, that we might arrive time enough on the coaft of *Mexico* to intercept the *Manila* galeon. But the wind being contrary detained us a night, and the next day, when we got into the offing (which we did through the fame channel by which we entered) we were obliged to keep hovering about the Ifland, in hopes of getting fight of the *Gloucefter*, who, as I have in the laft chapter mentioned, was feparated from us on our firft arrival. It was the 9th of *December*, in the morning, when we put to fea, and continuing to the fouthward of the Ifland, looking out for the *Gloucefter*, we, on the 10th, at five in the afternoon, difcerned a fmall fail to the northward of us, to which we gave chace, and coming up with her took her. She proved to be a bark from *Panama*, bound to *Cheripe*, an inconfiderable village on the Continent, and was called the *Jefu Nazareno*. She had nothing on board but fome oakum, about a tun of rock falt, and between 30 and 40 *l.* in fpecie, moft of it confifting of fmall filver money, intended for purchafing a cargoe of provifions at *Cheripe*.

And on occafion of this prize I cannot but obferve, for the ufe of future cruifers, that had we been in want of provifions, we had by this capture an obvious method of fupplying ourfelves. For at *Cheripe*, whither fhe was bound, there is a conftant ftore of provifions prepared for the veffels who go thither every week from *Panama*, the market of *Panama* being chiefly fupplied from thence: So that by putting a few of our hands on board our prize, we might eafily have feized a large ftore without any hazard, fince *Cheripe* is a place of no ftrength. And as provifions are the ftaple commodity of that place and of its neighbourhood, the knowledge of this circumftance may be of great ufe to fuch cruifers, as find their provifions grow fcant, and yet are defirous of continuing on that coaft as long as poffible. But to return:

On

On the 12th of *December* we were at laſt relieved from the perplexity we had ſuffered, by the ſeparation of the *Glouceſter* ; for on
that day ſhe joined us, and informed us, that in tacking to the
ſouthward, on our firſt arrival, ſhe had ſprung her fore-top-maſt,
which had diſabled her from working to windward, and prevented
her from joining us ſooner. And now we ſcuttled and ſunk the
Jeſu Nazareno, the prize we took laſt ; and having the greateſt
impatience to get into a proper ſtation for the galeon, we ſtood all together to the weſtward, leaving the Iſland of *Quibo* (notwithſtanding all the impediments we met with) in about nine days after our
firſt coming in ſight of it.

CHAP.

CHAP. IX.

From *Quibo* to the coaft of *Mexico*.

ON the 12th of *December* we ftood from *Quibo* to the weftward, and the fame day the Commodore delivered frefh inftructions to the Captains of the men of war, and the commanders of our prizes, appointing them the rendezvoufes they were
to make, and the courfes they were to fteer in cafe of a feparation.
And firft, they were directed to ufe all poffible difpatch in getting
to the northward of the harbour of *Acapulco*, where they were to
endeavour to fall in with the land, between the latitudes of 18
and 19 degrees; from thence, they were to beat up the coaft at
eight or ten leagues diftance from the fhore, till they came a-breaft
of Cape *Corientes*, in the latitude of 20°: 20. When they arrived there, they were to continue cruifing on that ftation till the
14th of *February*; and then they were to proceed to the middle
Ifland of the *Tres Marias*, in the latitude of 21°: 25′, bearing
from Cape *Corientes* N. W. by N, twenty-five leagues diftant. And
if at this Ifland they did not meet the Commodore, they were there
to recruit their wood and water, and then to make the beft of their
way to the Ifland of *Macao*, on the coaft of *China*. Thefe orders being diftributed to all the fhips, we had little doubt of arriving foon upon our intended ftation, as we expected, upon the
encreafing our offing from *Quibo*, to fall in with the regular trade-
wind. But, to our extream vexation, we were baffled for near a month,
either with tempeftuous weather from the weftern quarter, or with
dead calms and heavy rains, attended with a fultry air; fo that it
was the 25th of *December* before we got a fight of the Ifland of
Cocos, which by our reckoning was only a hundred leagues from
the Continent; and we had the mortification to make fo little way,

that

that we did not lofe fight of it again in five days. This Ifland we found to be in the latitude of 5° : 20′ North. It has a high hummock towards the weftern part, which defcends gradually, and at laft terminates in a low point to the eaftward. From the Ifland of *Cocos* we ftood W. by N, and were till the 9th of *January* in running an hundred leagues more. We had at firft flattered ourfelves, that the uncertain weather and weftern gales we met with were owing to the neighbourhood of the Continent, from which, as we got more diftant, we expected every day to be relieved, by falling in with the eaftern trade-wind: But as our hopes were fo long baffled, and our patience quite exhaufted, we began at length to defpair of fucceeding in the great purpofe we had in view, that of intercepting the *Manila* galeon; and this produced a general dejection amongft us, as we had at firft confidered this project as almoft infallible, and had indulged ourfelves in the moft boundlefs hopes of the advantages we fhould thence receive. However, our defpondency was at laft fomewhat alleviated, by a favourable change of the wind; for, on the 9th of *January*, a gale for the firft time fprung up from the N. E, and on this we took the *Carmelo* in tow, as the *Gloucefter* did the *Carmin*, making all the fail we could to improve the advantage, for we ftill fufpected that it was only a temporary gale, which would not laft long; but the next day we had the fatisfaction to find, that the wind did not only continue in the fame quarter, but blew with fo much brifknefs and fteadinefs, that we now no longer doubted of its being the true trade-wind. And as we advanced a-pace towards our ftation, our hopes began to revive, and our former defpair by degrees gave place to more fanguine prejudices: For though the cuftomary feafon of the arrival of the galeon at *Acapulco* was already elapfed, yet we were by this time unreafonable enough to flatter ourfelves, that fome accidental delay might, for our advantage, lengthen out her paffage beyond its ufual limits.

When we got into the trade-wind, we found no alteration in it till the 17th of *January*, when we were advanced to the latitude

G g

of

of 12° : 50′, but on that day it fhifted to the weftward of the North : This change we imputed to our having haled up too foon, though we then efteemed our felves full feventy·leagues from the coaft, which plainly fhows, that the trade-wind doth not take place, but at a confiderable diftance from the Continent. After this, the wind was not fo favourable to us as it had been : However, we ftill continued to advance, and, on the 26th of *January*, being then to the northward of *Acapulco*, we tacked and ftood to the eaftward, with a view of making the land.

In the preceding fortnight we caught fome turtle on the furface of the water, and feveral dolphins, bonitos, and albicores. One day, as one of the fail-makers mates was fifhing from the end of the gib-boom, he loft his hold and dropped into the fea; and the fhip, which was then going at the rate of fix or feven knots, went directly over him : But as we had the *Carmelo* in tow, we inftantly called out to the people on board her, who threw him over feveral ends of ropes, one of which he fortunately caught hold of, and twifting it round his arm, they haled him into the fhip, without his having received any other injury than a wrench in his arm, of which he foon recovered.

When, on the 26th of *January*, we ftood to the eaftward, we expected, by our reckonings, to have fallen in with the land on the 28th; but though the weather was perfectly clear, we had no fight of it at fun-fet, and therefore we continued on our courfe, not doubting but we fhould fee it by the next morning. About ten at night we difcovered a light on the larboard-bow, bearing from us N. N. E. The *Tryal*'s prize too, who was about a mile a head of us, made a fignal at the fame time for feeing a fail ; and as we had none of us any doubt but what we faw was a fhip's light; we were all extremely animated with a firm perfuafion, that it was the *Manila* galeon, which had been fo long the object of our wifhes : And what added to our alacrity, was our expectation of meeting with two of them inftead of one, for we took it for granted, that the light in view was carried in the top of one fhip for a

direction

direction to her confort. We immediately caft off the *Carmelo*, and preffed forward with all our canvafs, making a fignal for the *Gloucefter* to do the fame. Thus we chafed the light, keeping all our hands at their refpective quarters, under an expectation of engaging in the next half hour, as we fometimes conceived the chace to be about a mile diftant, and at other times to be within reach of our guns ; and fome on board us pofitively averred, that befides the light, they could plainly difcern her fails. The Commodore himfelf was fo fully perfuaded that we fhould be foon along fide of her, that he fent for his firft Lieutenant, who commanded between decks, and directed him to fee all the great guns loaded with two round-fhot for the firft broadfide, and after that with one round-fhot and one grape, ftrictly charging him, at the fame time, not to fuffer a gun to be fired, till he, the Commodore, fhould give orders, which he informed the Lieutenant would not be till we arrived within piftol-fhot of the enemy. In this conftant and eager attention we continued all night, always prefuming that another quarter of an hour would bring us up with this *Manila* fhip, whofe wealth, with that of her fuppofed confort, we now eftimated by round millions. But when the morning broke, and day-light came on, we were moft ftrangely and vexatioufly difappointed, by finding that the light which had occafioned all this buftle and expectancy, was only a fire on the fhore. Indeed the circumftances of this deception are fo extraordinary as to be fcarcely credible ; for, by our run during the night and the diftance of the land in the morning, there was no doubt to be made but this fire, when we firft difcovered it, was above twenty-five leagues from us: And yet I believe there was no perfon on board, who doubted of its being a fhip's light, or of its being near at hand. It was indeed upon a very high mountain, and continued burning for feveral days afterwards ; it was not a vulcano, but rather, as I fuppofe, ftubble or heath fet on fire for fome purpofe of agriculture.

At fun-rifing, after this mortifying delufion, we found ourfelves about nine leagues off the land, which extended from the N. W.

to

to E. ½ N. On this land we obferved two remarkable hummocks, fuch as are ufually called paps, which bore North from us: Thefe a *Spanifh* Pilot and two *Indians*, who were the only perfons amongft us that pretended to have traded in this part of the world, affirmed to be over the harbour of *Acapulco*. Indeed, we very much doubted their knowledge of the coaft ; for we found thefe paps to be in the latitude of 17°: 56, whereas thofe over *Acapulco* are faid to be in 17 degrees only ; and we afterwards found our fufpicions of their fkill to be well grounded: However, they were very confident, and affured us, that the height of the mountains was itfelf an infallible mark of the harbour; the coaft, as they pretended (though falfly) being generally low to the eaftward and weftward of it.

And now being in the track of the *Manila* galeon, it was a great doubt with us (as it was near the end of *January*) whether fhe was or was not arrived: But examining our prifoners about it, they affured us, that fhe was fometimes known to come in after the middle of *February* ; and they endeavoured to perfuade us, that the fire we had feen on fhore was a proof that fhe was as yet at fea, it being cuftomary, as they faid, to make ufe of thefe fires as fignals for her direction, when fhe continued longer out than ordinary. On this information, ftrengthened by our propenfity to believe them in a matter which fo pleafingly flattered our wifhes, we refolved to cruife for her for fome days ; and we accordingly fpread our fhips at the diftance of twelve leagues from the coaft, in fuch a manner, that it was impoffible fhe fhould pafs us unobferved : However, not feeing her foon, we were at intervals inclined to fufpect that fhe had gained her port already ; and as we now began to want a harbour to refrefh our people, the uncertainty of our prefent fituation gave us great uneafinefs, and we were very folicitous to get fome pofitive intelligence, which might either fet us at liberty to confult our neceffities, if the galeon was arrived, or might animate us to continue on our prefent cruife with chearfulnefs, if fhe was not. With this view the Commodore, after examining our pri-

foners

foners very particularly, refolved to fend a boat, under colour of the night, into the harbour of *Acapulco*, to fee if the *Manila* fhip was there or not, one of the *Indians* being very pofitive that this might be done without the boat itfelf being difcovered. To execute this project, the barge was difpatched the 6th of *February*, with a fufficient crew and two officers, who took with them a *Spanifh* Pilot, and the *Indian* who had infifted on the practicability of this meafure, and had undertaken to conduct it. Our barge did not return to us again till the eleventh, when the officers acquainted Mr. *Anfon*, that, agreeable to our fufpicion, there was nothing like a harbour in the place where the *Spanifh* Pilots had at firft afferted *Acapulco* to lie; that when they had fatisfied themfelves in this particular, they fteered to the eaftward, in hopes of difcovering it, and had coafted along fhore thirty-two leagues; that in this whole range they met chiefly with fandy beaches of a great length, over which the fea broke with fo much violence, that it was impoffible for a boat to land; that at the end of their run they could juft difcover two paps at a very great diftance to the eaftward, which from their appearance and their latitude, they concluded to be thofe in the neighbourhood of *Acapulco*; but that not having a fufficient quantity of frefh water and provifion for their paffage thither and back again, they were obliged to return to the Commodore, to acquaint him with their difappointment. On this intelligence we all made fail to the eaftward, in order to get into the neighbourhood of that port, the Commodore refolving to fend the barge a fecond time upon the fame enterprize, when we were arrived within a moderate diftance. And the next day, which was the 12th of *February*, we being by that time confiderably advanced, the barge was again difpatched, and particular inftructions given to the officers to preferve themfelves from being feen from the fhore. On the 13th we efpied a high land to the eaftward, which we firft imagined to be that over the harbour of *Acapulco*; but we afterwards found that it was the high land of *Seguateneio*, where there is a fmall harbour, of which we fhall have occafion to make more ample mention hereafter.

after. And now, having waited fix days without any news of our barge, we began to be uneafy for her fafety; but, on the 7th day, that is, on the 19th of *February*, fhe returned. The officers informed the Commodore, that they had difcovered the harbour of *Acapulco*, which they efteemed to bear from us E. S. E. at leaft fifty leagues diftant: That on the 17th, about two in the morning, they were got within the Ifland that lies at the mouth of the harbour, and yet neither the *Spanifh* Pilot, nor the *Indian* who were with them, could give them any information where they then were; but that while they were lying upon their oars in fufpence what to do, being ignorant that they were then at the very place they fought for, they difcerned a fmall light upon the furface of the water, on which they inftantly plied their paddles, and moving as filently as poffible towards it, they found it to be in a fifhing canoe, which they furprized, with three Negroes that belonged to it. It feems the Negroes at firft attempted to jump overboard; and being fo near the land they would eafily have fwam on fhore; but they were prevented by prefenting a piece at them, on which they readily fubmitted, and were taken into the barge. The officers further added, that they had immediately turned the canoe adrift againft the face of a rock, where it would inevitably be dafhed to pieces by the fury of the fea: This they did to deceive thofe who perhaps might be fent from the town to fearch after the canoe; for upon feeing feveral pieces of a wreck, they would immediately conclude that the people on board her had been drowned, and would have no fufpicion of their having fallen into our hands. When the crew of the barge had taken this precaution, they exerted their utmoft ftrength in pulling out to fea, and by dawn of day had gained fuch an offing, as rendered it impoffible for them to be feen from the coaft.

And now having gotten the three Negroes in our poffeffion, who were not ignorant of the tranfactions at *Acapulco*, we were foon fatiffied about the moft material points which had long kept us in fufpence: And on examination we found, that we were indeed difappointed in our expectation of intercepting the galeon before her arri-

val

val at *Acapulco*; but we learnt other circumſtances which ſtill revived our hopes, and which, we then conceived, would more than balance the opportunity we had already loſt: For tho' our Negroe priſoners informed us that the galeon arrived at *Acapulco* on our 9th of *January*, which was about twenty days before we fell in with this coaſt, yet they at the ſame time told us, that the galeon had delivered her cargoe, and was taking in water and proviſions for her return, and that the Viceroy of *Mexico* had by proclamation, fixed her depārture from *Acapulco* to the 14th of *March*, N. S. This laſt news was moſt joyfully received by us, as we had no doubt but ſhe muſt certainly fall into our hands, and as it was much more eligible to ſeize her on her return, than it would have been to have taken her before her arrival, as the ſpecies for which ſhe had ſold her cargoe and which ſhe would now have on board, would be prodigiouſly more to be eſteemed by us than the cargoe itſelf; great part of which would have periſhed on our hands, and no part of it could have been diſpoſed of by us at ſo advantageous a mart as *Acapulco*.

Thus we were a ſecond time engaged in an eager expectation of meeting with this *Manila* ſhip, which, by the fame of its wealth, we had been taught to conſider as the moſt deſirable prize that was to be met with in any part of the globe. As all our future projects will be in ſome ſort regulated with a view to the poſſeſſion of this celebrated galeon, and as the commerce which is carried on by means of theſe veſſels between the city of *Manila* and the port of *Acapulco* is perhaps the moſt valuable, in proportion to its quantity, of any in the known world, I ſhall endeavour, in the enſuing chapter, to give as diſtinct an account as I can of all the particulars relating thereto, both as it is a matter in which I conceive the public to be in ſome degree intereſted, and as I flatter myſelf, that from the materials which have fallen into my hands, I am enabled to deſcribe it with more diſtinctneſs than has hitherto been done, at leaſt in our language.

CHAP.

C H A P. X.

An account of the commerce carried on between the city of *Manila* on the Ifland of *Luconia*, and the port of *Acapulco* on the Coaft of *Mexico*.

ABOUT the end of the 15th Century and the beginning of the 16th, the difcovery of new countries and of new branches of commerce was the reigning paffion of feveral of the *European* Princes. But thofe who engaged moft deeply and fortunately in thefe purfuits were the Kings of *Spain* and *Portugal*; the firft of thefe having difcovered the immenfe and opulent Continent of *America* and its adjacent Iflands, whilft the other, by doubling the Cape of *Good Hope*, had opened to his fleets a paffage to the fouthern coaft of *Afia*, ufually called the *Eaft-Indies*, and by his fettlements in that part of the globe, became poffeffed of many of the manufactures and natural productions with which it abounded, and which, for fome ages, had been the wonder and delight of the more polifhed and luxurious part of mankind.

In the mean time, thefe two Nations of *Spain* and *Portugal*, who were thus profecuting the fame views, though in different quarters of the world, grew extremely jealous of each other, and became apprehenfive of mutual encroachments. And therefore to quiet their jealoufies, and to enable them with more tranquillity to purfue the propagation of the Catholick Faith in thefe diftant countries, (they having both of them given diftinguifhed marks of their zeal for their mother church, by their butchery of innocent Pagans) Pope *Alexander* VI. granted to the *Spanifh* Crown the property and dominion of all places, either already difcovered, or that fhould be difcovered an hundred leagues to the weftward of the Iflands of *Azores*, leaving all the unknown countries to the eaftward of this

limit,

limit, to the induſtry and future diſquiſition of the *Portugueſe:* And this boundary being afterwards removed two hundred and fifty leagues more to the weſtward, by the agreement of both Nations, it was imagined that by this regulation all the ſeeds of future conteſts were ſuppreſſed. For the *Spaniards* preſumed, that the *Portugueſe* would be hereby prevented from meddling with their colonies in *America:* And the *Portugueſe* ſuppoſed that their *Eaſt-Indian* ſettlements, and particularly the ſpice Iſlands, which they had then newly diſcovered, were ſecured from any future attempts of the *Spaniſh* Nation.

But it ſeems the infallibility of the Holy Father had, on this occaſion, deſerted him, and for want of being more converſant in geography, he had not foreſeen that the *Spaniards,* by puſhing their diſcoveries to the Weſt, and the *Portugueſe* to the Eaſt, might at laſt meet with each other, and be again embroiled; as it actually happened within a few years afterwards. For *Frederick Magellan,* who was an officer in the King of *Portugal*'s ſervice, having received ſome diſguſt from that Court, either by the defalcation of his pay, or by having his parts, as he conceived, too cheaply conſidered, he entered into the ſervice of the King of *Spain*; and being as it appears a man of ability, he was very deſirous of ſignalizing his talents by ſome enterprize, which might prove extremely vexatious to his former Maſters, and might teach them to eſtimate his worth by the greatneſs of the miſchief he brought upon them, this being the moſt obvious and natural turn of all fugitives, and more eſpecially of thoſe, who, being really men of capacity, have quitted their country by reaſon of the ſmall account that has been made of them. *Magellan,* in purſuance of theſe vindictive views, knowing that the *Portugueſe* Court conſidered their poſſeſſion of the ſpice iſlands as their moſt important acquiſition in the *Eaſt-Indies,* reſolved with himſelf to inſtigate the Court of *Spain* to an enterprize, which, by ſtill puſhing their diſcoveries, would give them a right to interfere both in the property and commerce of thoſe renowned *Portugueſe* ſettlements; and the King of *Spain*

H h   approving

approving of this project, *Magellan*, in the year 1519, set sail from the port of *Sevil*, in order to carry this enterprize into execution. He had with him a considerable force, consisting of five ships and two hundred and thirty-four men, with which he stood for the coast of South *America*, and ranging along shore, he at last, towards the end of *October* 1520, had the good fortune to discover those Streights, which have since been denominated from him, and which opened him a passage into the *Pacific* Ocean. And this first part of his scheme being thus happily accomplished, he, after some stay on the coast of *Peru*, set sail again to the westward, with a view of falling in with the spice islands. In this extensive run he first discovered the *Ladrones* or *Marian* Islands; and continuing on his course, he at length reached the *Philippine* Islands, which are the most eastern part of *Asia*, where, venturing on shore in an hostile manner, and skirmishing with the *Indians*, he was slain.

By the death of *Magellan*, the original project of securing some of the spice islands was defeated; for those who were left in command contented themselves with ranging through them, and purchasing some spices from the natives; after which they returned home round the Cape of *Good Hope*, being the first ships which had ever surrounded this terraqueous globe; and thereby demonstrated, by a palpable experiment obvious to the grossest and most vulgar capacity, the reality of its long disputed spherical figure.

But though *Spain* did not hereby acquire the property of any of the spice islands, yet the discovery made in this expedition of the *Philippine* Islands, was thought too considerable to be neglected; for these were not far distant from those places which produced spices, and were very well situated for the *Chinese* trade, and for the commerce of other parts of *India*; and therefore a communication was soon established, and carefully supported between these Islands and the *Spanish* colonies on the coast of *Peru*: So that the city of *Manila*, (which was built on the Island of *Luconia*, the chief of the *Philippines*) soon became the mart for all *Indian* commodities, which were bought up by the inhabitants, and were annually sent

to

to the *South-Seas* to be there vended on their account; and the returns of this commerce to *Manila* being principally made in silver, the place by degrees grew extremely opulent and confiderable, and its trade fo far encreafed, as to engage the attention of the Court of *Spain*, and to be frequently controlled and regulated by royal edicts.

In the infancy of this trade, it was carried on from the port of *Callao* to the city of *Manila*, in which voyage the trade-wind continually favoured them; fo that notwithftanding thefe places were diftant between three and four thoufand leagues, yet the voyage was often made in little more than two months: But then the return from *Manila* was extremely troublefome and tedious, and is faid to have fometimes taken them up above a twelve month, which, if they pretended to ply up within the limits of the trade-wind, is not at all to be wondered at; and it is afferted, that in their firft voyages they were fo imprudent and unfkilful as to attempt this courfe. However, that route was foon laid afide by the advice, as it is faid, of a Jefuit, who perfuaded them to fteer to the northward till they got clear of the trade-winds, and then by the favour of the wefterly winds, which generally prevail in high latitudes, to ftretch away for the coaft of *California*. This has been the practice for at leaft a hundred and fixty years paft: For Sir *Thomas Cavendifh*, in the year 1586, engaged off the South end of *California* a veffel bound from *Manila* to the *American* coaft. And it was in compliance with this new plan of navigation, and to fhorten the run both backwards and forwards, that the ftaple of this commerce to and from *Manila* was removed from *Callao* on the coaft of *Peru*, to the port of *Acapulco* on the coaft of *Mexico*, where it continues fixed at this time.

Such was the commencement, and fuch were the early regulations of this commerce; but its prefent condition being a much more interefting fubject, I muft beg leave to dwell longer on this head, and to be indulged in a more particular narration, beginning

H h 2   with

with a defcription of the Ifland of *Luconia*, and of the port and bay of *Manila*.

The Ifland of *Luconia*, though fituated in the latitude of 15° North, is efteemed to be in general extremely healthy, and the water, that is found upon it, is faid to be the beft in the world : It produces all the fruits of the warm climates, and abounds in a moft excellent breed of horfes, fuppofed to be carried thither firft from *Spain* : It is very well fituated for the *Indian* and *Chinefe* trade ; and the bay and port of *Manila*, which lies on its weftern fide, is perhaps the moft remarkable on the whole globe, the bay being a large circular bafon, near ten leagues in diameter, and great part of it entirely land-locked. On the eaft fide of this bay ftands the city of *Manila*, which is very large and populous ; and which, at the beginning of this war, was only an open place, its principal defence confifting in a fmall fort, which was in great meafure furrounded on every fide by houfes ; but they have lately made confiderable additions to its fortifications, though I have not yet learnt in what manner. The port, peculiar to the city, is called *Cabite*, and lies near two leagues to the fouthward ; and in this port all the fhips employed in the *Acapulco* trade are ufually ftationed. As I have never feen but one engraved plan of this bay, and that in a very fcarce book, I have hereafter added, towards the beginning of the third book, a plan which fell into my hands, and which differs confiderably from that already publifhed : But I cannot pretend to decide which of the two is moft to be relied on.

The city of *Manila* itfelf is in a very healthy fituation, is well watered, and is in the neighbourhood of a very fruitful and plentiful country ; but as the principal bufinefs of this place is its trade to *Acapulco*, it lies under fome difadvantage, from the difficulty there is in getting to fea to the eaftward : For the paffage is among iflands and through channels where the *Spaniards*, by reafon of their unfkilfulnefs in marine affairs, wafte much time, and are often in great danger. Thefe difficulties will be better apprehended by the reader

by

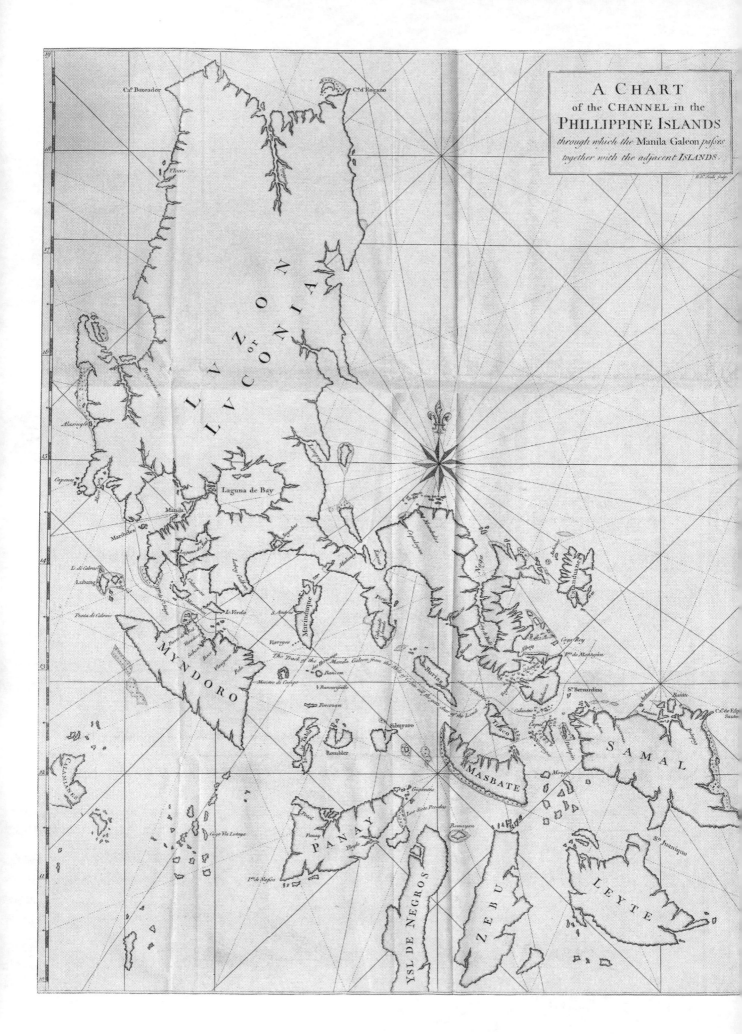

The material originally positioned here is too large for reproduction in this reissue. A PDF can be downloaded from the web address given on page iv of this book, by clicking on 'Resources Available'.

by the annexed draught of the Ifland of *Luconia*, and of its neighbouring ifles, which was taken from the enemy, and had been newly drawn and corrected but a fhort time before.

The trade carried on from this place to *China* and different parts of *India*, is principally for fuch commodities as are intended to fupply the Kingdoms of *Mexico* and *Peru*. Thefe are fpices, all forts of *Chinefe* filks and manufactures; particularly filk ftockings, of which I have heard that no lefs than fifty thoufand pair were the ufual number fhipped on board the annual fhip; vaft quantities of *Indian* ftuffs, as callicoes and chints, which are much worn in *America*, together with other minuter articles, as goldfmiths work, &c. which is principally done at the city of *Manila* itfelf by the *Chinefe*; for it is faid there are at leaft twenty thoufand *Chinefe* who conftantly refide there, either as fervants, manufacturers, or brokers. All thefe different commodities are collected at *Manila*, thence to be tranfported annually in one or more fhips, to the port of *Acapulco*, in the Kingdom of *Mexico*.

But this trade to *Acapulco* is not laid open to all the inhabitants of *Manila*, but is confined by very particular regulations, fomewhat analogous to thofe by which the trade of the regifter fhips from *Cadiz* to the *Weft-Indies* is reftrained. The fhips employed herein are found by the King of *Spain*, who pays the officers and crew; and the tunnage is divided into a certain number of bales, all of the fame fize: Thefe are diftributed amongft the Convents at *Manila*, but principally to the Jefuits, as a donation for the fupport of their miffions for the propagation of the Catholick Faith; and thefe Convents have hereby a right to embark fuch a quantity of goods on board the *Manila* fhip, as the tunnage of their bales amounts to; or if they chufe not to be concerned in trade themfelves, they have the power of felling this privilege to others; and as the Merchants to whom they grant their fhares are often unprovided of a ftock, it is ufual for the Convents to lend them confiderable fums of money on bottomry.

The

The trade is by the royal edicts limited to a certain value, which the annual cargoe ought not to exceed. Some *Spanish* manuscripts, I have seen, mention this limitation to be 600,000 dollars ; but the annual cargoe does certainly surpass this sum ; and though it may be difficult to fix its exact value, yet from many comparisons I conclude, that the return cannot be greatly short of three millions of dollars.

It is sufficiently obvious, that the greatest part of the treasure, returned from *Acapulco* to *Manila*, does not remain in that place, but is again dispersed into different parts of *India*. And as all *European* Nations have generally esteemed it good policy to keep their *American* settlements in an immediate dependence on their mother country, without permitting them to carry on directly any gainful traffick with other powers, these considerations have occasioned many remonstrances to be presented to the Court of *Spain* against the *Indian* trade, hereby allowed to the Kingdom of *Peru* and *Mexico*; it having been urged, that the silk manufactures of *Valencia* and other parts of *Spain* are hereby greatly prejudiced, and the linnens carried from *Cadiz* are much injured in their sale : Since the *Chinese* silks coming almost directly to *Acapulco*, can be afforded much cheaper there than any *European* manufactures of equal goodness ; and the cottons from the *Coromondel* coast, make the *European* linnens almost useless. So that the *Manila* trade renders both *Mexico* and *Peru* less dependent upon *Spain* for a supply of their necessities than they ought to be ; and exhausts those countries of a considerable quantity of silver, the greatest part of which, were this trade prohibited, would center in *Spain*, either in payment for *Spanish* commodities, or in gains to the *Spanish* Merchant ; whereas now the only advantage which arises from it is, the enriching the Jesuits and a few particular persons besides, at the other extremity of the world. These arguments did so far influence Don *Joseph Patinho*, who was then prime Minister, but an enemy to the Jesuits, that about the year 1725, he had resolved to abolish this trade, and to have permitted no *Indian* commodities to be introduced

into

into any of the *Spanish* ports in the *West-Indies*, but what were carried there in the register ships from *Europe*. But the powerful intrigues of the Jesuits prevented this regulation from taking place.

This trade from *Manila* to *Acapulco* and back again, is usually carried on in one or at most two annual ships, which set sail from *Manila* about *July*, and arrive at *Acapulco* in the *December*, *January*, or *February* following, and having there disposed of their effects, return for *Manila* sometime in *March*, where they generally arrive in *June*; so that the whole voyage takes up very near an entire year: For this reason, though there is often no more than one ship employed at a time, yet there is always one ready for the sea when the other arrives; and therefore the commerce at *Manila* are provided with three or four stout ships, that, in case of any accident, the trade may not be suspended. The largest of these ships, whose name I have not learnt, is described as little less than one of our first rate men of war, and indeed she must be of an enormous size; for it is known, that when she was employed with other ships from the same port, to cruise for our *China* trade, she had no less than twelve hundred men on board. Their other ships, though far inferior in bulk to this, are yet stout large vessels, of the burthen of twelve hundred tun and upwards, and usually carry from three hundred and fifty to six hundred hands, passengers included, with fifty odd guns. As these are all King's ships commissioned and paid by him, there is usually one of the Captains, who is stiled the General, and who carries the royal standard of *Spain* at the maintop gallant mast-head, as we shall more particularly observe hereafter.

And now having described the port of *Manila* and the shipping they employ, it is necessary to give a more circumstantial detail of their navigation. The ship having received her cargoe on board, and being fitted for the sea, generally weighs from the mole of *Cabite* about the middle of *July*, taking the advantage of the westerly monsoon, which then sets in, to carry them to sea. It appears by the chart already inserted, that the getting through the *Boccadero* to

the

the eaftward muft be a troublefome navigation, and in fact, it is
fometimes the end of *Auguft* before they get clear of the land.
When they have got through this paffage, and are clear of the
Iflands, they ftand to the northward of the eaft, in order to get
into the latitude of 30 odd degrees, where they expect to meet with
wefterly winds, before which they run away for the coaft of *California*. To give a better idea of the track which they hold in this
navigation, I have inferted, towards the latter end of the third book,
the copy of a manufcript chart, which was taken on board one of
thefe fhips, containing all that Ocean between the *Philippine* Iflands
and the coaft of *Mexico*, in which I have laid down the particular
route of this veffel, both in her paffage from *Manila* to *Acapulco*,
and from *Acapulco* back again. In this chart (as it was drawn for
the ufe of the *Spanifh* General) there are contained all the difcoveries which the *Manila* fhips have at any time made in traverfing
this vaft Ocean ; whence it appears what minute and inconfiderable
fragments of land are difperfed in that prodigious fea; and it is moft
remarkable, that by the concurrent teftimony of all the *Spanifh* Navigators, there is not one port, nor even a tolerable road as yet found
out betwixt the *Philippine* Iflands and the coaft of *California* and
Mexico; fo that from the time the *Manila* fhip firft lofes fight of
land, fhe never lets go her anchor till fhe arrives on the coaft of
California, and very often not till fhe gets to its fouthermoft extremity : And therefore as this voyage is rarely of lefs than fix months
continuance, and the fhip is deep laden with merchandize and
crowded with people, it may appear wonderful how they can be
fupplied with a ftock of frefh water for fo long a time ; and indeed
their method of procuring it is extremely fingular, and deferves a
very particular recital.

It is well known to thofe who are acquainted with the *Spanifh*
cuftoms in the *South-Seas*, that their water is preferved on fhip-
board not in cafks but in earthern jars, which in fome fort refemble
the large oil jars we often fee in *Europe*. When the *Manila* fhip
firft puts to fea, they take on board a much greater quantity of wa-

ter

ter than can be ftowed between decks, and the jars which contain
it are hung all about the fhrouds and ftays, fo as to exhibit at a diftance a very odd appearance. And though it is one convenience of
their jars that they are much more manageable than cafks, and are
liable to no leekage, unlefs they are broken, yet it is fufficiently obvious, that a fixth, or even a three months ftore of water could
never be ftowed in a fhip fo loaded, by any management whatever; and therefore without fome other fupply, this navigation
could not be performed : A fupply indeed they have, but the reliance upon it feems at firft fight fo extremely precarious, that it is
wonderful fuch numbers fhould rifque the perifhing by the moft
dreadful of all deaths, on the expectation of fo cafual a circumftance. In fhort, their only method of recruiting their water is by
the rains, which they meet with between the latitudes of 30 and
40° North, and which they are always prepared to catch : For this
purpofe they take to fea with them a great number of mats, which
they place flopingly againft the gunwale, whenever the rain defcends;
thefe mats extend from one end of the fhip to the other, and their
lower edges reft on a large fplit bamboe, fo that all the water which
falls on the mats drains into the bamboe, and by this, as a trough,
is conveyed into a jar; and this method of fupplying their water,
however accidental and extraordinary it may at firft fight appear,
hath never been known to fail them, fo that it is common for them,
when their voyage is a little longer than ufual, to fill all their water
jars feveral times over.

However, though their diftreffes for frefh water are much fhort
of what might be expected in fo tedious a navigation, yet there are
other inconveniencies generally attendant upon a long continuance at
fea, from which they are not exempted. The principal of thefe
is the fcurvy, which fometimes rages with extreme violence, and
deftroys great numbers of the people; but at other times their paffage to *Acapulco* (of which alone I would be here underftood to
fpeak) is performed with little lofs.

The

The length of time employed in this paſſage, ſo much beyond what uſually occurs in any other known navigation, is perhaps in part to be imputed to the indolence and unſkilfulneſs of the *Spaniſh* ſailors, and to an unneceſſary degree of caution and concern for ſo rich a veſſel : For it is ſaid, that they never ſet their main ſail in the night, and often lie by unneceſſarily. And indeed the inſtructions given to their Captains (which I have ſeen) ſeem to have been drawn up by ſuch as were more apprehenſive of too ſtrong a gale, though favourable, than of the inconveniencies and mortality attending a lingring and tedious voyage ; for the Captain is particularly ordered to make his paſſage in the latitude of 30 degrees if poſſible, and to be extremely careful to ſtand no farther to the northward than is abſolutely neceſſary for the getting a weſterly wind. This, according to our conceptions, appears to be a very abſurd reſtriction ; ſince it can ſcarcely be doubted, but that in the higher latitudes the weſterly winds are much ſteadier and briſker than in the latitude of 30 degrees : So that the whole conduct of this navigation ſeems liable to very great cenſure. For if inſtead of ſteering E. N. E. into the latitude of 30 odd degrees, they at firſt ſtood N. E, or even ſtill more northerly, into the latitude of 40 or 45 degrees, in part of which courſe the trade-winds would greatly aſſiſt them, I doubt not but by this management they might conſiderably contract their voyage, and perhaps perform it in half the time, which is now allotted for it ; for in the journals I have ſeen of theſe voyages it appears, that they are often a month or ſix weeks after their laying the land, before they get into the latitude of 30 degrees ; whereas, with a more northerly courſe, it might eaſily be done in a fourth part of the time ; and when they were once well advanced to the northward, the weſterly winds would ſoon blow them over to the coaſt of *California*, and they would be thereby freed from the other embaraſſments, to which they are now ſubjected, only at the expence of a rough ſea and a ſtiff gale. And this is not merely matter of ſpeculation ; for I am credibly informed, that about the year 1721, a *French* ſhip, by purſuing this courſe,

ran

ran from the coaſt of *China* to the valley of *Vanderas* on the coaſt of *Mexico*, in leſs than fifty days: But it was ſaid that this ſhip, notwithſtanding the ſhortneſs of her paſſage, ſuffered prodigiouſly by the ſcurvy, ſo that ſhe had only four or five of her crew left when ſhe arrived in *America*.

However, I ſhall deſcant no longer on the probability of performing this voyage in a much ſhorter time, but ſhall content myſelf with reciting the actual occurrences of the preſent navigation. The *Manila* ſhip having ſtood ſo far to the northward as to meet with a weſterly wind, ſtretches away nearly in the ſame latitude for the coaſt of *California*: And when ſhe has run into the longitude of 96 degrees from Cape *Eſpiritu Santo*, ſhe generally meets with a plant floating on the ſea, which, being called *Porra* by the *Spaniards*, is, I preſume, a ſpecies of ſea-leek. On the ſight of this plant they eſteem themſelves ſufficiently near the *Californian* ſhore, and immediately ſtand to the ſouthward ; and they rely ſo much on this circumſtance, that on the firſt diſcovery of the plant the whole ſhip's company chaunt a ſolemn *Te Deum*, eſteeming the difficulties and hazards of their paſſage to be now at an end ; and they conſtantly correct their longitude thereby, without ever coming within ſight of land. After falling in with theſe SIGNS, as they denominate them, they ſteer to the ſouthward, without endeavouring to fall in with the coaſt, till they have run into a lower latitude; for as there are many iſlands, and ſome ſhoals adjacent to *California*, the extreme caution of the *Spaniſh* Navigators makes them very apprehenſive of being engaged with the land ; however, when they draw near its ſouthern extremity, they venture to hale in, both for the ſake of making Cape St. *Lucas* to aſcertain their reckoning, and alſo to receive intelligence from the *Indian* inhabitants, whether or no there are any enemies on the coaſt ; and this laſt circumſtance, which is a particular article in the Captain's inſtructions, makes it neceſſary to mention the late proceedings of the Jeſuits amongſt the *Californian Indians*.

Since

Since the firſt diſcovery of *California*, there have been various wandring Miſſionaries who have viſited it at different times, though to little purpoſe ; but of late years the Jeſuits, encouraged and ſupported by a large-donation from the Marquis *de Valero*, a moſt munificent bigot, have fixed themſelves upon the place, and have eſtabliſhed a very conſiderable miſſion. Their principal ſettlement lies juſt within Cape St. *Lucas*, where they have collected a great number of ſavages, and have endeavoured to inure them to agriculture and other mechanic arts : And their efforts have not been altogether ineffectual ; for they have planted vines at their ſettlements with very good ſucceſs, ſo that they already make a conſiderable quantity of wine, reſembling in flavour the inferior ſorts of *Madera*, which begins to be eſteemed in the neighbouring kingdom of *Mexico*.

The Jeſuits then being thus firmly rooted on *California*, they have already extended their juriſdiction quite acroſs the country from ſea to ſea, and are endeavouring to ſpread their influence farther to the northward ; with which view they have made ſeveral expeditions up the gulf between *California* and *Mexico*, in order to diſcover the nature of the adjacent countries, all which they hope hereafter to bring under their power. And being thus occupied in advancing the intereſts of their ſociety, it is no wonder if ſome ſhare of attention is engaged about the ſecurity of the *Manila* ſhip, in which their Convents at *Manila* are ſo deeply concerned. For this purpoſe there are refreſhments, as fruits, wine, water, &c. conſtantly kept in readineſs for her ; and there is beſides care taken at Cape St. *Lucas*, to look out for any ſhip of the enemy, which might be cruiſing there to intercept her ; this being a ſtation where ſhe is conſtantly expected, and where ſhe has been often waited for and fought with, though generally with little ſucceſs. In conſequence then of the meaſures mutually ſettled between the Jeſuits of *Manila* and their brethren at *California*, the Captain of the galeon is ordered to fall in with the land to the northward of Cape St. *Lucas*, where the inhabitants are directed, on ſight of the veſ-

ſel,

fel, to make the proper fignals with fires ; and on difcovering thefe fires, the Captain is to fend his launch on fhore with twenty men, well armed, who are to carry with them the letters from the Convents at *Manila* to the *Californian* Miffionaries, and are to bring back the refrefhments which will be prepared for them, and likewife intelligence whether or no there are any enemies on the coaft. And if the Captain finds, from the account which is fent him, that he has nothing to fear, he is directed to proceed for Cape St. *Lucas*, and thence to Cape *Corientes*, after which he is to coaft it along for the port of *Acapulco*.

The moft ufual time of the arrival of the galeon at *Acapulco* is towards the middle of *January* : But this navigation is fo uncertain, that fhe fometimes gets in a month fooner, and at other times has been detained at fea above a month longer. The port of *Acapulco* is by much the fecureft and fineft in all the northern parts of the *Pacific* Ocean, being, as it were, a bafon furrounded by very high mountains : But the town is a moft wretched place, and extremely unhealthy, for the air about it is fo pent up by the hills, that it has fcarcely any circulation. The place is befides deftitute of frefh water, except what is brought from a confiderable diftance; and is in all refpects fo inconvenient, that except at the time of the mart, whilft the *Manila* galeon is in the port, it is almoft deferted. To compenfate in fome meafure for the fhortnefs of this defcription, I have added in the third book, in the fame plate with the bay of *Manila* abovementioned, a plan of this place and of its port and citadel, in which are likewife drawn the new works which were added on their firft intelligence of the equipment of our fquadron. As this plan was taken from the *Spaniards*, I cannot anfwer for its accuracy ; but having feen two or three other *Spanifh* draughts of the place, I conceive, by comparing them together, that this I have here inferted is not very diftant from the truth.

When

When the galeon arrives in this port, fhe is generally moored on its weftern fide to the two trees marked in the plan, and her cargoe is delivered with all poffible expedition. And now the town of *Acapulco*, from almoft a folitude, is immediately thronged with Merchants from all parts of the kingdom of *Mexico*. The cargoe being landed and difpofed of, the filver and the goods intended for *Manila* are taken on board, together with provifions and water, and the fhip prepares to put to fea with the utmoft expedition. There is indeed no time to be loft ; for it is an exprefs order to the Captain to be out of the port of *Acapulco* on his return, before the firft day of *April, N. S.*

And having mentioned the goods intended for *Manila*, I muft obferve, that the principal return is always made in filver, and confequently the reft of the cargoe is but of little account, the other articles, befides the filver, being fome cochineal and a few fweetmeats, the produce of the *American* fettlements, together with *European* millinery ware for the women at *Manila*, and fome *Spanifh* wines, fuch as tent and fherry, which are intended for the ufe of their Priefts in the adminiftration of the Sacrament.

And this difference in the cargoe of the fhip to and from *Manila*, occafions a very remarkable variety in the manner of equipping the fhip for thefe two different voyages. For the galeon, when fhe fets fail from *Manila*, being deep laden with a variety of bulky goods, fhe has not the conveniency of mounting her lower tire of guns, but carries them in her hold, till fhe draws near Cape St. *Lucas*, and is apprehenfive of an enemy. Her hands too are as few as is confiftent with the fafety of the fhip, that fhe may be lefs peftered with the ftowage of provifions. But on her return from *Acapulco*, as her cargoe lies in lefs room, her lower tire is (or ought to be) always mounted before fhe leaves the port, and her crew is augmented with a fupply of failors, and with one or two companies of foot, which are intended to reinforce the garrifon at *Manila*. And there being befides many Merchants who take their paffage to *Manila*

nila

nila on board the galeon, her whole number of hands on her return is usually little short of six hundred, all which are easily provided for, by reason of the small stowage necessary for the silver.

The galeon being thus fitted for her return, the Captain, on leaving the port of *Acapulco*, steers for the latitude of 13° or 14°, and runs on that parallel, till he gets sight of the Island of *Guam*, one of the *Ladrones*. In this run the Captain is particularly directed to be careful of the shoals of St. *Bartholomew*, and of the Island of *Gasparico*. He is also told in his instructions, that to prevent his passing the *Ladrones* in the dark, there are orders given that, through all the month of *June*, fires shall be lighted every night on the highest part of *Guam* and *Rota*, and kept in till the morning.

At *Guam* there is a small *Spanish* garrison, (as will be more particularly mentioned hereafter) purposely intended to secure that place for the refreshment of the galeon, and to yield her all the assistance in their power. However, the danger of the road at *Guam* is so great, that though the galeon is ordered to call there, yet she rarely stays above a day or two, but getting her water and refreshments on board as soon as possible, she steers away directly for Cape *Espiritu Santo*, on the Island of *Samal*. Here the Captain is again ordered to look out for signals ; and he is told, that centinels will be posted not only on that Cape, but likewise in *Catanduanas*, *Butusan*, *Birriborongo*, and on the Island of *Batan*. These centinels are instructed to make a fire when they discover the ship, which the Captain is carefully to observe : For if, after this first fire is extinguished, he perceives that four or more are lighted up again, he is then to conclude that there are enemies on the coast ; and on this he is immediately to endeavour to speak with the centinel on shore, and to procure from him more particular intelligence of their force, and of the station they cruise in ; pursuant to which, he is to regulate his conduct, and to endeavour to gain some secure port amongst those Islands, without coming in sight of the enemy ; and in case

he

he ſhould be diſcovered when in port, and ſhould be apprehenſive
of an attack, he is then to land his treaſure, and to take ſome
of his artillery on ſhore for its defence, not neglecting to ſend frequent and particular accounts to the city of *Manila* of all that
paſſes. But if, after the firſt fire on ſhore, the Captain obſerves
that two others only are made by the centinels, he is then to conclude, that there is nothing to fear : And he is to purſue his courſe
without interruption, and to make the beſt of his way to the port
of *Cabité*, which is the port to the city of *Manila*, and the conſtant ſtation for all the ſhips employed in this commerce to
Acapulco.

C H A P.

C H A P. XI.

Our cruife off the port of *Acapulco* for the *Manila* fhip.

I HAVE already mentioned, in the ninth chapter, that the return of our barge from the port of *Acapulco*, where fhe had furprized three Negro fifhermen, gave us inexpreffible fatisfaction, as we learnt from our prifoners, that the galeon was then preparing to put to fea, and that her departure was fixed, by an edict of the Viceroy of *Mexico*, to the 14th of *March*, *N. S.* that is, to the 3d of *March*, according to our reckoning.

What related to this *Manila* fhip being the matter to which we were moft attentive, it was neceffarily the firft article of our examination ; but having fatisfied ourfelves upon this head, we then indulged our curiofity in enquiring after other news ; when the prifoners informed us, that they had received intelligence at *Acapulco*, of our having plundered and burnt the town of *Paita* ; and that, on this occafion, the Governor of *Acapulco* had augmented the fortifications of the place, and had taken feveral precautions to prevent us from forcing our way into the harbour ; that in particular, he had placed a guard on the Ifland which lies at the harbour's mouth, and that this guard had been withdrawn but two nights before the arrival of our barge : So that had the barge fucceeded in her firft attempt, or had fhe arrived at the port the fecond time two days fooner, fhe could fcarcely have avoided being feized on, or if fhe had efcaped, it muft have been with the lofs of the greateft part of her crew, as fhe would have been under the fire of the guard, before fhe had known her danger.

The

The withdrawing of this guard was a circumstance that greatly encouraged us, as it seemed to demonstrate, not only that the enemy had not as yet discovered us, but likewise that they had now no farther apprehensions of our visiting their coast. Indeed the prisoners assured us, that they had no knowledge of our being in those seas, and that they had therefore flattered themselves, that, in the long interval since our taking of *Paita*, we had steered another course. But we did not consider the opinion of these Negro prisoners as so authentick a proof of our being hitherto concealed, as the withdrawing of the guard from the harbour's mouth; for this being the action of the Governor, was of all arguments the most convincing, as he might be supposed to have intelligence, with which the rest of the inhabitants were unacquainted.

Satisfied therefore that we were undiscovered, and that the time was fixed for the departure of the galeon from *Acapulco*, we made all necessar preparations, and waited with the utmost impatience for the important day. As this was the 3d of *March*, and it was the 10th of *February* when the barge returned and brought us our intelligence, the Commodore resolved to continue the greatest part of the intermediate time on his present station, to the westward of *Acapulco*, conceiving that in this situation there would be less danger of his being seen from the shore, which was the only circumstance that could deprive us of the immense treasure, on which we had at present so eagerly fixed our thoughts. During this interval, we were employed in scrubbing and cleansing our ships bottoms, in bringing them into their most advantageous trim, and in regulating the orders, signals and stations to be observed when we should arrive off *Acapulco*, and the time of the departure of the galeon should draw nigh.

And now, on the first of *March*, we made the high lands, usually called the paps over *Acapulco*, and got with all possible expedition into the situation prescribed by the Commodore's orders. The distribution of our squadron on this occasion, both for the inter-

cepting

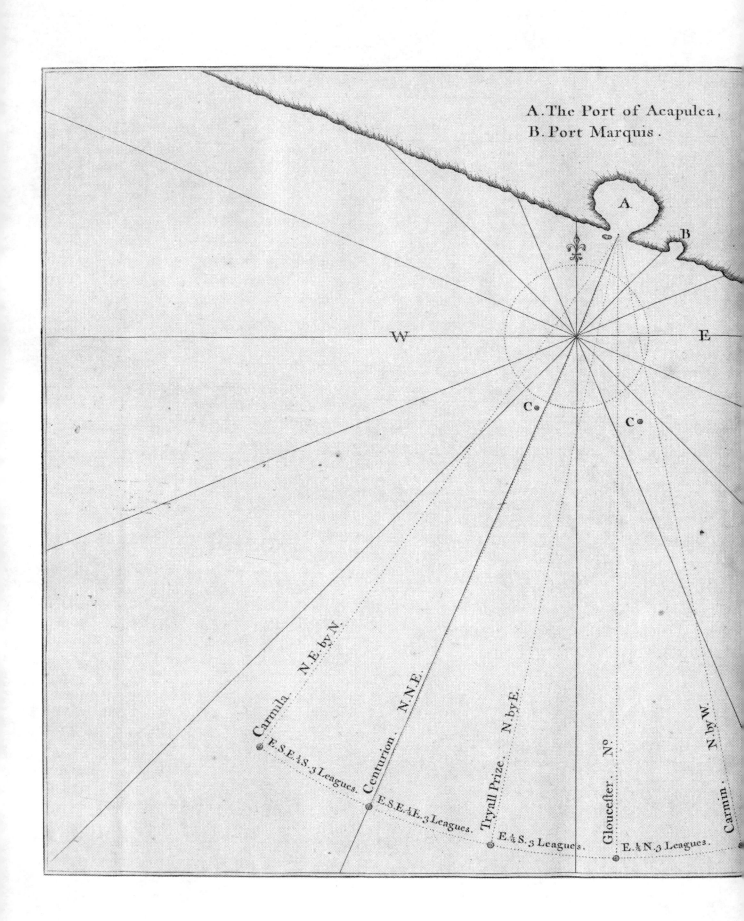

A. The Port of Acapulca,
B. Port Marquis.

W

E

A

B

C

C

Carmila. N.E. by N.

Centurion. N.N.E.

Tryall Prize. N. by E.

Glouceſter. N?

Carmin. N. by W.

E.S.E.¼S. 3 Leagues.

E.S.E.½E. 3 Leagues.

E.¼S. 3 Leagues.

E.¼N. 3 Leagues.

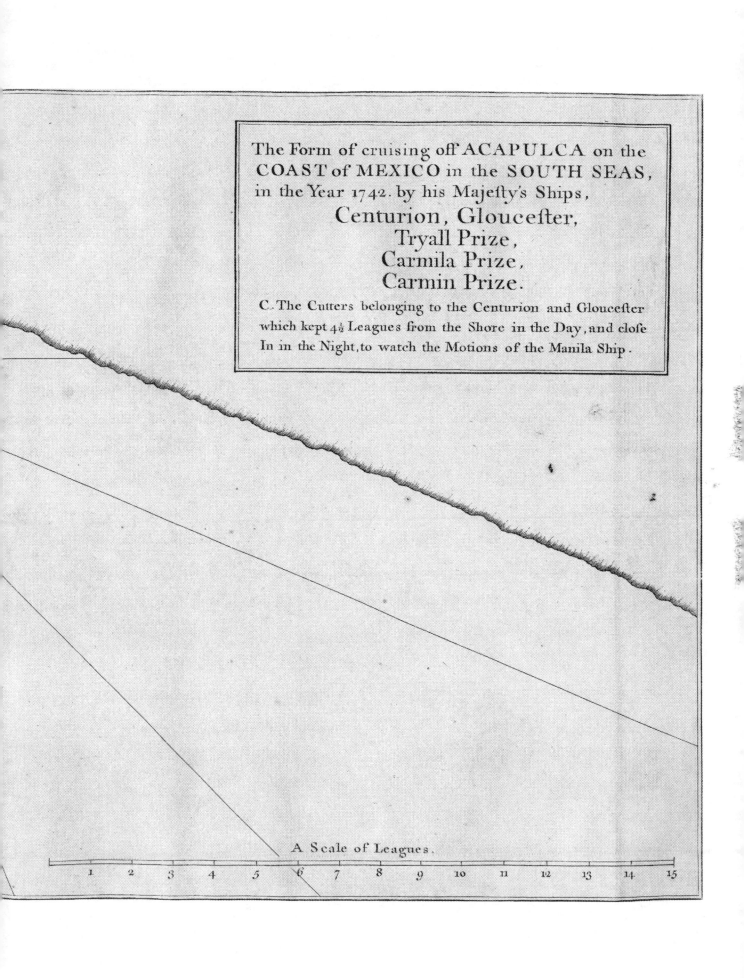

The Form of cruising off ACAPULCA on the
COAST of MEXICO in the SOUTH SEAS,
in the Year 1742. by his Majesty's Ships,
Centurion, Gloucester,
Tryall Prize,
Carmila Prize,
Carmin Prize.

C. The Cutters belonging to the Centurion and Gloucester
which kept 4½ Leagues from the Shore in the Day, and close
In in the Night, to watch the Motions of the Manila Ship.

A Scale of Leagues.

1   2   3   4   5   6   7   8   9   10   11   12   13   14   15

cepting the galeon, and for the avoiding a difcovery from the fhore, was fo very judicious, that it well merits to be diftinctly defcribed. The order of it was thus:

The *Centurion* brought the paps over the harbour to bear N.N.E, at fifteen leagues diftance, which was a fufficient offing to prevent our being feen by the enemy. To the weftward of the *Centurion* there was ftationed the *Carmelo*, and to the eaftward were the *Tryal* Prize, the *Gloucefter*, and the *Carmin*: Thefe were all ranged in a circular line, and each fhip was three leagues diftant from the next; fo that the *Carmelo* and the *Carmin*, which were the two extremes, were twelve leagues diftant from each other: And as the galeon could, without doubt, be difcerned at fix leagues diftance from either extremity, the whole fweep of our fquadron, within which nothing could pafs undifcovered, was at leaft twenty-four leagues in extent; and yet we were fo connected by our fignals, as to be eafily and fpeedily informed of what was feen in any part of the line: And to render this difpofition ftill more compleat, and to prevent even the poffibility of the galeon's efcaping us in the night, the two Cutters belonging to the *Centurion* and the *Gloucefter* were both manned and fent in fhore, and were ordered to lie all day at the diftance of four or five leagues from the entrance of the port, where, by reafon of their fmallnefs, they could not poffibly be difcovered; but in the night they were directed to ftand nearer to the harbour's mouth, and as the light of the morning came on, they were to return back again to their day-pofts. When the Cutters fhould firft difcover the *Manila* fhip, one of them was ordered to return to the fquadron, and to make a fignal, whether the galeon ftood to the eaftward or to the weftward; whilft the other was to follow the galeon at a diftance, and if it grew dark, was to direct the fquadron in their chace, by fhewing falfe fires. The particular fituation of each fhip and of the Cutters, and the bearings from each other, which they were to obferve in order to keep their ftations, will be better underftood by the delineation exhibited in the

K k 2

annexed

annexed plate; a draught of which was delivered to each of the Commanders at the same time with their orders.

Besides the care we had taken to prevent the galeon from passing by us unobserved, we had not been inattentive to the means of engaging her to advantage, when we came up with her : For considering the thinness of our hands, and the vaunting accounts given by the *Spaniards* of her size, her guns, and her strength, this was a consideration not to be neglected. As we supposed that none of our ships but the *Centurion* and the *Gloucester* were capable of lying along side of her, we took on board the *Centurion* all the hands belonging to the *Carmelo* and the *Carmin*, except what were just sufficient to navigate those ships; and Captain *Saunders* was ordered to send from the *Tryal* Prize ten *Englishmen*, and as many Negroes, to reinforce the crew of the *Gloucester* : And for the encouragement of our Negroes, of which we had a considerable number on board, we promised them, that on their good behaviour they should all have their freedom; and as they had been almost every day trained to the management of the great guns for the two preceding months, they were very well qualified to be of service to us; and from their hopes of liberty, and in return for the usage they had met with amongst us, they seemed disposed to exert themselves to the utmost of their power.

And now being thus prepared for the reception of the galeon, we expected, with the utmost impatience, the so often mentioned 3d of *March*, the day fixed for her departure. And on that day we were all of us most eagerly engaged in looking out towards *Acapulco*; and we were so strangely prepossessed with the certainty of our intelligence, and with an assurance of her coming out of port, that some or other on board us were constantly imagining that they discovered one of our Cutters returning with a signal. But to our extreme vexation, both this day and the succeeding night passed over, without any news of the galeon : However, we did not yet despair, but were all heartily disposed to flatter ourselves, that some

unforeseen

unforeseen accident had intervened, which might have put off her departure for a few days; and suggestions of this kind occurred in plenty, as we knew that the time fixed by the Viceroy for her sailing, was often prolonged on the petition of the Merchants of *Mexico*. Thus we kept up our hopes, and did not abate of our vigilance; and as the 7th of *March* was *Sunday* the beginning of Passion week, which is observed by the Papists with great strictness, and a total cessation from all kinds of labour, so that no ship is permitted to stir out of port during the whole week, this quieted our apprehensions for some days, and disposed us not to expect the galeon till the week following. On the *Friday* in this week our Cutters returned to us, and the officers on board them were very confident that the galeon was still in port, for that she could not possibly have come out but they must have seen her. On the *Monday* morning succeeding passion week, that is, on the 15th of *March*, the Cutters were again dispatched to their old station, and our hopes were once more indulged in as sanguine prepossessions as before; but in a week's time our eagerness was greatly abated, and a general dejection and despondency took place in its room. It is true, there were some few amongst us who still kept up their spirits, and were very ingenious in finding out reasons to satisfy themselves, that the disappointment we had hitherto met with had only been occasioned by a casual delay of the galeon, which a few days would remove, and not by a total suspension of her departure for the whole season: But these speculations were not relished by the generality of our people; for they were persuaded that the enemy had, by some accident, discovered our being upon the coast, and had therefore laid an embargo on the galeon till the next year. And indeed this persuasion was but too well founded; for we afterwards learnt, that our barge, when sent on the discovery of the port of *Acapulco*, had been seen from the shore; and that this circumstance (no embarkations but canoes ever frequenting that coast) was to them a sufficient proof of the neighbourhood of our squadron; on which, they stopped the galeon till the succeeding year.

The

The Commodore himſelf, though he declared not his opinion, was yet in his own thoughts very apprehenſive that we were diſcovered, and that the departure of the galeon was put off; and he had, in conſequence of this opinion, formed a plan for poſſeſſing himſelf of *Acapulco*; for he had no doubt but the treaſure as yet remained in the town, even though the orders for the diſpatching of the galeon were countermanded. Indeed the place was too well defended to be carried by an open attempt; for beſides the garriſon and the crew of the galeon, there were in it at leaſt a thouſand men well armed, who had marched thither as guards to the treaſure, when it was brought down from the city of *Mexico*: For the roads thereabouts are ſo much infeſted either by independent *Indians* or fugitives, that the *Spaniards* never truſt the ſilver without an armed force to protect it. And beſides, had the ſtrength of the place been leſs conſiderable, and ſuch as might have appeared not ſuperior to the efforts of our ſquadron, yet a declared attack would have prevented us from receiving any advantages from its ſucceſs; ſince upon the firſt diſcovery of our ſquadron, all the treaſure would have been ordered into the country, and in a few hours would have been out of our reach; ſo that our conqueſt would have been only a deſolate town, where we ſhould have found nothing that could have been of the leaſt conſequence to us.

For theſe reaſons, the ſurpriſal of the place was the only method that could at all anſwer our purpoſe; and therefore the manner in which Mr. *Anſon* propoſed to conduct this enterprize was, by ſetting ſail with the ſquadron in the evening, time enough to arrive at the port in the night; and as there is no danger on that coaſt, he would have ſtood boldly for the harbour's mouth, where he expected to arrive, and might perhaps have entered it, before the *Spaniards* were acquainted with his deſigns: Aſſoon as he had run into the harbour, he intended to have puſht two hundred of his men on ſhore in his boats, who were immediately to attempt the fort markt (D) in the plan mentioned in the preceding chapter, and inſerted towards the beginning of the third book; whilſt he, the

Com-

Commodore, with his ships, was employed in firing upon the town, and the other batteries. And these different operations, which would have been executed with great regularity, could hardly have failed of succeeding against an enemy, who would have been prevented by the suddenness of the attack, and by the want of day-light, from concerting any measures for their defence; so that it was extremely probable that we should have carried the fort by storm; and then the other batteries, being open behind, must have been soon abandoned; after which, the town, and its Inhabitants, and all the treasure, must necessarily have fallen into our hands; for the place is so cooped up with mountains, that it is scarcely possible to escape out of it, but by the great road, markt (I. I.) in the plan, which passes under the fort. This was the project which the Commodore had settled in general in his thoughts; but when he began to inquire into such circumstances as were necessary to be considered in order to regulate the particulars of its execution, he found there was a difficulty, which, being insuperable, occasioned the enterprize to be laid aside: For on examining the prisoners about the winds which prevail near the shore, he learnt (and it was afterwards confirmed by the officers of our cutters) that nearer in shore there was always a dead calm for the greatest part of the night, and that towards morning, when a gale sprung up, it constantly blew off the land; so that the setting sail from our present station in the evening, and arriving at *Acapulco* before day-light, was impossible.

This scheme, as hath been said, was formed by the Commodore, upon a supposition that the galeon was detained till the next year: But as this was a matter of opinion only, and not founded on intelligence, and there was a possibility that she might still put to sea in a short time, the Commodore thought it prudent to continue his cruise upon this station, as long as the necessary attention to his stores of wood and water, and to the convenient season for his future passage to *China*, would give him leave; and therefore, as the Cutters had been ordered to remain before *Acapulco* till the 23d of *March*, the squadron did not change its position till that day; when

the

the Cutters not appearing, we were in some pain for them, appre-
hending they might have suffered either from the enemy or the
weather; but we were relieved from our concern the next morning, when we discovered them, though at a great distance and to
the leeward of the squadron: We bore down to them and took
them up, and were informed by them, that, conformable to their
orders, they had left their station the day before, without having
seen any thing of the galeon; and we found, that the reason of
their being so far to the leeward of us was a strong current, .which
had driven the whole squadron to windward.

And here it is necessary to mention, that, by information which
was afterwards received, it appeared that this prolongation of our
cruise was a very prudent measure, and afforded us no contemptible
chance of seizing the treasure, on which we had so long fixed our
thoughts. For it seems, after the embargo was laid on the galeon,
as is before mentioned, the persons principally interested in the cargoe sent several expresses to *Mexico*, to beg that she might still be
permitted to depart : For as they knew, by the accounts sent from
Paita, that we had not more than three hundred men in all, they
insisted that there was nothing to be feared from us; for that the
galeon (carrying above twice as many hands as our whole squadron)
would be greatly an overmatch for us. And though the Viceroy
was inflexible, yet, on the account of their representation, she was
kept ready for the sea for near three weeks after the first order came
to detain her.

When we had taken up the Cutters, all the ships being joined,
the Commodore made a signal to speak with their Commanders;
and upon enquiry into the stock of fresh water remaining on board
the squadron, it was found to be so very slender, that we were under a necessity of quitting our station to procure a fresh supply : And
consulting what place was the properest for this purpose, it was
agreed, that the harbour of *Seguataneo* or *Chequetan* being the nearest to us, was, on that account, the most eligible; and it was therefore immediately resolved to make the best of our way thither : And
that

that, even while we were recruiting our water, we might not totally abandon our views upon the galeon, which perhaps, upon certain intelligence of our being employed at *Chequetan*, might venture to flip out to fea, our Cutter, under the command of Mr. *Hughes*, the Lieutenant of the *Tryal* Prize, was ordered to cruife off the port of *Acapulco* for twenty-four days; that if the galeon fhould fet fail in that interval, we might be fpeedily informed of it. In purfuance of thefe refolutions we endeavoured to ply to the weftward, to gain our intended port, but were often interrupted in our progrefs by calms and adverfe currents: In thefe intervals we employed ourfelves in taking out the moft valuable part of the cargoes of the *Carmelo* and *Carmin* prizes, which two fhips we intended to deftroy as foon as we had tolerably cleared them. By the firft of *April* we were fo far advanced towards *Seguataneo*, that we thought it expedient to fend out two boats, that they might range along the coaft, and difcover the watering place; they were gone fome days, and our water being now very fhort, it was a particular felicity to us that we met with daily fupplies of turtle, for had we been entirely confined to falt provifions, we muft have fuffered extremely in fo warm a climate. Indeed our prefent circumftances were fufficiently alarming, and gave the moft confiderate amongft us as much concern as any of the numerous perils we had hitherto encountered; for our boats, as we conceived by their not returning, had not as yet difcovered a place proper to water at, and by the leakage of our cafk and other accidents, we had not ten days water on board the whole fquadron: So that from the known difficulty of procuring water on this coaft, and the little reliance we had on the Buccaneer writers (the only guides we had to truft to) we were apprehenfive of being foon expofed to a calamity, the moft terrible of any in the long difheartning catalogue of the diftreffes of a feafaring life.

But thefe gloomy fuggeftions were foon happily ended; for our boats returned on the 5th of *April*, having difcovered a place proper for our purpofe, about feven miles to the weftward of the rocks

L l   of

of *Seguataneo*, which, by the defcription they gave of it, appeared to be the port, called by *Dampier* the harbour of *Chequetan*. The fuccefs of our boats was highly agreeable to us, and they were ordered out again the next day, to found the harbour and its entrance, which they had reprefented as very narrow. At their return they reported the place to be free from any danger; fo that on the 7th we ftood in, and that evening came to an anchor in eleven fathom. The *Gloucefter* came to an anchor at the fame time with us; but the *Carmelo* and the *Carmin* having fallen to leeward, the *Tryal* Prize was ordered to join them, and to bring them in, which in two or three days fhe effected.

Thus, after a four months continuance at fea from the leaving of *Quibo*, and having but fix days water on board, we arrived in the harbour of *Chequetan*, the defcription of which, and of the adjacent coaft, fhall be the bufinefs of the enfuing chapter.

CHAP.

C H A P. XII.

Defcription of the harbour of *Chequetan*, and of the adjacent coaft and country.

THE harbour of *Chequetan*, which we here propofe to de-fcribe, lies in the latitude of 17°: 36' North, and is about thirty leagues to the weftward of *Acapulco*. It is eafy to be difcovered by any fhip that will keep well in with the land, ef-pecially by fuch as range down coaft from *Acapulco*, and will attend to the following particulars.

There is a beach of fand, which extends eighteen leagues from the harbour of *Acapulco* to the weftward, againft which the fea breaks with fuch violence, that it is impoffible to land in any part of it: But yet the ground is fo clean, that fhips, in the fair fea-fon, may anchor in great fafety, at the diftance of a mile or two from the fhore. The land adjacent to this beach is generally low, full of villages, and planted with a great number of trees; and on the tops of fome fmall eminencies there are feveral look-out towers; fo that the face of the country affords a very agreeable profpect: For the cultivated part, which is the part here defcribed, extends fome leagues back from the fhore, and there appears to be bounded by the chain of mountains, which ftretch to a confiderable diftance on either fide of *Acapulco*. It is a moft remarkable particularity, that in this whole extent, being, as hath been mentioned, eighteen leagues, and containing, in appearance, the moft populous and beft planted diftrict of the whole coaft, there fhould be neither canoes, boats, nor any other embarkations either for fifhing, coafting, or for pleafure.

The

The beach here defcribed is the fureft guide for finding the harbour of *Chequetan*; for five miles to the weftward of the extremity of this beach there appears a hummock, which at firft makes like an ifland, and is in fhape not very unlike the hill of *Petaplan* hereafter mentioned, though much fmaller. Three miles to the weftward of this hummock is a white rock lying near the fhore, which cannot eafily be paffed by unobferved: It is about two cables length from the land, and lies in a large bay about nine leagues over. The weftward point of this bay is the hill of *Petaplan*, which is reprefented in the fame plate with the view of the Ifland of *Quicara* and *Quibo*, and is here inferted. This hill too, like the forementioned hummock, may be at firft miftaken for an ifland, though it be, in reality, a peninfula, which is joined to the Continent by a low and narrow Ifthmus, covered over with fhrubs and fmall trees. The bay of *Seguataneo* extends from this hill a great way to the weftward ; and it appears, by a plan of the bay of *Petaplan*, which is part of that of *Seguatanco*, and is here annexed, that at a fmall diftance from the hill, and oppofite to the entrance of the bay, there is an affemblage of rocks, which are white from the excrements of boobies and tropical birds. Four of thefe rocks are high and large, and, together with feveral fmaller ones, are by the help of a little imagination, pretended to refemble the form of a crofs, and are called the *White Friars*. Thefe rocks, as appears by the plan, bear W. by N. from *Petaplan*; and about feven miles to the weftward of them lies the harbour of *Chequetan*, which is ftill more minutely diftinguifhed by a large and fingle rock, that rifes out of the water a mile and an half diftant from its entrance, and bears S $\frac{1}{2}$ W. from the middle of it. The appearance of the entrance of this harbour is very accurately reprefented in the annexed plate, where (*e*) is the Eaft point of the harbour, and (*d*) the Weft, the forementioned rock being marked (*f*). In the fame view (*a*) is a large fandy bay, but where there is no landing ; (*b*) are four remarkable white rocks ; and from the ifland (*c*) there runs a large bay to the weftward.

Thefe

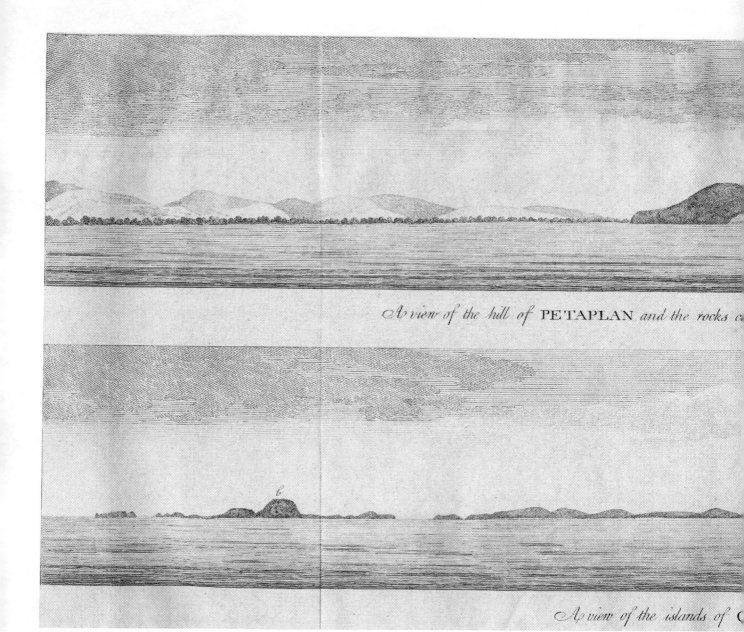

A view of the hill of PETAPLAN *and the rocks c...*

A view of the islands of C...

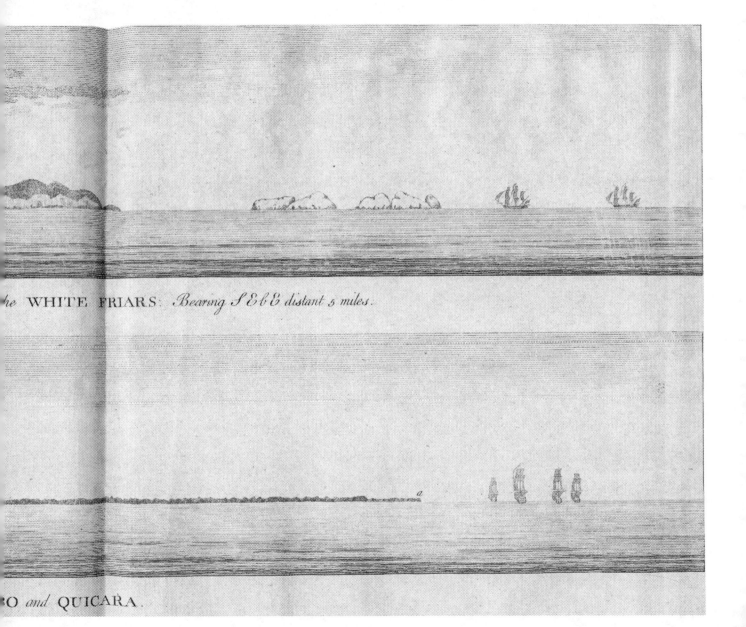

he WHITE FRIARS, *Bearing S E b E distant 5 miles.*

O *and* QUICARA.

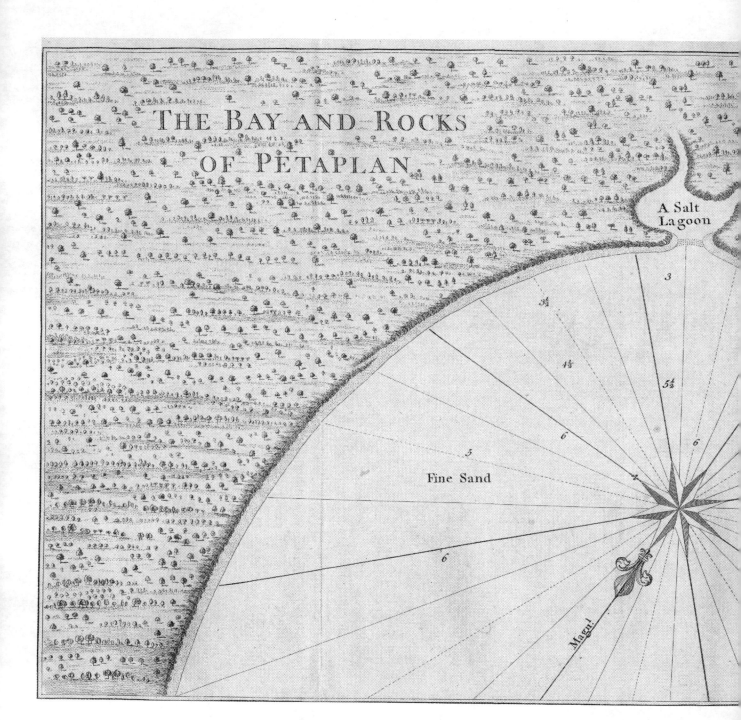

THE BAY AND ROCKS
OF PETAPLAN

A Salt
Lagoon

3

3½

4½

5½

6

5

Fine Sand

6

6

Magn.ᵗ

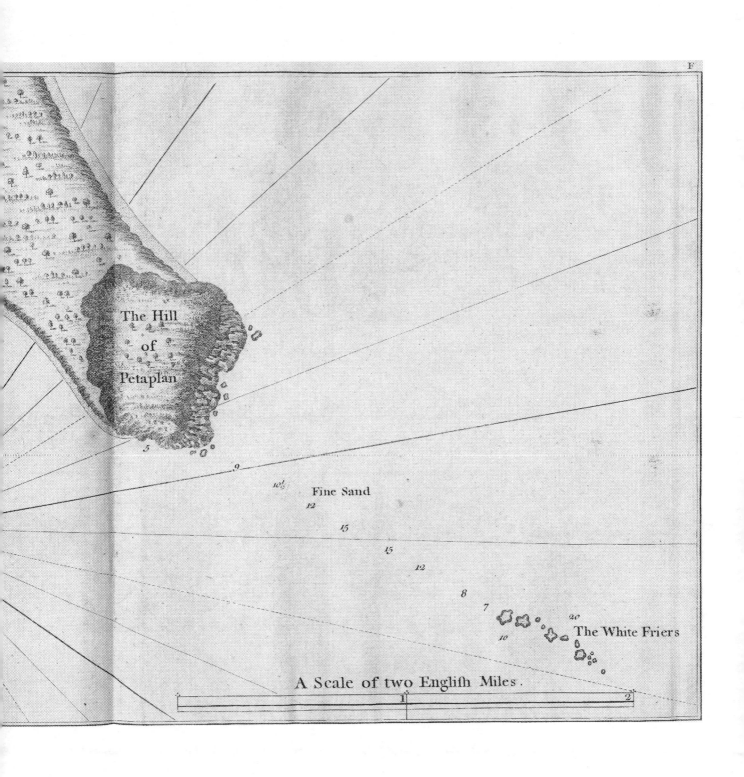

The Hill
of
Petaplan

Fine Sand

The White Friers

A Scale of two English Miles.

F

A View of the Entrance of CHEQUETAN

ZEGUATANEO *bearing N.E. distant five Miles.*

IS. Müller sculp.

A PLAN

OF THE HARBOUR

OF CHEQUETAN

OR SEGUATANEO

Lying in the Lat.d of 17.d 36.m N°

Fresh Water

Brackish Water

A Brackish Lagoon

Fine Sand

Coarse Sand and Shells

Gravel

Magnet!

Fine Sand

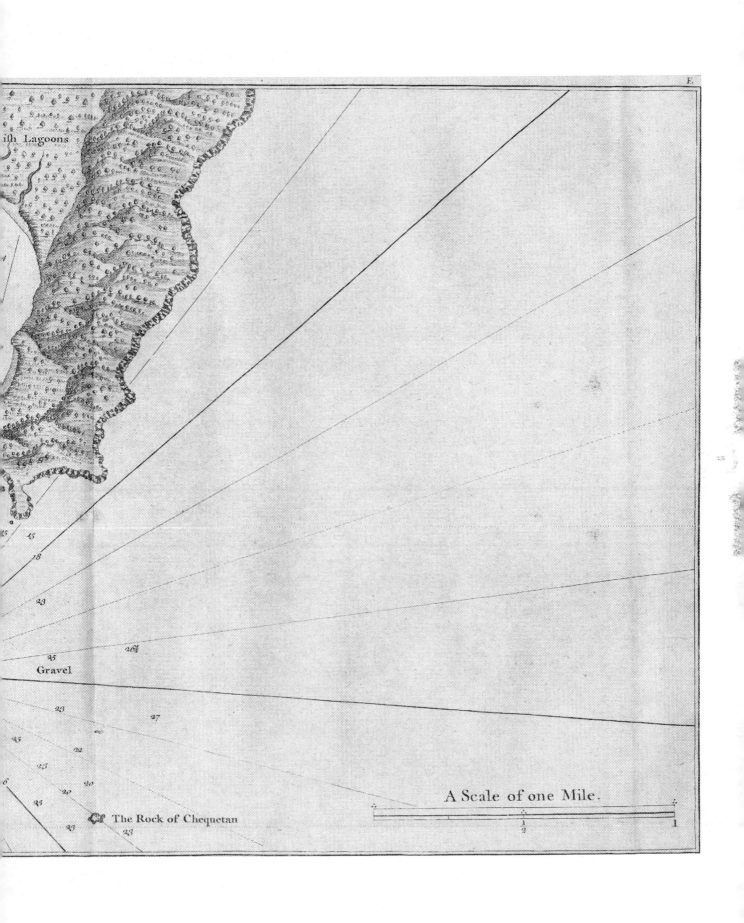

ith Lagoons

E

45

45

18

23

25 26¾

Gravel

23

23 27

45

22

23

6 20

20

45

A Scale of one Mile.

𝔊 The Rock of Chequetan

½   1

27   23

Thefe are the infallible marks by which the harbour of *Cheque-tan* may be known to thofe who keep well in with the land; and I muft add, that the coaft is no ways to be dreaded from the middle of *October* to the beginning of *May*, nor is there then any danger from the winds: Though in the remaining part of the year there are frequent and violent tornadoes, heavy rains, and hard gales in all directions of the compafs. But as to thofe who keep at any confiderable diftance from the coaft, there is no other method to be taken by them for finding this harbour, than that of making it by its latitude: For there are fo many ranges of mountains rifing one upon the back of another within land, that no drawings of the appearance of the coaft can be at all depended on, when off at fea; for every little change of diftance or variation of pofition brings new mountains in view, and produces an infinity of different profpects, which would render all attempts of delineating the afpect of the coaft impoffible.

This may fuffice as to the methods of difcovering the harbour of *Chequetan*. A plan of the harbour itfelf is reprefented in the annexed plate; where it appears, that its entrance is but about half a mile broad; the two points which form it, and which are faced with rocks that are almoft perpendicular, bear from each other S. E. and N. W. The harbour is invironed on all fides, except to the weftward, with high mountains overfpread with trees. The paffage into it is very fafe on either fide of the rock that lies off the mouth of it, though we, both in coming in and going out, left it to the eaftward. The ground without the harbour is gravel mixed with ftones, but within it is a foft mud: And it muft be remembered, that in coming to an anchor a good allowance fhould be made for a large fwell, which frequently caufes a great fend of the fea; as likewife, for the ebbing and flowing of the tide, which we obferved to be about five feet, and that it fet nearly E. and W.

The watering place is fituated in that part of the harbour, which is taken notice of in the plan for frefh water, This, during the whole time of our ftay, had the appearance of a large ftanding lake,

without

without any vifible outlet into the fea, from which it is feparated by a part of the ftrand. The origin of this lake is a fpring; that bubbles out of the ground near half a mile within the country. We found the water a little brackifh, but more confiderably fo towards the fea-fide; for the nearer we advanced towards the fpring-head the fofter and frefher it proved: This laid us under a neceffity of filling all our cafks from the furtheft part of the lake, and occafioned us fome trouble; and would have proved ftill more difficult, had it not been for our particular management, which for the conveniency of it deferves to be recommended to all who fhall hereafter water at this place. Our method confifted in making ufe of canoes which drew but little water; for, loading them with a number of fmall cafk, they eafily got up the lake to the fpring-head, and the fmall cafk being there filled were in the fame manner tranfported back again to the beach, where fome of our hands always attended to ftart them into other cafks of a larger fize.

Though this lake, during our continuance there, appeared to have no outlet into the fea, yet there is reafon to fuppofe that in the wet feafon it overflows the ftrand, and communicates with the ocean; for *Dampier*, who was formerly here, fpeaks of it as a large river. Indeed there muft be a very great body of water amaffed before the lake can rife high enough to overflow the ftrand; for the neighbouring country is fo low, that great part of it muft be covered with water, before it can run out over the beach.

As the country in the neighbourhood, particularly the tract which we have already defcribed, appeared to be well peopled, and cultivated, we hoped thence to have procured frefh provifion and other refrefhments which we ftood in need of. With this view, the morning after we came to an anchor, the Commodore ordered a party of forty men, well armed, to march into the country, and to endeavour to difcover fome town or village, where they were to attempt to fet on foot a correfpondence with the inhabitants; for we doubted not, if we could have any intercourfe with them, but that by prefents of fome of the coarfe merchandife, with which our prizes abounded

abounded (which, though of little confequence to us, would to them be extremely valuable) we fhould allure them to furnifh us with whatever fruits or frefh provifions were in their power. Our people were directed on this occafion to proceed with the greateft circumfpection, and to make as little oftentation of hoftility as poffible; for we were fenfible, that we could meet with no wealth here worth our notice, and that what neceffaries we really wanted, we fhould in all probability be better fupplied with by an open amicable traffic, than by violence and force of arms. But this endeavour of opening an intercourfe with the inhabitants proved ineffectual ; for towards evening, the party which had been ordered to march into the country, returned greatly fatigued with their unufual exercife, and fome of them fo far fpent as to have fainted by the way, and to be obliged to be brought back upon the fhoulders of their companions. They had marched in all, as they conceived, about ten miles, in a beaten road, where they often faw the frefh dung of horfes or mules. When they had got about five miles from the harbour, the road divided between the mountains into two branches, one running to the Eaft, and the other to the Weft : After fome deliberation about the courfe they fhould take, they agreed to purfue the eaftern road, which, when they had followed for fome time, led them at once into a large plain or Savannah ; on one fide of which they difcovered a centinel on horfeback with a piftol in his hand : It was fuppofed that when they firft faw him he was afleep, but his horfe ftartled at the glittering of their arms, and turning round fuddenly rode off with his mafter, who was very near being unhorfed in the furprize, but he recovered his feat, and efcaped with the lofs only of his hat and his piftol which he dropped on the ground. Our people ran after him, in hopes of difcovering fome village or habitation which he would retreat to, but as he had the advantage of being on horfeback, he foon loft fight of them. However, they were unwilling to come back without making fome difcovery, and therefore ftill followed the track they were in ; but the heat of the day encreafing, and finding no water to quench their

thirft,

thirft, they were firft obliged to halt, and then refolved to return; for as they faw no figns of plantations or cultivated land, they had no reafon to believe that there was any village or fettlement near them : But to leave no means untried of procuring fome intercourfe with the people, the officers ftuck up feveral poles in the road, to which were affixed declarations, written in *Spanifh*, encouraging the inhabitants to come down to the harbour, and to traffic with us, giving the ftrongeft affurances of a kind reception, and faithful payment for any provifions they fhould bring us. This was doubtlefs a very prudent meafure, but yet it produced no effect; for we never faw any of them during the whole time of our continuance at this port of *Chequetan*. But had our men upon the divifion of the path, taken the weftern road inftead of the eaftern, it would foon have led them to a village or town, which in fome *Spanifh* manufcripts is mentioned as being in the neighbourhood of this port, and which we afterwards learnt was not above two miles from that turning.

And on this occafion I cannot help mentioning another adventure, which happened to fome of our people in the bay of *Petaplan*, as it may help to give the reader a juft idea of the temper of the inhabitants of this part of the world. Sometime after our arrival at *Chequetan*, Lieutenant *Brett* was fent by the Commodore, with two of our boats under his command, to examine the coaft to the eaftward, particularly to make obfervations on the bay and watering place of *Petaplan*, a plan of which has been already inferted in this chapter. As Mr. *Brett* with one of the boats was preparing to go on fhore towards the hill of *Petaplan*, he, accidentally looking acrofs the bay, perceived, on the oppofite ftrand, three fmall fquadrons of horfe parading upon the beach, and feeming to advance towards the place where he propofed to land. On fight of this he immediately put off the boat, though he had but fixteen men with him, and ftood over the bay towards them : And he foon came near enough to perceive that they were mounted on very fightly horfes, and were armed with carbines and lances. On

feeing

feeing him make towards them, they formed upon the beach, and feemed refolved to difpute his landing, firing feveral diftant fhot at him as he drew near; till at laft the boat being arrived within a reafonable diftance of the moft advanced fquadron, Mr. *Brett* ordered his people to fire, upon which this refolute cavalry inftantly ran in great confufion into the wood, through a fmall opening which appears in the plan. In this precipitate flight one of their horfes fell down and threw his rider; but, whether he was wounded or not, we could not learn, for both man and horfe foon got up again, and followed the reft into the wood. In the mean time the other two fquadrons, who were drawn up at a great diftance behind, out of the reach of our fhot, were calm fpectators of the rout of their comrades; for they had halted on our firft approach, and never advanced afterwards. It was doubtlefs fortunate for our people that the enemy acted with fo little prudence, and exerted fo little fpirit; for had they concealed themfelves till our men had landed, it is fcarcely poffible but the whole boat's crew muft have fallen into their hands; fince the *Spaniards* were not much fhort of two hundred, and the whole number with Mr. *Brett*, as hath been already mentioned, only amounted to fixteen. However, the difcovery of fo confiderable a force, collected in this bay of *Petaplan*, obliged us conftantly to keep a boat or two before it: For we were apprehenfive that the Cutter, which we had left to cruife off *Acapulco*, might, on her return, be furprized by the enemy, if fhe did not receive timely information of her danger. But now to proceed with the account of the harbour of *Chequetan*.

After our unfuccefsful attempt to engage the people of the country, to furnifh us with the neceffaries we wanted, we defifted from any more endeavours of the fame nature, and were obliged to be contented with what we could procure for ourfelves in the neighbourhood of the port. We caught fifh here in tolerable quantities, efpecially when the fmoothnefs of the water permitted us to hale the Seyne. Amongft the reft, we got here cavallies, breams, mullets, foles, fiddle-fifh, fea eggs, and lobfters: And we here, and in no

M m   other

other place, met with that extraordinary fish called the *Torpedo*, or numbing fish, which is in shape very like the fiddle-fish, and is not to be known from it but by a brown circular spot of about the bigness of a crown piece near the center of its back ; perhaps its figure will be better understood, when I say it is a flat fish, much resembling the thorn-back. This fish, the *Torpedo*, is indeed of a most singular nature, productive of the strangest effects on the human body : For whoever handles it, or happens even to set his foot upon it, is presently seized with a numbness all over him ; but which is more distinguishable, in that limb which was in immediate contact with it. The same effect too will be in some degree produced by touching the fish with any thing held in the hand ; for I myself had a considerable degree of numbness conveyed to my right arm, through a walking cane which I rested on the body of the fish for some time ; and I make no doubt but I should have been much more sensibly affected, had not the fish been near expiring when I made the experiment : For it is observable that this influence acts with most vigour when the fish is first taken out of the water, and entirely ceases when it is dead, so that it may be then handled or even eaten without any inconvenience. I shall only add that the numbness of my arm on this occasion did not go off on a sudden, as the accounts of some Naturalists gave me reason to expect, but diminished gradually, so that I had some sensation of it remaining till the next day.

To the account given of the fish we met with here, I must add, that though turtle now grew scarce, and we met with none in this harbour of *Chequetan*, yet our boats, which, as I have mentioned, were stationed off *Petaplan* , often supplied us therewith ; and though this was a food that we had now been so long as it were confined to, (for it was the only fresh provisions which we had tasted for near six months) yet we were far from being cloyed with it, or from finding that the relish we had of it at all diminished.

The

The animals we met with on shore were principally guanos, with which the country abounds, and which are by some reckoned delicious food. We saw no beasts of prey here, except we should esteem that amphibious animal, the alligator, as such, several of which our people discovered, but none of them very large. However, we were satisfied that there were great numbers of tygers in the woods, though none of them came in sight; for we every morning found the beach near the watering place imprinted very thick with their footsteps: But we never apprehended any mischief from them; for they are by no means so fierce as the *Asiatic* or *African* tyger, and are rarely, if ever known, to attack mankind. Birds were here in sufficient plenty; for we had abundance of pheasants of different kinds, some of them of an uncommon size, but they were very dry and tasteless food. And besides these we had a variety of smaller birds, particularly parrots, which we often killed for food.

The fruits and vegetable refreshments at this place were neither plentiful, nor of the best kinds: There were, it is true, a few bushes scattered about the woods, which supplied us with limes, but we scarcely could procure enough for our present use; and these, with a small plumb of an agreeable acid, called in *Jamaica* the *Hog-Plumb*, together with another fruit called a *Papah*, were the only fruits to be found in the woods. Nor is there any other useful vegetable here worth mentioning, except brook-lime: This indeed grew in great quantities near the fresh-water banks; and, as it was esteemed an antiscorbutic, we fed upon it frequently, though its extreme bitterness made it very unpalatable.

These are the articles most worthy of notice in this harbour of *Chequetan*. I shall only mention a particular of the coast lying to the westward of it, that to the eastward having been already described. As Mr. *Anson* was always attentive to whatever might be of consequence to those who might frequent these seas hereafter; and, as we had observed, that there was no double land to the westward of *Chequetan*, which stretched out to a considerable di-

stance,

ſtance, with a kind of opening, which appeared not unlike the in-
let to ſome harbour, the Commodore, ſoon after we came to an
anchor, ſent a boat to diſcover it more accurately, and it was found,
on a nearer examination, that the two hills, which formed the double land, were joined together by a valley, and that there was no
harbour nor ſhelter between them.

By all that hath been ſaid it will appear, that the conveniencies
of this port of *Chequetan*, particularly in the articles of refreſhment,
are not altogether ſuch as might be deſired : But yet, upon the
whole, it is a place of conſiderable conſequence, and the knowledge of it may be of great import to future cruiſers. For it is the
only ſecure harbour in a vaſt extent of coaſt, except *Acapulco*,
which is in the hands of the enemy. It lies at a proper diſtance
from *Acapulco* for the convenience of ſuch ſhips as may have any
deſigns on the *Manila* galeon ; and it is a place, where wood and
water may be taken in with great ſecurity, in deſpight of the efforts of the inhabitants of the adjacent diſtrict : For there is but
one narrow path which leads through the woods into the country,
and this is eaſily to be ſecured by a very ſmall party, againſt all the
ſtrength the *Spaniards* in that neighbourhood can muſter. After
this account of *Chequetan*, and the coaſt contiguous to it, we ſhall
return to the recital of our own proceedings.

CHAP.

C H A P. XIII.

Our proceedings at *Chequetan* and on the adjacent coaft, till our fetting fail for *Afia*.

THE next morning, after our coming to an anchor in the harbour of *Chequetan*, we fent about ninety of our men well armed on fhore, forty of whom were ordered to march into the country, as hath been mentioned, and the remaining fifty were employed to cover the watering place, and to prevent any interruption from the natives.

Here we compleated the unloading of the *Carmelo* and *Carmin*, which we had begun at fea; at leaft, we took out of them the indico, cacao, and cochineal, with fome iron for ballaft, which were all the goods we intended to preferve, though they did not amount to a tenth of their cargoes. Here too it was agreed, after a mature confultation, to deftroy the *Tryal*'s Prize, as well as the *Carmelo* and *Carmin*, whofe fate had been before refolved on. Indeed the fhip was in good repair and fit for the fea; but as the whole numbers on board our fquadron did not amount to the complement of a fourth rate man of war, we found it was impoffible to divide them into three fhips, without rendering them incapable of navigating in fafety in the tempeftuous weather we had reafon to expect on the coaft of *China*, where we fuppofed we fhould arrive about the time of the change of the monfoons. Thefe confiderations determined the Commodore to deftroy the *Tryal* Prize, and to reinforce the *Gloucefter* with the greateft part of her crew. And in confequence of this refolve, all the ftores on board the *Tryal* Prize were removed into the other fhips, and the Prize herfelf, with the *Carmelo* and *Carmin*, were prepared for fcuttling with all the expedition we were mafters of; but the greateft difficulties we were under in laying in a

ftore

ftore of water (which have been already touched on) together with the neceffary repairs of our rigging and other unavoidable occupations, took us up fo much time, and found us fuch unexpected employment, that it was near the end of *April* before we were in a condition to leave the place.

During our ftay here, there happened an incident, which, as it proved the means of convincing our friends in *England* of our fafety, which for fome time they had defpaired of, and were then in doubt about, I fhall beg leave particularly to recite. I have obferved, in the preceding chapter, that from this harbour of *Chequetan* there was but one path-way which led through the woods into the country. This we found much beaten, and were thence convinced, that it was well known to the inhabitants. As it paffed by the fpring-head, and was the only avenue by which the *Spaniards* could approach us, we, at fome diftance beyond the fpring-head, felled feveral large trees, and laid them one upon the other acrofs the path ; and at this barricadoe we conftantly kept a guard : And we befides ordered our men employed in watering to have their arms ready, and, in cafe of any alarm, to march inftantly to this poft. And though our principal intention was to prevent our being difturbed by any fudden attack of the enemy's horfe, yet it anfwered another purpofe, which was not in itfelf lefs important; this was to hinder our own people from ftraggling fingly into the country, where we had reafon to believe they would be furprized by the *Spaniards*, who would doubtlefs be extremely folicitous to pick up fome of them, in hopes of getting intelligence of our future defigns. To avoid this inconvenience, the ftricteft orders were given to the centinels, to let no perfon whatever pafs beyond their poft : But notwithftanding this precaution, we miffed one *Lewis Leger*, who was the Commodore's Cook ; and as he was a *Frenchman*, and fufpected to be a Papift, it was by fome imagined that he had deferted, with a view of betraying all that he knew to the enemy; but this appeared, by the event, to be an ill-grounded furmife; for it was afterwards known, that he had been taken by fome *Indians*,

who

who carried him prifoner to *Acapulco*, from whence he was transferred to *Mexico*, and then to *Vera Cruz*, where he was fhipped on board a veffel bound to *Old Spain :* And the veffel being obliged by fome accident to put into *Lifbon*, *Leger* efcaped on fhore, and was by the *Britifh* Conful fent from thence to *England*; where he brought the firft authentick account of the fafety of the Commodore, and of what he had done in the *South-Seas.* The relation he gave of his own feizure was, that he rambled into the woods at fome diftance from the barricadoe, where he had firft attempted to pafs, but had been ftopped and threatned to be punifhed; that his principal view was to get a quantity of limes for his Mafter's ftore; and that in this occupation he was furprized unawares by four *Indians*, who ftripped him naked, and carried him in that condition to *Acapulco*, expofed to the fcorching heat of the fun, which at that time of the year fhone with its greateft violence : And afterwards at *Mexico* his treatment in prifon was fufficiently fevere, and the whole courfe of his captivity was a continued inftance of the hatred, which the *Spaniards* bear to all thofe who endeavour to difturb them in the peaceable poffeffion of the coafts of the *South-Seas.* Indeed *Leger's* fortune was, upon the whole, extremely fingular; for after the hazards he had run in the Commodore's fquadron, and the feverities he had fuffered in his long confinement amongft the enemy, a more fatal difafter attended him on his return to *England* : For though, when he arrived in *London*, fome of Mr. *Anfon's* friends interefted themfelves in relieving him from the poverty to which his captivity had reduced him; yet he did not long enjoy the benefit of their humanity, for he was killed in an infignificant night brawl, the caufe of which could fcarcely be difcovered.

And here I muft obferve, that though the enemy never appeared in fight during our ftay in this harbour, yet we perceived that there were large parties of them incamped in the woods about us; for we could fee their fmokes, and could thence determine that they were pofted in a circular line furrounding us at a diftance; and

juft

juft before our coming away they feemed, by the increafe of their fires, to have received a confiderable reinforcement. But to return:

Towards the latter end of *April*, the unloading of our three prizes, our wooding and watering, and, in fhort, all our propofed employments at the harbour of *Chequetan*, were compleated: So that, on the 27th of *April*, the *Tryal*'s Prize, the *Carmelo* and the *Carmin*, all which we here intended to deftroy, were towed on fhore and fcuttled, and a quantity of combuftible materials were diftributed in their upper works; and the next morning the *Centurion* and the *Gloucefter* weighed anchor, but as there was but little wind, and that not in their favour, they were obliged to warp out of the harbour. When they had reached the offing, one of the boats was difpatched back again to fet fire to our prizes, which was accordingly executed. And a canoe was left fixed to a grapnel in the middle of the harbour, with a bottle in it well corked, inclofing a letter to Mr. *Hughes*, who commanded the Cutter, which was ordered to cruife before the port of *Acapulco*, when we came off that ftation. And on this occafion I muft mention more particularly than I have yet done, the views of the Commodore in leaving the Cutter before that port.

When we were neceffitated to make for *Chequetan* to take in our water, Mr. *Anfon* confidered that our being in that harbour would foon be known at *Acapulco*; and therefore he hoped, that on the intelligence of our being employed in port, the galeon might put to fea, efpecially as *Chequetan* is fo very remote from the courfe generally fteered by the galeon: He therefore ordered the Cutter to cruife twenty-four days off the port of *Acapulco*, and her Commander was directed, on perceiving the galeon under fail, to make the beft of his way to the Commodore at *Chequetan*. As the *Centurion* was doubtlefs a much better failor than the galeon, Mr. *Anfon*, in this cafe, refolved to have got to fea as foon as poffible, and to have purfued the galeon acrofs the *Pacific* Ocean: And fuppofing he fhould not have met with her in his paffage (which confidering

that

that he would have kept nearly the same parallel, was not very improbable) yet he was certain of arriving off Cape *Espiritu Santo*, on the Island of *Samal*, before her; and that being the first land she makes on her return to the *Philippines*, we could not have failed to have fallen in with her, by cruising a few days in that station. But the Viceroy of *Mexico* ruined this project, by keeping the galeon in the port of *Acapulco* all that year.

The letter left in the canoe for Mr. *Hughes*, the Commander of the Cutter, (the time of whose return was now considerably elapsed) directed him to go back immediately to his former station before *Acapulco*, where he would find Mr. *Anson*, who resolved to cruise for him here for a certain number of days; after which it was added, that the Commodore would return to the southward to join the rest of the squadron. This last article was inserted to deceive the *Spaniards*, if they got possession of the canoe, (as we afterwards learnt they did) but could not impose on Mr. *Hughes*, who well knew that the Commodore had no squadron to join, nor any intention of steering back to *Peru*.

Being now in the offing of *Chequetan*, bound cross the vast *Pacific* Ocean in our way to *China*, we were impatient to run off the coast as soon as possible; for as the stormy season was approaching apace, and as we had no further views in the *American* seas, we had hoped that nothing would have prevented us from standing to the westward, the moment we got out of the harbour of *Chequetan*: And it was no small mortification to us, that our necessary employment there had detained us so much longer than we expected; and now we were farther detained by the absence of the Cutter, and the standing towards *Acapulco* in search of her. Indeed, as the time of her cruise had been expired for near a fortnight, we suspected that she had been discovered from the shore; and that the Governor of *Acapulco* had thereupon sent out a force to seize her, which, as she carried but six hands, was no very difficult enterprize. However, this being only conjecture, the Commodore, as soon as he was got clear of the harbour of *Chequetan*, stood along the coast

N n

to the eaftward in fearch of her: And to prevent her from paffing by us in the dark, we brought to every night; and the *Gloucefter*, whofe ftation was a league within us towards the fhore, carried a light, which the Cutter could not but perceive, if fhe kept along fhore, as we fuppofed fhe would do; and as a farther fecurity, the *Centurion* and the *Gloucefter* alternately fhowed two falfe fires every half hour. Indeed, had fhe efcaped us, fhe would have found orders in the canoe to have returned immediately before *Acapulco*, where Mr. *Anfon* propofed to cruife for her fome days.

By *Sunday*, the 2d of *May*, we were advanced within three leagues of *Acapulco*, and having feen nothing of our boat, we gave her over for loft, which, befides the compaffionate concern for our fhipmates, and for what it was apprehended they might have fuffered, was in itfelf a misfortune, which, in our prefent fcarcity of hands, we were all greatly interefted in: For the crew of the Cutter, confifting of fix men and the Lieutenant, were the very flower of our people, purpofely pickt out for this fervice, and known to be every one of them of tried and approved refolution, and as fkilful feamen as ever trod a deck. However, as it was the general belief among us that they were taken and carried into *Acapulco*, the Commodore's prudence fuggefted a project which we hoped would recover them. This was founded on our having many *Spanifh* and *Indian* prifoners in our poffeffion, and a number of fick Negroes, who could be of no fervice to us in the navigating of the fhip. The Commodore therefore wrote a letter the fame day to the Governor of *Acapulco*, telling him, that he would releafe them all, provided the Governor returned the Cutter's crew; and the letter was difpatched the fame afternoon by a *Spanifh* officer, of whofe honour we had a good opinion, and who was furnifhed with a launch belonging to one of our prizes, and a crew of fix other prifoners who all gave their parole for their return. The officer too, befides the Commodore's letter, carried with him a joint petition figned by all the reft of the prifoners, befeeching his Excellence to acquiefce in the terms propofed for their liberty. From a confideration of the number of our pri-

4

foners

foners, and the quality of fome of them, we did not doubt but the Governor would readily comply with Mr *Anfon*'s propofal, and therefore we kept plying on and off the whole night, intending to keep well in with the land, that we might receive an anfwer at the limited time, which was the next day, being *Monday :* But both on the *Monday* and *Tuefday* we were driven fo far off fhore, that we could not hope to receive any anfwer; and on the *Wednef-day* morning we found ourfelves fourteen leagues from the harbour of *Acapulco* ; but as the wind was now favourable, we preffed forwards with all our fail, and did not doubt of getting in with the land in a few hours. Whilft we were thus ftanding in, the man at the maft-head called out that he faw a boat under fail at a confiderable diftance to the South eaftward : This we took for granted was the anfwer of the Governor to the Commodore's meffage, and we inftantly edged towards it ; but when we drew nearer, we found to our unfpeakable joy that it was our own Cutter. While fhe was ftill at a diftance we imagined that fhe had been difcharged out of the port of *Acapulco* by the Governor ; but when fhe drew nearer, the wan and meager countenances of the crew, the length of their beards, and the feeble and hollow tone of their voices, convinced us that they had fuffered much greater hardfhips than could be expected from even the feverities of a *Spanifh* prifon. They were obliged to be helped into the fhip, and were immediately put to bed, and with reft, and nourifhing diet, which they were plentifully fupplied with from the Commodore's table, they recovered their health and vigour apace : And now we learnt that they had kept the fea the whole time of their abfence, which was above fix weeks, that when they had finifhed their cruife before *Acapulco*, and had juft begun to ply to the weftward in order to join the fquadron, a ftrong adverfe current had forced them down the coaft to the eaftward in fpight of all their efforts ; that at length their water being all expended, they were obliged to fearch the coaft farther on to the eaftward, in queft of fome convenient landing-place, where they might get a frefh fupply ; that in this diftrefs they ran upwards of

eighty

eighty leagues to leeward, and found every where fo large a furf, that there was not the leaft poffibility of their landing ; that they paffed fome days in this dreadful fituation, without water, and having no other means left them to allay their thirft than fucking the blood of the turtle, which they caught ; and at laft, giving up all hopes of relief, the heat of the climate too augmenting their neceffities, and rendring their fufferings infupportable, they abandoned them-felves to defpair, fully perfuaded that they fhould perifh by the moft terrible of all deaths ; but that they were foon after happily relieved by a moft unexpected incident, for there fell fo heavy a rain, that by fpreading their fails horizontally, and by putting bullets in the cen-ters of them to draw them to a point, they caught as much water, as filled all their cafk ; that immediately upon this fortunate fupply they ftood to the weftward in queft of the Commodore ; and being now luckily favoured by a ftrong current, they joined us in lefs than fifty hours, from the time they ftood to the weftward, after having been abfent from us full forty-three days. Thofe who have an idea of the inconfiderable fize of a Cutter belonging to a fixty gun fhip, (being only an open boat about twenty-two feet in length) and who will attend to the various accidents to which fhe was expofed during a fix weeks continuance alone, in the open ocean, on fo im-practicable and dangerous a coaft, will readily own, that her return to us at laft, after all the difficulties which fhe actually experienced, and the hazards to which fhe was each hour expofed, may be con-fidered as little fhort of miraculous.

I cannot finifh the article of this Cutter, without remarking how little reliance Navigators ought to have on the accounts of the Buc-caneer writers : For though in this run of hers, eighty leagues to the eaftward of *Acapulco*, fhe found no place where it was poffible for a boat to land, yet thofe writers have not been afhamed to feign harbours and convenient watering places within thefe limits, thereby expofing fuch as fhould confide in their relations, to the rifque of being deftroyed by thirft.

I muft farther add on this occafion, that when we ftood near

4   the

A View of the Entranc

Plate XXXII.

the Port of ACAPULCO.

the port of *Acapulco*, in order to fend our meffage to the Governor,
and to receive his anfwer, Mr. *Brett* took that opportunity of delineating a view of the entrance of the port, and of the neighbouring coaft, which, added to the plan of the place formerly mentioned,
may be of confiderable ufe hereafter, and is therefore annexed. In
this plate (*q*) is the weft point of the harbour called the *Griffo*,
being in the latitude of 16° : 45′; (*b c*) is the Ifland bearing from
the obferver N. by E, three leagues diftant; (*d*) is the eaft point
of the harbour; (*e*) port *Marquis*; (*f*) *Sierra di Brea*; (*h*) a
white rock in the harbour, and (*g*) watch towers.

And now having received our Cutter, the fole object of our
coming a fecond time before *Acapulco*, the Commodore refolved
not to lofe a moment's time longer, but to run off the coaft with
the utmoft expedition, both as the ftormy feafon on the coaft of
Mexico was now approaching apace, and as we were apprehenfive
of having the wefterly monfoon to ftruggle with when we came
upon the coaft of *China*; and therefore he no longer ftood towards *Acapulco*, as he now wanted no anfwer from the Governor;
but yet he refolved not to deprive his prifoners of the liberty, which
he had promifed them; fo that they were all immediately embarked in two launches which belonged to our prizes, thofe from the
Centurion in one launch, and thofe from the *Gloucefter* in the other.
The launches were well equipped with mafts, fails and oars; and
leaft the wind might prove unfavourable, they had a ftock of water and provifions put on board them fufficient for fourteen days.
There were difcharged thirty-nine perfons from on board the *Centurion*, and eighteen from the *Gloucefter*, the greateft part of them
Spaniards, the reft *Indians* and fick Negroes: But as our crews
were very weak, we kept the Mulattoes and fome of the ftouteft of the Negroes, with a few *Indians* to affift us; but we difmif-
fed every *Spanifh* prifoner whatever. We have fince learnt, that
thefe two launches arrived fafe at *Acapulco*, where the prifoners
could not enough extol the humanity with which they had been
treated; and that the Governor, before their arrival, had returned

a very

a very obliging anfwer to the Commodore's letter, and had attended it with a prefent of two boats laden with the choiceft refrefh-ments and provifions which were to be got at *Acapulco*; but that thefe boats not having found our fhips, were at length obliged to put back again, after having thrown all their provifions over-board in a ftorm which threatened their deftruction.

The fending away our prifoners was our laft tranfaction on the *American* coaft; for no fooner had we parted with them, than we and the *Gloucefter* made fail to the S. W, propofing to get a good offing from the land, where we hoped, in a few days, to meet with the regular trade-wind, which the accounts of former Navigators had reprefented as much brifker and fteadier in this ocean, than in any other part of the globe : For it has been efteemed no uncommon paffage, to run from hence to the eaftermoft parts of *Afia* in two months; and we flattered ourfelves that we were as capable of making an expeditious paffage, as any fhips that had ever run this courfe before us : So that we hoped foon to gain the coaft of *China*, for which we were now bound. And conformable to the general idea of this navigation given by former Voyagers, we confidered it as free from all kinds of embarraffment of bad weather, fatigue, or ficknefs ; and confequently we undertook it with alacrity, efpecially as it was no contemptible ftep towards our arrival at our native country, for which many of us by this time began to have great longings. Thus, on the 6th of *May*, we, for the laft time, loft fight of the mountains of *Mexico*, perfuaded, that in a few weeks we fhould arrive at the river of *Canton* in *China*, where we expected to meet with many *Englifh* fhips, and numbers of our countrymen ; and hoped to enjoy the advantages of an amicable well frequented port, inhabited by a polifhed people, and abounding with the conveniencies and indulgencies of a civilized life; bleffings, which now for near twenty months had never been once in our power. But there yet remains (before we take our leave of *America*) the confideration of a matter well worthy of attention, the difcuffion of which fhall be referred to the enfuing chapter.

CHAP.

CHAP. XIV.

A brief account of what might have been expected from our squadron, had it arrived in the *South-Seas* in good time.

AFTER the recital of the transactions of the Commodore, and the ships under his command, on the coasts of *Peru* and *Mexico*, as contained in the preceding part of this book, it will be no useless digression to examine what the whole squadron might have been capable of atchieving, had it arrived in those seas in so good a plight, as it would probably have done, had the passage round Cape *Horn* been attempted in a more seasonable time of the year. This disquisition may be serviceable to those who shall hereafter form projects of the like nature for that part of the world, or may be entrusted with their execution. And therefore I propose, in this chapter, to consider as succinctly as I can, the numerous advantages which the Public might have received from the operations of the squadron, had it set sail from *England* a few months sooner.

And first, I must suppose, that in the summer time we might have got round Cape *Horn* with an inconsiderable loss, and without any damage to our ships or rigging. For the Duke and Duchess of *Bristol,* who between them had above three hundred men, buried no more than two, from the coast of *Brazil* to *Juan Fernandes* ; and out of a hundred and eighty-three hands which were on board the Duke, there were only twenty-one sick of the scurvy, when they arrived at that Island : Whence as men of war are much better provided with all conveniencies than privateers, we might doubtless have appeared before *Baldivia* in full strength, and in a condition of entering immediately on action ; and therefore, as that place

was

was in a very defenceless ftate, its cannon incapable of fervice, and its garrifon in great meafure unarmed, it was impoffible that it could have oppofed our force, or that its half ftarved inhabitants, moft of whom are convicts banifhed thither from other parts, could have had any other thoughts than that of fubmitting ; and *Baldivia*, which is a moft excellent port, being once taken, we fhould immediately have been terrible to the whole kingdom of *Chili*, and fhould doubtlefs have awed the moft diftant parts of the *Spanifh* Empire. Indeed, it is far from improbable that, by a prudent ufe of our advantages, we might have given a violent fhock to the authority of *Spain* on that whole Continent; and might have rendered fome, at leaft, of her provinces independent. This would doubtlefs have turned the whole attention of the *Spanifh* Miniftry to that part of the world, where the danger would have been fo preffing : And thence *Great-Britain*, and her Allies, might have been rid of the numerous embarraffments, which the wealth of the *Spanifh Indies*, operating in conjunction with the *Gallick* intrigues, have conftantly thrown in her way.

And that I may not be thought to over-rate the force of this fquadron, by afcribing to it a power of overturning the *Spanifh* Government in *America*, it is neceffary to premife a few obfervations on the condition of the provinces bordering on the *South-Seas*, and on the difpofition of the inhabitants, both *Spaniards* and *Indians*, at that time ; by which it will appear, that there was great diffenfion amongft the Governors, and difaffection among the *Creolians* ; that they were in want of arms and ftores, and had fallen into a total neglect of all military regulations in their garrifons ; and that as to the *Indians* on their frontier, they were univerfally difcontented, and feemed to be watching with impatience for the favourable moment, when they might take a fevere revenge for the barbarities they had groaned under for more than two ages ; fo that every circumftance concurred to facilitate the enterprizes of our fquadron. Of all thefe particulars we were amply informed by the letters

we

we took on board our prizes, none of these vessels, as I remember, having had the precaution to throw her papers over-board.

The ill blood amongst the Governors was greatly augmented by their apprehensions of our squadron ; for every one being willing to have it believed, that the bad condition of his Government was not the effect of negligence, there were continual demands and remonstrances amongst them, in order to throw the blame upon each other. Thus, for instance, the President of St. *Jago* in *Chili*, the President of *Panama*, and many other Governors, and military officers, were perpetually soliciting the Viceroy of *Peru* to furnish them with the necessary supplies of money for putting their provinces and places in a proper state of defence to oppose our designs : But the customary answer of the Viceroy to these representations was the emptiness of the royal chest at *Lima*, and the difficulties he was under to support the expences of his own Government ; and in one of his letters, (which we intercepted,) he mentioned his apprehensions that he might even be necessitated to stop the pay of the troops and of the garrison of *Callao*, the key of the whole kingdom of *Peru*. Indeed he did at times remit to these Governors some part of their demands ; but as what he sent them was greatly short of their wants, it rather tended to the raising jealousies and heartburnings amongst them, than contributed to the purposes for which it was intended.

And besides these mutual janglings amongst the Governors, the whole body of the people were extremely dissatisfied ; for they were fully persuaded that the affairs of *Spain* for many years before had been managed by the influence of a particular foreign interest, which was altogether detached from the advantages of the *Spanish* Nation : So that the inhabitants of these distant provinces believed themselves to be sacrificed to an ambition, which never considered their convenience or interests, or paid any regard to the reputation of their name, or the honour of their country. That this was the temper of the *Creolian Spaniards* at that time, might be evinced from a hundred instances ; but I shall content myself with one, which is indeed

O o

conclusive:

conclufive : This is the teftimony of the *French* Mathematicians fent into *America*, to meafure the magnitude of an equatorial degree of longitude. For in the relation of the murther of a furgeon belonging to their company in one of the cities of *Peru*, and of the popular tumult occafioned thereby, written by one of thofe aftronomers, the author confeffes, that the inhabitants, during the uproar, all joined in imprecations on their bad Governors, and beftowed the moft abufive language upon the *French*, detefting them, in all probability, more particularly as belonging to a nation, to whofe influence in the *Spanifh* Counfels the *Spaniards* imputed all their misfortunes.

And whilft the *Creolian Spaniards* were thus diffatisfied, it appears by the letters we intercepted, that the *Indians* on almoft every frontier, were ripe for a revolt, and would have taken up arms on the flighteft encouragement ; in particular, the *Indians* in the fouthern parts of *Peru* ; as likewife the *Arraucos*, and the reft of the *Chilian Indians*, the moft powerful and terrible to the *Spanifh* name of any on that Continent. For it feems, that in the difputes between the *Spaniards* and the *Indians*, which happened fome time before our arrival, the *Spaniards* had infulted the *Indians* with an account of the force, which they expected from *Old Spain* under the command of Admiral *Pizarro*, and had vaunted that he was coming thither to compleat the great work, which had been left unfinifhed by his anceftors. Thefe threats alarmed the *Indians*, and made them believe that their extirpation was refolved on : For the *Pizarro's* being the firft conquerors of that coaft, the *Peruvian Indians* held the name, and all that bore it, in execration ; not having forgot the deftruction of their Monarchy, the maffacre of their beloved *Inca*, *Atapalipa*, the extinction of their religion, and the flaughter of their anceftors ; all perpetrated by the family of the *Pizarro's*. The *Chilian Indians* too abhorred a Chief defcended from thofe, who, by their Lieutenants, had firft attempted to inflave them, and had neceffitated their Tribes, for more

I

than

than a Century, to be continually wafting their blood in defence of their independency.

And let it not be fuppofed, that among thofe barbarous nations the traditions of fuch diftant tranfactions could not be continued till the prefent times; for all who have been acquainted with that part of the world agree, that the *Indians*, in their publick feafts, and annual folemnities, conftantly revive the memory of thefe tragick incidents; and thofe who have been prefent at thefe fpectacles, have obferved, that all the recitals and reprefentations of this kind were received with an enthufiaftick rage, and with fuch vehement emotions, as plainly evinced how ftrongly the memory of their former wrongs was implanted in them, and how acceptable the means of revenge would at all times prove. To this account I muft add too, that the *Spanifh* Governors themfelves were fo fully informed of the difpofition of the *Indians*, and were fo apprehenfive of a general defection among them, that they employed all their induftry to reconcile the moft dangerous tribes, and to prevent them from immediately taking up arms: Among the reft, the Prefident of *Chili* in particular made large conceffions to the *Arraucos*, and the other *Chilian Indians*, by which, and by diftributing confiderable prefents to their leading men, he at loft got them to confent to a prolongation of the truce between the two nations. But thefe negotiations were not concluded at the time when we might have been in the *South-Seas*; and had they been compleated, yet the hatred of thefe *Indians* to the *Spaniards* was fo great, that it would have been impoffible for their Chiefs to have prevented their joining us.

Thus then it appears, that on our arrival in the *South-Sea* we might have found the whole coaft unprovided with troops, and deftitute even of arms: For we well knew from very particular intelligence, that there were not three hundred fire-arms, of which too the greateft part were matchlocks, in all the province of *Chili*. At the fame time, the *Indians* would have been ready to revolt, the *Spaniards* difpofed to mutiny, and the Governors enraged with each

other,

other, and each prepared to rejoice at the difgrace of his antagonift; whilft we, on the other hand, might have confifted of near two thoufand men, the greateft part in health and vigour, all well-armed, and united under a Chief, whofe enterprifing genius (as we have feen) could not be depreffed by a continued feries of the moft finifter events, and whofe equable and prudent turn of temper would have remained unvaried, in the midft of the greateft degree of good fuccefs; and who befides poffeffed, in a diftinguifhed manner, the two qualities, the moft neceffary in thefe uncommon undertakings; I mean, that of maintaining his authority, and preferving, at the fame time, the affections of his people. Our other officers too, of every rank, appear, by the experience the Public hath fince had of them, to have been equal to any enterprize they might have been charged with by their Commander: And our men (at all times brave if well conducted) in fuch a caufe where treafure was the object, and under fuch leaders, would doubtlefs have been prepared to rival the moft celebrated atchievements hitherto performed by *Britifh* Mariners.

It cannot then be contefted, but that *Baldivia* muft have furrendered on the appearance of our fquadron: After which, it may be prefumed, that the *Arraucos*, the *Pulches* and *Penguinches*, inhabiting the banks of the river *Imperial*, about twenty-five leagues to the northward of this place, would have immediately taken up arms, being difpofed as hath been already related, and encouraged by the arrival of fo confiderable a force in their neighbourhood. As thefe *Indians* can bring into the field near thirty thoufand men, the greateft part of them horfe, their firft ftep would doubtlefs have been the invading the province of *Chili*, which they would have found totally unprovided of ammunition and weapons; and as its inhabitants are a luxurious and effeminate race, they would have been incapable, on fuch an emergency, of giving any oppofition to this rugged enemy: So that it is no ftrained conjecture to imagine, that the *Indians* would have been foon mafters of the whole country. And the other *Indians* on the frontiers of *Peru* being

ing equally difpofed with the *Arraucos* to fhake off the *Spanifh* yoke, it is highly probable, that they likewife would have embraced the occafion, and that a general infurrection would have taken place through all the *Spanifh* territories in *South America*; in which cafe, the only refource left to the *Creolians* (diffatisfied as they were with the *Spanifh* Government) would have been to have made the beft terms they could with their *Indian* neighbours, and to have withdrawn themfelves from the obedience of a Mafter, who had fhown fo little regard to their fecurity. This laft fuppofition may perhaps appear chimerical to thofe, who meafure the poffibility of all events by the fcanty ftandard of their own experience; but the temper of the times, and the ftrong diflike of the natives to the meafures then purfued by the *Spanifh* Court, fufficiently evince at leaft its poffibility. But not to infift on the prefumption of a general revolt, it is fufficient for our purpofe to conclude, that the *Arraucos* would fcarcely have failed of taking arms on our appearance : For this alone would fo far have embarraffed the enemy, that they would no longer have thought of oppofing us; but would have turned all their care to the *Indian* affairs ; as they ftill remember, with the utmoft horror, the facking of their cities, the rifling of their convents, the captivity of their wives and daughters, and the defolation of their country by thefe refolute favages, in the laft war between the two nations. For it muft be remembered, that this tribe of *Indians* have been frequently fuccefsful againft the *Spaniards*, and poffefs at this time a large tract of country, which was formerly full of *Spanifh* towns and villages, whofe inhabitants were all either deftroyed, or carried into captivity by the *Arraucos* and the neighbouring *Indians*, who, in a war againft the *Spaniards*, never fail to join their forces.

But even, independent of an *Indian* revolt, there were but two places on all the coaft of the *South-Sea*, which could be fuppofed capable of refifting our fquadron; thefe were the cities of *Panama* and *Callao*: As to the firft of thefe, its fortifications were fo decayed, and it was fo much in want of powder, that the Governor

himfelf,

himfelf, in an intercepted letter, acknowledged it was incapable of being defended ; fo that I take it for granted, it would have given us but little trouble, efpecially if we had opened a communication acrofs the Ifthmus with our fleet on the other fide : And for the city and port of *Callao*, its condition was not much better than that of *Panama*; for its walls are built upon the plain ground, without either outwork or ditch before them, and confift only of very flender feeble mafonry, without any earth behind them ; fo that a battery of five or fix pieces of cannon, raifed any where within four or five hundred paces of the place, would have had a full view of the whole rampart, and would have opened it in a fhort time ; and the breach hereby formed, as the walls are fo extremely thin, could not have been difficult of afcent ; for the ruins would have been but little higher than the furface of the ground ; and it would have yielded this particular advantage to the affailants, that the bullets, which grazed upon it, would have driven before them fuch fhivers of brick and ftone, as would have prevented the garrifon from forming behind it, fuppofing that the troops employed in the defence of the place, fhould have fo far furpaffed the ufual limits of *Creolian* bravery, as to refolve to ftand a general affault : Indeed, fuch a refolution cannot be imputed to them; for the garrifon and people were in general diffatisfied with the Viceroy's behaviour, and were never expected to act a vigorous part. The Viceroy himfelf greatly apprehended that the Commodore would make him a vifit at *Lima*, the capitol of the kingdom of *Peru* ; to prevent which, if poffible, he had ordered twelve gallies to be built at *Guaiaquil* and other places, which were intended to oppofe the landing of our boats, and to hinder us from pufhing our men on fhore. But this was an impracticable project, and proceeded on the fuppofition that our fhips, when we fhould land our men, would keep at fuch a diftance, that thefe gallies, by drawing little water, would have been out of the reach of their guns; whereas the Commodore, before he had made fuch an attempt, would doubtlefs have been poffeffed of feveral prize fhips, which he would not have hefitated to

have

have run on shore for the protection of his boats ; and besides there were many places on that coast, and one in particular in the neighbourhood of *Callao*, where there was good anchoring, though a great depth of water, within a cable's length of the shore ; so that the cannon of the men of war would have swept all the coast to above a mile's distance from the water's edge, and would have effectually prevented any force from assembling, to oppose the landing and forming of our men : And the place had this additional advantage, that it was but two leagues distant from the city of *Lima* ; so that we might have been at that city within four hours after we should have been first discovered from the shore. The place I have here in view is about two leagues South of *Callao*, and just to the northward of the head-land called, in *Frezier*'s draught of that coast, *Morro Solar*. Here there is seventy or eighty fathom of water, within two cables length of the shore ; and the *Spaniards* themselves were so apprehensive of our attempting to land there, that they had projected to build a fort close to the water ; but there being no money in the royal chests, they could not go on with that work, and therefore they contented themselves with keeping a guard of an hundred horse there, that they might be sure to receive early notice of our appearance on that coast. Indeed some of them (as we were told) conceiving our management at sea to be as pusillanimous as their own, pretended that the Commodore would never dare to bring in his ships there, for fear that in so great a depth of water their anchors could not hold them.

And here let it not be imagined, that I am proceeding upon groundless and extravagant presumptions, when I conclude, that fifteen hundred or a thousand of our people, well conducted, should have been an over-match for any numbers the *Spaniards* could muster in *South America*. For not to mention the experience we had of them at *Paita* and *Petaplan*, it must be remembered, that our Commodore was extremely solicitous to have all his men trained to the dexterous use of their fire-arms ; whereas the *Spaniards*, in this part of the world, were in great want of arms, and were very

awkard

awkard in the management of the few they had : And though, on their repeated reprefentations, the Court of *Spain* had ordered feveral thoufand firelocks to be put on board *Pizarro*'s fquadron, yet thofe, it is evident, could not have been in *America* time enough to have been employed againft us ; fo that by our arms, and our readinefs in the ufe of them (not to infift on the timidity and foftnefs of our enemy) we fhould in fome degree have had the fame advantages, which the *Spaniards* themfelves had, in the firft difcovery of this country, againft its naked and unarmed inhabitants.

And now let it be confidered what were the events which we had to fear, or what were the circumftances which could have prevented us from giving law to all the coaft of *South America*, and thereby cutting off from *Spain* the refources which fhe drew from thofe immenfe provinces. By fea there was no force capable of oppofing us ; for how foon foever we had failed, *Pizarro*'s fquadron could not have failed fooner than it did, and therefore could not have avoided the fate it met with : As we fhould have been mafters of the ports of *Chili*, we could there have fupplied ourfelves with the provifions we wanted in the greateft plenty ; and from *Baldivia* to the equinoctial we ran no rifque of lofing our men by ficknefs, (that being of all climates the moft temperate and healthy) nor of having our fhips difabled by bad weather ; and had we wanted hands to affift in the navigating our fquadron, whilft a confiderable part of our men were employed on fhore, we could not have failed of getting whatever numbers we pleafed in the ports we fhould have taken, and the prizes which would have fallen into our hands ; and I muft obferve that the *Indians*, who are the principal failors in that part of the world, are extremely docile, and dexterous, and though they are not fit to ftruggle with the inclemencies of a cold climate, yet in temperate feas they are moft ufeful and laborious feamen.

Thus then it appears, what important revolutions might have been brought about by our fquadron, had it departed from *England* as early as it ought to have done : And from hence it is eafy to conclude, what immenfe advantages might have thence accrued to the

Public.

public. For, as on our fuccefs it would have been impoffible for the kingdom of *Spain* to have received any treafure from the provinces bordering on the *South-Seas*, or even to have had any communication with them, it is certain that the whole attention of that Monarchy muft have been immediately employed in regaining the poffeffion of thefe ineftimable territories, either by force or compact. By the firft of thefe methods it was fcarcely poffible they could fucceed; for it muft have been at leaft a twelvemonth from our arrival, before any fhips from *Spain* could get into the *South-Seas*, and thofe perhaps feparated, difabled, and fickly; and by that time they would have had no port in their poffeffion, either to rendezvous at or to refit; whilft we might have been fupplied acrofs the Ifthmus with whatever neceffaries, ftores, or even men we wanted, and might thereby have maintained our fquadron in as good a plight, as when it firft fet fail from St. *Helens*. In fhort, it required but little prudence in the conduct of this bufinefs to have rendered all the efforts of *Spain*, feconded by the power of *France*, ineffectual, and to have maintained our conquefts in defiance of them both: So that they muft either have refolved to have left *Great-Britain* mafters of the wealth of *South America*, (the principal fupport of all their deftructive projects) or they muft have fubmitted to her terms, and have been contented to receive thefe provinces back again, as an equivalent for thofe reftrictions to their future ambition, which her prudence fhould have dictated to them. Having thus difcuffed the prodigious weight which the operations of our Squadron might have added to the national influence of this kingdom, I fhall here end this fecond book, referring to the next, the paffage of the fhattered remains of our force acrofs the *Pacific* Ocean, and all their future tranfactions till the Commodore's arrival in *England*.

END of BOOK II.

P p   A VOYAGE

A
VOYAGE
ROUND THE
WORLD, &c.

BOOK III.
CHAP. I.

The run from the coaſt of *Mexico* to the *Ladrones* or *Marian* Iſlands.

W HEN, on the 6th of *May* 1742, we left the coaſt of *America*, we ſtood to the S. W. with a view of meeting with the N. E. trade-wind, which the accounts of former writers made us expeƈt at ſeventy or eighty leagues diſtance from the land : We had beſides another reaſon for ſtanding to the ſouthward, which was the getting into the latitude of 13 or 14° North ; that being the parallel where the *Pacific* Ocean is moſt uſually croſſed, and conſequently where the navigation is eſteemed the ſafeſt : This laſt purpoſe we had ſoon anſwered, being in a day or two ſufficiently advanced to the South. At the ſame time we were alſo farther from the ſhore, than we had preſumed was neceſſary for the falling in with the trade-wind : But in this

particular

particular we were moſt grievouſly diſappointed; for the wind ſtill continued to the weſtward, or at beſt variable. As the getting into the N. E. trade was to us a matter of the laſt conſequence, we ſtood more to the ſouthward, and made many experiments to meet with it; but all our efforts were for a long time unſucceſsful: So that it was ſeven weeks, from our leaving the coaſt, before we got into the true trade-wind. This was an interval, in which we believed we ſhould well nigh have reached the eaſtermoſt parts of *Aſia*: But we were ſo baffled with the contrary and variable winds, which for all that time perplexed us, that we were not as yet advanced a-bove a fourth part of the way. The delay alone would have been a ſufficient mortification; but there were other circumſtances at-tending it, which rendered this ſituation not leſs terrible, and our apprehenſions perhaps ſtill greater than in any of our paſt diſtreſſes. For our two ſhips were by this time extremely crazy; and many days had not paſſed, before we diſcovered a ſpring in the foremaſt of the *Centurion*, which rounded about twenty-ſix inches of its circum-ference, and which was judged to be at leaſt four inches deep: And no ſooner had our Carpenters ſecured this with fiſhing it, but the *Glouceſter* made a ſignal of diſtreſs; and we learnt that ſhe had a dangerous ſpring in her main-maſt, twelve feet below the truſſel-trees; ſo that ſhe could not carry any ſail upon it. Our Carpen-ters, on a ſtrict examination of this maſt, found it ſo very rotten and decayed, that they judged it neceſſary to cut it down as low as it appeared to have been injured; and by this it was re-duced to nothing but a ſtump, which ſerved only as a ſtep to the top-maſt. Theſe accidents augmented our delay, and occaſioned us great anxiety about our future ſecurity: For on our leaving the coaſt of *Mexico*, the ſcurvy had begun to make its appearance again amongſt our people; though from our departure from *Juan Fernandes* we had till then enjoyed a moſt uninterrupted ſtate of health. We too well knew the effects of this diſeaſe, from our former fatal experience, to ſuppoſe that any thing but a ſpeedy paſ-ſage could ſecure the greater part of our crew from periſhing by it:

And

And as, after being feven weeks at fea, there did not appear any reafons that could perfuade us, we were nearer the trade-wind, than when we firft fet out, there was no ground for us to fuppofe, but our paffage would prove at leaft three times as long as we at firft expected; and confequently we had the melancholy profpect, either of dying by the fcurvy, or perifhing with the fhip for want of hands to navigate her. Indeed, fome amongft us were at firft willing to believe, that in this warm climate, fo different from what we felt in paffing round Cape *Horn*, the violence of this difeafe, and its fatality, might be in fome degree mitigated; as it had not been unufual to fuppofe that its particular virulence in that paffage was in a great meafure owing to the feverity of the weather: But the havock of the diftemper, in our prefent circumftances, foon convinced us of the falfity of this fpeculation; as it likewife exploded fome other opinions, which ufually pafs current about the caufe and nature of this difeafe.

For it has been generally prefumed, that plenty of frefh provifions, and of water are effectual preventives of this malady; but it happened that in the prefent inftance we had a confiderable ftock of frefh provifions on board, as hogs and fowls, which were taken at *Paita*; and we befides almoft every day caught great abundance of bonito's, dolphins, and albicores; and the unfettled feafon, which deprived us of the benefit of the trade-wind, proved extremely rainy; fo that we were enabled to fill up our water cafk, almoft as faft as they were empty; and each man had five pints of water allowed him every day, during the paffage. But notwithftanding this plenty of water, and that the frefh provifions were diftributed amongft the fick, and the whole crew often fed upon fifh, yet neither were the fick hereby relieved, nor the progrefs and advancement of the difeafe retarded: Nor was it in thefe inftances only that we found ourfelves difappointed; for though it has been ufually efteemed a neceffary piece of management to keep all fhips, where the crews are large, as clean and airy between decks as poffible; and it hath been believed by many, that this particular, if well attended

to,

to, would prevent the appearance of the fcurvy, or at leaft, mitigate its effects; yet we obferved, during the latter part of our run, that though we kept all our ports open, and took uncommon pains in cleanfing and fweetning the fhips, yet neither the progrefs, nor the virulence of the difeafe were thereby fenfibly abated.

However, I would not be underftood to affert, that frefh provifions, plenty of water, and a conftant frefh fupply of fweet air between decks, are matters of no moment : I am, on the contrary, well fatisfied, that they are all of them articles of great importance, and are doubtlefs extremely conducive to the health and vigour of a crew, and may in many cafes prevent the fatal malady we are now fpeaking of from taking place. All I have aimed at, in what I have advanced, is only to fhew that in fome inftances, both the cure, and prevention of this difeafe, is impoffible to be effected by any management, or by the application of any remedies, which can be made ufe of at fea. Indeed, I am myfelf fully perfuaded, that when it has once got to a certain head, there are no other means in nature for relieving the difeafed, but carrying them on fhore, or at leaft bringing them into the neighbourhood of land. Perhaps a diftinct and adequate knowledge of the fource of this difeafe may never be difcovered; but in general, there is no difficulty in conceiving, that as a continued fupply of frefh air is neceffary to all animal life, and as this air is fo particular a fluid, that without lofing its elafticity, or any of its obvious properties, it may be rendered unfit for this purpofe, by the mixing with it fome very fubtle and otherwife imperceptible effluvia; it may be conceived, I fay, that the fteams arifing from the ocean may have a tendency to render the air they are fpread through lefs properly adapted to the fupport of the life of terreftrial animals, unlefs thefe fteams are corrected by effluvia of another kind, and which perhaps the land alone can fupply.

To what hath been already faid in relation to this difeafe, I fhall add, that our furgeon (who during our paffage round Cape *Horn*, had afcribed the mortality we fuffered to the feverity of the climate)

I   exerted

exerted himſelf in the preſent run to the utmoſt, and at laſt declared, that all his meaſures were totally ineffectual, and did not in the leaſt avail his patients: On which it was reſolved by the Commodore to try the effects of two medicines, which, juſt before his departure from *England*, were the ſubject of much diſcourſe, I mean the pill and drop of Mr. *Ward*. For however violent the effects of theſe medicines are ſaid to have ſometimes proved, yet in the preſent inſtance, where deſtruction ſeemed inevitable without ſome remedy, the experiment at leaſt was thought adviſeable: And therefore, one or both of them, at different times, were given to perſons in every ſtage of the diſtemper. Out of the numbers that took them, one, ſoon after ſwallowing the pill, was ſeized with a violent bleeding at the noſe: He was before given over by the ſurgeon, and lay almoſt at the point of death; but he immediately found himſelf much better, and continued to recover, though ſlowly, till we arrived on ſhore, which was near a fortnight after. A few others too were relieved for ſome days, but the diſeaſe returned again with as much violence as ever; though neither did theſe, nor the reſt, who received no benefit, appear to be reduced to a worſe condition than they would have been if they had taken nothing. The moſt remarkable property of theſe medicines, and what was obvious in almoſt every one that took them, was, that they operated in proportion to the vigour of the patient; ſo that thoſe who were within two or thee days of dying were ſcarcely affected; and as the patient was differently advanced in the diſeaſe, the operation was either a gentle perſpiration, an eaſy vomit, or a moderate purge: But if they were taken by one in full ſtrength, they then produced all the beforementioned effects with conſiderable violence, which ſometimes continued for ſix or eight hours together with little intermiſſion. But to return to the proſecution of our voyage.

I have already obſerved, that, a few days after our running off the coaſt of *Mexico*, the *Glouceſter* had her main-maſt cut down to a ſtump, and we were obliged to fiſh our fore-maſt; and that theſe misfortunes were greatly aggravated, by our meeting with contrary

and

and variable winds for near feven weeks. I fhall now add, that when we reached the trade-wind, and it fettled between the North and the Eaft, yet it feldom blew with fo much ftrength, but the *Centurion* might have carried all her fmall fails abroad with the greateft fafety; fo that now had we been a fingle fhip, we might have run down our longitude apace, and have reached the *Ladrones* foon enough to have recovered great numbers of our men, who af-terwards perifhed. But the *Gloucefter*, by the lofs of her main-maft, failed fo very heavily, that we had feldom any more than our top-fails fet, and yet were frequently obliged to lie too for her: And, I conceive, that in the whole we loft little lefs than a month by our attendance upon her, in confequence of the various mif-chances fhe encountered. In all this run it was remarkable, that we were rarely many days together, without feeing great numbers of birds; which is a proof that there are many iflands, or at leaft rocks, fcattered all along, at no very confiderable diftance from our track. Some indeed there are marked in the *Spanifh* chart, hereafter infert-ed; but the frequency of the birds feem to evince, that there are many more than have been hitherto difcovered: For the greateft part of the birds, we obferved, were fuch as are known to rooft on fhore; and the manner of their appearance fufficiently made out, that they came from fome diftant haunt every morning, and re-turned thither again in the evening; for we never faw them early or late; and the hour of their arrival and departure gradually va-ried, which we fuppofed was occafioned by our running nearer their haunts, or getting farther from them.

The trade-wind continued to favour us without any fluctuation, from the end of *June* till towards the end of *July*. But on the 26th of *July*, being then, as we efteemed, about three hundred leagues diftant from the *Ladrones*, we met with a wefterly wind, which did not come about again to the eaftward in four days time. This was a moft difpiriting incident, as it at once damped all our hopes of fpeedy relief, efpecially too as it was attended with a vex-atious accident to the *Gloucefter*: For in one part of thefe four days

2 the

the wind flatted to a calm, and the ſhips rolled very deep; by which means the *Gloucefter*'s forecap ſplit, and her top-maſt came by the board, and broke her fore-yard directly in the ſlings. As ſhe was hereby rendered incapable of making any ſail for ſome time, we were obliged as ſoon as a gale ſprung up, to take her in tow; and near twenty of the healthieſt and ableſt of our ſeamen were taken from the buſineſs of our own ſhip, and were employed for eight or ten days together on board the *Gloucefter* in repairing her damages: But theſe things, mortifying as we thought them, were but the beginning of our diſafters; for ſcarce had our people fi-niſhed their buſineſs in the *Gloucefter*, before we met with a moſt violent ſtorm in the weſtern board, which obliged us to lie to. In the beginning of this ſtorm our ſhip ſprung a leak, and let in ſo much water, that all our people, officers included, were employed continually in working the pumps: And the next day we had the vexation to ſee the *Gloucefter*, with her top-maſt once more by the board; and whilſt we were viewing her with great concern for this new diſtreſs, we ſaw her main-top maſt, which had hitherto ſerved her as a jury main-maſt, ſhare the ſame fate. This compleat-ed our misfortunes, and rendered them without reſource; for we knew the *Gloucefter*'s crew were ſo few and feeble, that without our aſſiſtance they could not be relieved: And our ſick were now ſo far encreaſed, and thoſe that remained in health ſo continually fa-tigued with the additional duty of our pumps, that it was impoſſi-ble for us to lend them any aid. Indeed we were not as yet fully apprized of the deplorable ſituation of the *Gloucefter*'s crew; for when the ſtorm abated, (which during its continuance prevented all communication with them) the *Gloucefter* bore up under our ſtern; and Captain *Mitchel* informed the Commodore, that beſides the loſs of his maſts, which was all that had appeared to us, the ſhip had then no leſs than ſeven feet of water in her hold, although his officers and men had been kept conſtantly at the pump for the laſt twenty-four hours.

Q q

This

This laft circumftance was indeed a moft terrible accumulation to the other extraordinary diftreffes of the *Gloucefter*, and required, if poffible, the moft fpeedy and vigorous affiftance ; which Captain *Mitchel* begged the Commodore to fend him : But the debility of our people, and our own immediate prefervation, rendered it impoffible for the Commodore to comply with his requeft. All that could be done was to fend our boat on board for a more particular condition of the fhip ; and it was foon fufpected that the taking her people on board us, and then deftroying her, was the only meafure that could be profecuted in the prefent emergency, both for the fecurity of their lives and of our own.

Our boat foon returned with a reprefentation of the ftate of the *Gloucefter*, and of her feveral defects, figned by Captain *Mitchel* and all his officers ; by which it appeared, that fhe had fprung a leak by the ftern poft being loofe, and working with every roll of the fhip, and by two beams a midfhips being broken in the orlope ; no part of which the Carpenters reported was poffible to be repaired at fea : That both officers and men had worked twenty-four hours at the pump without intermiffion, and were at length fo fatigued, that they could continue their labour no longer ; but had been forced to defift, with feven feet of water in the hold, which covered their cafk, fo that they could neither come at frefh water, nor provifion : That they had no maft ftanding, except the fore-maft, the mizen-maft, and the mizen top-maft, nor had they any fpare mafts to get up in the room of thofe they had loft : That the fhip was befides extremely decayed in every part, for her knees and clamps were all worked quite loofe, and her upper works in general were fo loofe, that the quarter-deck was ready to drop down : And that her crew was greatly reduced, for there remained alive on board her no more than feventy-feven men, eighteen boys, and two prifoners, officers included ; and that of this whole number, only fixteen men, and eleven boys were capable of keeping the deck, and feveral of thefe very infirm.

I

The

The Commodore, on the perufal of this melancholy reprefenta-tion, prefently ordered them a fupply of water and provifions, of which they feemed to be in immediate want, and at the fame time fent his own Carpenter on board them, to examine into the truth of every particular ; and it being found, on the ftricteft enquiry, that the preceding account was in no inftance exaggerated, it plainly appeared, that there was no poffibility of preferving the *Gloucefter* any longer, as her leaks were irreparable, and the united hands on board both fhips, capable of working, would not be able to free her, even if our own fhip fhould not employ any part of them. What then could be refolved on, when it was the utmoft we our-felves could do to manage our own pumps ? Indeed there was no room for deliberation ; the only ftep to be taken was, the faving the lives of the few that remained on board the *Gloucefter*, and getting out of her as much as was poffible before fhe was deftroyed. And therefore the Commodore immediately fent an order to Captain *Mitchel*, as the weather was now calm and favourable, to fend his people on board the *Centurion*, as expeditioufly as he could; and to take out fuch ftores as he could get at, whilft the fhip could be kept above water. And as our leak required lefs attention, whilft the prefent eafy weather continued, we fent our boats with as many men as we could fpare, to Captain *Mitchel*'s affiftance.

The removing the *Gloucefter*'s people on board us, and the get-ting out fuch ftores as could moft eafily be come at, gave us full employment for two days. Mr. *Anfon* was extremely defirous to have gotten two of her cables and an anchor, but the fhip rolled fo much, and the men were fo exceffively fatigued, that they were incapable of effecting it ; nay, it was even with the greateft diffi-culty that the prize money, which the *Gloucefter* had taken in the *South-Seas*, was fecured, and fent on board the *Centurion:* How-ever, the prize goods on board her, which amounted to feveral thoufand pounds in value, and were principally the *Centurion*'s pro-perty, were entirely loft ; nor could any more provifion be got out than five cafk of flower, three of which were fpoiled by the falt-

water.

water. Their sick men amounting to near seventy, were removed into the boats with as much care as the circumstances of that time would permit; but three or four of them expired as they were hoisting them into the *Centurion*.

It was the 15th of *August*, in the evening, before the *Gloucester* was cleared of every thing that was proposed to be removed; and though the hold was now almost full of water, yet, as the Carpenters were of opinion that she might still swim for some time, if the calm should continue, and the water become smooth, she was set on fire; for we knew not how near we might now be to the Island of *Guam*, which was in the possession of our enemies, and the wreck of such a ship would have been to them no contemptible acquisition. When she was set on fire, Captain *Mitchel* and his officers left her, and came on board the *Centurion*: And we immediately stood from the wreck, not without some apprehensions (as we had now only a light breeze) that if she blew up soon, the concussion of the air might damage our rigging; but she fortunately burnt, though very fiercely, the whole night, her guns firing successively, as the flames reached them. And it was six in the morning, when we were about four leagues distant, before she blew up; the report she made upon this occasion was but a small one, but there was an exceeding black pillar of smoke, which shot up into the air to a very considerable height.

Thus perished his Majesty's ship the *Gloucester*. And now it might have been expected, that being freed from the embarrassments which her frequent disasters had involved us in, we might proceed on our way much brisker than we had hitherto done, especially as we had received some small addition to our strength, by the taking on board the *Gloucester*'s crew; but our anxieties were not yet to be relieved; for, notwithstanding all that we had hitherto suffered, there remained much greater distresses, which we were still to struggle with. For the late storm, which had proved so fatal to the *Gloucester*, had driven us to the northward of our intended course; and the current setting the same way, after

the

the weather abated, had forced us still a degree or two farther, so
that we were now in 17° ¼ of North latitude, instead of being in
13° ½, which was the parallel we proposed to keep, in order to
reach the Island of *Guam*: And as it had been a perfect calm for
some days since the cessation of the storm, and we were ignorant
how near we were to the meridian of the *Ladrones*, and supposed
ourselves not to be far from it, we apprehended that we might be
driven to the leeward of them by the current, without discovering
them: In this case, the only land we could make would be some
of the eastern parts of *Asia*, where, if we could arrive, we should
find the western monsoon in its full force, so that it would be impossible for the stoutest best manned ship to get in. And this coast
being removed between four and five hundred leagues farther, we,
in our languishing circumstances, could expect no other than to be
destroyed by the scurvy, long before the most favourable gale could
carry us to such a distance: For our deaths were now extremely
alarming, no day passing in which we did not bury eight or ten,
and sometimes twelve of our men; and those, who had hitherto
continued healthy, began to fall down apace. Indeed we made the
best use we could of the present calm, by employing our Carpenters in searching after the leak, which was now considerable notwithstanding the little wind we had: The Carpenters at length discovered it to be in the Gunner's fore store-room, where the water
rushed in under the breast-hook, on each side of the stem; but
though they found where it was, they agreed that it was impossible
to stop it, till we should get into port, and till they could come at
it on the outside: However, they did the best they could within
board, and were fortunate enough to reduce it, which was a considerable relief to us.

We had hitherto considered the calm which succeeded the storm,
and which continued for some days, as a very great misfortune;
since the currents were driving us to the northward of our parallel,
and we thereby risqued the missing of the *Ladrones*, which we
now conceived ourselves to be very near. But when a gale sprung

up,

up, our condition was still worse; for it blew from the S. W, and consequently was directly opposed to the course we wanted to steer: And though it soon veered to the N. E, yet this served only to tantalize us, for it returned back again in a very short time to its old quarter. However, on the 22d of *August* we had the satisfaction to find that the current was shifted; and had set us to the southward: And the 23d, at day-break, we were cheered with the discovery of two Islands in the western board: This gave us all great joy, and raised our drooping spirits; for before this an universal dejection had seized us, and we almost despaired of ever seeing land again: The nearest of these Islands we afterwards found to be *Anatacan*; we judged it to be full fifteen leagues from us, and it seemed to be high land, though of an indifferent length: The other was the Island of *Serigan*; and had rather the appearance of a high rock, than a place we could hope to anchor at. The view of these Islands is inserted at the top of the annexed plan. We were extremely impatient to get in with the nearest Island, where we expected to meet with anchoring ground, and an opportunity of refreshing our sick: But the wind proved so variable all day, and there was so little of it, that we advanced towards it but slowly; however, by the next morning we were got so far to the westward, that we were in view of a third Island which was that of *Paxaros*, though marked in the chart only as a rock. This was small and very low land, and we had passed within less than a mile of it, in the night, without seeing it: And now at noon, being within four miles of the Island of *Anatacan*, the boat was sent away to examine the anchoring ground and the produce of the place; and we were not a little solicitous for her return, as we then conceived our fate to depend upon the report we should receive: For the other two Islands were obviously enough incapable of furnishing us with any assistance, and we knew not then that there were any others which we could reach. In the evening the boat came back, and the crew informed us that there was no place for a ship to anchor, the bottom being every where foul ground, and all except one small spot, not less than fifty fathom in depth; that

on

a. Anatacan W. by S. dist.ᵗ 13 Leagues.

A view of 2 of t...

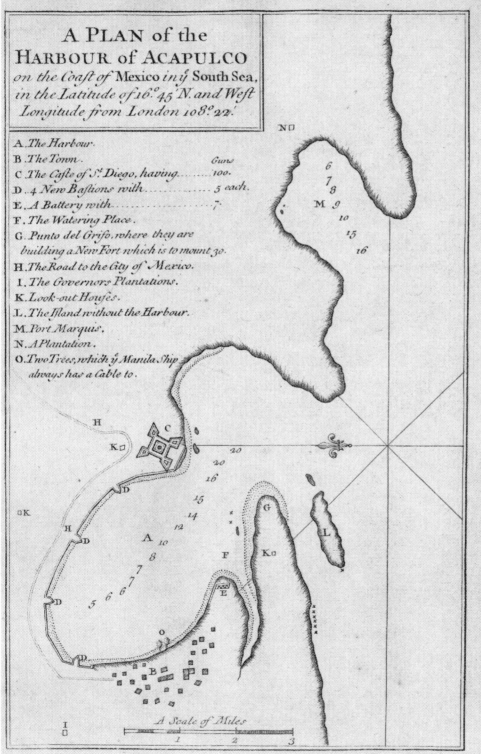

A PLAN of the
HARBOUR of ACAPULCO
on the Coast of Mexico in ȳ South Sea,
in the Latitude of 16.° 45' N. and West
Longitude from London 108.° 22'.

A. The Harbour.
B. The Town.
C. The Castle of St. Diego, having .... Guns 100.
D. 4 New Bastions with .... 5 each.
E. A Battery with .... 7.
F. The Watering Place.
G. Punto del Griso, where they are
   building a New Fort which is to mount 30.
H. The Road to the City of Mexico.
I. The Governors Plantations.
K. Look-out Houses.
L. The Island without the Harbour.
M. Port Marquis.
N. A Plantation.
O. Two Trees, which ȳ Manila Ship
   always has a Cable to.

A Scale of Miles

Lenbobe

Bulo Can

Pampanga

Arangnay
Puerto de Silit
Maquannaye
a las Can
Sabrivaneal
Baton
Silit
Ma...

Caliba Siba

P.ᵗᵒ de Capones

Plate XXXIII.

Ladrone Islands.   *b. Serigan W. by N. diftant 13 Leagues.*

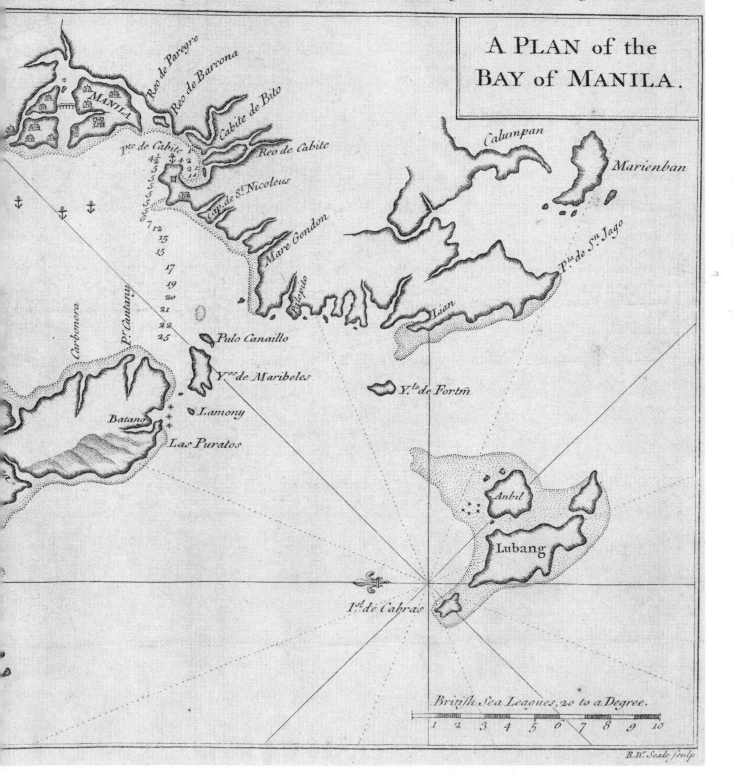

A PLAN of the
BAY of MANILA.

Reo de Paregre

Reo de Barrona

MANILA

Cabite de Bito

Pto de Cabite

Reo de Cabite

4½   4   2
5   2   2
5   2   1
5   2
5

Cap de St Nicoleus

Calumpan

Marienban

6
6
7   12
     13
     15
     17
     19
     20
     21
     22
     25

Mare Gondon

Pto de Sn Jago

Palopulo

Lian

Carbomera

Pt Cantary

Pulo Canaillo

Yas de Maribeles

Yla de Fortñ

Lamony

Batang

Las Puratos

Anbil

Lubang

Ist de Cabras

Britifh Sea Leagues, 20 to a Degree.

1   2   3   4   5   6   7   8   9   10

R.W. Seale fculp

on that spot there was thirty fathom, though not above half a mile from the shore; and that the bank was steep to, and could not be depended on : They farther told us, that they had landed on the Island, but with some difficulty on account of the greatness of the swell; that they found the ground was every where covered with a kind of wild cane, or rush; but that they met with no water, and did not believe the place to be inhabited; though the soil was good, and abounded with groves of coco-nut-trees.

This account of the impossibility of anchoring at this Island occasioned a general melancholy on board; for we considered it as little less than the prelude to our destruction; and our despondency was encreased by a disappointment we met with the succeeding night; for, as we were plying under top-sails, with an intention of getting nearer to the Island, and of sending our boat on shore to load with coco-nuts for the refreshment of our sick, the wind proved squally, and blew so strong off shore, that we were driven so far to the southward, that we dared not to send off our boat. And now the only possible circumstance, that could secure the few which remained alive from perishing, was the accidental falling in with some other of the *Ladrone* Islands, better prepared for our accommodation; and as our knowledge of these Islands was extremely imperfect, we were to trust entirely to chance for our guidance; only as they are all of them usually laid down near the same meridian, and we had conceived those we had already seen to be part of them, we concluded to stand to the southward, as the most probable means of falling in with the next. Thus, with the most gloomy perswasion of our approaching destruction, we stood from the Island of *Anatacan*, having all of us the strongest apprehensions (and those not ill founded) either of dying of the scurvy, or of perishing with the ship, which, for want of hands to work her pumps, might in a short time be expected to founder.

C H A P.

CHAP. II.

Our arrival at *Tinian*, and an account of the Ifland, and of our proceedings there, till the *Centurion* drove out to fea.

IT was the 26th of *Auguft* 1742, in the morning, when we loft fight of *Anatacan*. The next morning we difcovered three other Iflands to the eaftward, which were from ten to fourteen leagues from us. Thefe were, as we afterwards learnt, the Iflands of *Saypan*, *Tinian*, and *Aguigan*. We immediately fteered towards *Tinian*, which was the middlemoft of the three, but had fo much of calms and light airs, that tho' we were helped forwards by the currents, yet next day, at day-break, we were at leaft five leagues diftant from it. However, we kept on our courfe, and about ten in the morning we perceived a proa under fail to the fouthward, between *Tinian* and *Aguigan*. As we imagined from hence that thefe Iflands were inhabited, and knew that the *Spaniards* had always a force at *Guam*, we took the neceffary precautions for our own fecurity, and for preventing the enemy from taking advantage of our prefent wretched circumftances, of which they would be fufficiently informed by the manner of our working the fhip; we therefore muftered all our hands, who were capable of ftanding to their arms, and loaded our upper and quarter-deck guns with grape-fhot; and that we might the more readily procure fome intelligence of the ftate of thefe Iflands, we fhowed *Spanifh* colours, and hoifted a red flag at the fore top-maft-head, to give our fhip the appearance of the *Manila* galeon, hoping thereby to decoy fome of the inhabitants on board us. Thus preparing ourfelves, and ftanding towards the land, we were near enough, at three in the afternoon, to fend the Cutter in fhore, to find out a proper birth for the fhip;

and

and we foon perceived that a proa came off the fhore to meet the
Cutter, fully perfuaded, as we afterwards found, that we were the
Manila fhip. As we faw the Cutter returning back with the proa
in tow, we immediately fent the Pinnace to receive the proa and
the prifoners, and to bring them on board, that the Cutter might proceed on her errand. The Pinnace came back with a *Spaniard* and
four *Indians*, which were the people taken in the proa. The *Spaniard* was immediately examined as to the produce and circumftances of this Ifland of *Tinian*, and his account of it furpaffed
even our moft fanguine hopes; for he informed us that it was uninhabited, which, in our prefent defencelefs condition, was an advantage not to be defpifed, efpecially as it wanted but few of the
conveniencies that could be expected in the moft cultivated country;
for he affured us, that there was great plenty of very good water,
and that there were an incredible number of cattle, hogs, and poultry running wild on the Ifland, all of them excellent in their kind;
that the woods produced fweet and fower oranges, limes, lemons
and coco-nuts in great plenty, befides a fruit peculiar to thefe
Iflands (called by *Dampier*, *Bread-fruit*); that from the quantity and
goodnefs of the provifions produced here, the *Spaniards* at *Guam*
made ufe of it as a ftore for fupplying the garrifon; that he himfelf was a Serjeant of that garrifon, and was fent here with twenty-
two *Indians* to jerk beef, which he was to load for *Guam* on board
a fmall bark of about fifteen tun, which lay at anchor near the
fhore.

This account was received by us with inexpreffible joy: Part of
it we were ourfelves able to verify on the fpot, as we were by this
time near enough to difcover feveral numerous herds of cattle feeding in different places of the Ifland; and we did not any ways
doubt the reft of his relation, as the appearance of the fhore prejudiced us greatly in its favour, and made us hope, that not only
our neceffities might be there fully relieved, and our difeafed recovered, but that, amidft thofe pleafing fcenes which were then in
view, we might procure ourfelves fome amufement and relaxation,

after

after the numerous fatigues we had undergone : For the profpect of the country did by no means refemble that of an uninhabited and uncultivated place, but had much more the air of a magnificent plantation, where large lawns and ftately woods had been laid out together with great fkill, and where the whole had been fo artfully combined, and fo judicioufly adapted to the flopes of the hills, and the inequalities of the ground, as to produce a moft ftriking effect, and to do honour to the invention of the contriver. Thus, (an event not unlike what we had already feen) we were forced upon the moft defirable and falutary meafures by accidents, which at firft fight we confidered as the greateft of misfortunes; for had we not been driven by the contrary winds and currents to the northward of our courfe, (a circumftance, which at that time gave us the moft terrible apprehenfions) we fhould, in all probability, never have arrived at this delightful Ifland, and confequently we fhould have miffed of that place, where alone all our wants could be moft amply relieved, our fick recovered, and our enfeebled crew once more refrefhed, and enabled to put again to fea.

The *Spanifh* Serjeant, from whom we received the account of the Ifland, having informed us that there were fome *Indians* on fhore under his command, employed in jerking beef, and that there was a bark at anchor to take it on board, we were defirous, if poffible, to prevent the *Indians* from efcaping, who doubtlefs would have given the Governor of *Guam* intelligence of our arrival; and we therefore immediately difpatched the Pinnace to fecure the bark, which the Serjeant told us was the only imbarkation on the place; and then, about eight in the evening, we let go our anchor in twenty-two fathom; and though it was almoft calm, and whatever vigour and fpirit was to be found on board was doubtlefs exerted to the utmoft on this pleafing occafion, when, after having kept the fea for fome months, we were going to take poffeffion of this little paradife, yet we were full five hours in furling our fails : It is true, we were fomewhat weakened by the crews of the Cutter and Pinnace which were fent on fhore ; but it is not lefs true, that, includ-

I

ing

ing thofe abfent with the boats and fome Negroe and *Indian* prifoners, all the hands we could mufter capable of ftanding at a gun
amounted to no more than feventy-one, moft of which number too
were incapable of duty ; but on the greateft emergencies this was
all the force we could colle&, in our prefent enfeebled condition,
from the united crews of the *Centurion*, the *Gloucefter*, and the *Tryal*,
which, when we departed from *England*, confifted all together of
near a thoufand hands.

When we had furled our fails, the remaining part of the night
was allowed to our people for their repofe, to recover them from
the fatigue they had undergone; and in the morning a party was
fent on fhore well armed, of which I myfelf was one, to make ourfelves mafters of the landing place, as we were not certain what oppofition might be made by the *Indians* on the Ifland : We landed
without difficulty, for the *Indians* having perceived, by our feizure of
the bark the night before, that we were enemies, they immediately
fled into the woody parts of the Ifland. We found on fhore many
huts which they had inhabited, and which faved us both the time
and trouble of ere&ing tents ; one of thefe huts which the *Indians*
made ufe of for a ftore-houfe was very large, being twenty yards long,
and fifteen broad ; this we immediately cleared of fome bales of jerked beef, which we found in it, and converted it into an hofpital for
our fick, who affoon as the place was ready to receive them were
brought on fhore, being in all a hundred and twenty-eight : Numbers of thefe were fo very helplefs, that we were obliged to carry
them from the boats to the hofpital upon our fhoulders, in which
humane employment (as before at *Juan Fernandes*) the Commodore himfelf, and every one of his officers, were engaged without
diftin&ion ; and, notwithftanding the great debility and the dying
afpe&s of the greateft part of our fick, it is almoft incredible how
foon they began to feel the falutary influence of the land ; for,
though we buried twenty-one men on this and the preceeding day,
yet we did not loofe above ten men more during our whole two
months ftay here ; and in general, our difeafed received fo much

benefit

benefit from the fruits of the Iſland, particularly the fruits of the acid kind, that, in a week's time, there were but few who were not ſo far recovered, as to be able to move about without help.

And now being in ſome ſort eſtabliſhed at this place, we were enabled more particularly to examine its qualities and productions; and that the reader may the better judge of our manner of life here, and future Navigators be better apprized of the conveniencies we met with, I ſhall, before I proceed any farther in the hiſtory of our own adventures, throw together the moſt intereſting particulars that came to our knowledge, in relation to the ſituation, ſoil, produce, and conveniencies of this Iſland of *Tinian*.

This Iſland lies in the latitude of 15° : 8 North, and longitude from *Acapulco* 114° : 50 Weſt. Its length is about twelve miles, and its breadth about half as much ; it extending from the S. S. W. to N. N. E. The ſoil is every where dry and healthy, and ſomewhat ſandy, which being leſs diſpoſed than other ſoils to a rank and over luxuriant vegetation, occaſions the meadows and the bottoms of the woods, to be much neater and ſmoother than is cuſtomary in hot climates. The land riſes by eaſy ſlopes, from the very beach where we watered to the middle of the Iſland ; tho' the general courſe of its aſcent is often interrupted and traverſed by gentle deſcents and vallies ; and the inequalities, that are formed by the different combinations of theſe gradual ſwellings of the ground, are moſt beautifully diverſified with large lawns, which are covered with a very fine trefoil, intermixed with a variety of flowers, and are ſkirted by woods of tall and well-ſpread trees, moſt of them celebrated either for their aſpect or their fruit. The turf of the lawns is quite clean and even, and the bottoms of the woods in many places clear of all buſhes and underwoods ; and the woods themſelves uſually terminate on the lawns with a regular outline, not broken, nor confuſed with ſtraggling trees, but appearing as uniform, as if laid out by art. Hence aroſe a great variety of the moſt elegant and entertaining proſpects, formed by the mixture of theſe woods and lawns, and their various interſections with each other, as

they

they fpread themfelves differently through the vallies, and over the flopes and declivities with which the place abounds. The fortunate animals too, which for the greateft part of the year are the fole lords of this happy foil, partake in fome meafure of the romantic caft of the Ifland, and are no fmall addition to its wonderful fcenary: For the cattle, of which it is not uncommon to fee herds of fome thoufands feeding together in a large meadow, are certainly the moft remarkable in the world; for they are all of them milk-white, except their ears, which are generally black. And though there are no inhabitants here, yet the clamour and frequent parading of domeftic poultry, which range the woods in great numbers, perpetually excite the ideas of the neighbourhood of farms and villages, and greatly contribute to the chearfulnefs and beauty of the place. The cattle on the Ifland we computed were at leaft ten thoufand; and we had no difficulty in getting near them, as they were not fhy of us. Our firft method of killing them was fhooting them; but at laft, when, by accidents to be hereafter recited, we were obliged to hufband our ammunition, our men ran them down with eafe. Their flefh was extremely well tafted, and was believed by us to be much more eafily digefted, than any we had ever met with. The fowls too were exceeding good, and were likewife run down with little trouble; for they could fcarce fly further than an hundred yards at a flight, and even that fatigued them fo much, that they could not readily rife again; fo that, aided by the opennefs of the woods, we could at all times furnifh ourfelves with whatever number we wanted. Befides the cattle and the poultry, we found here abundance of wild hogs: Thefe were moft excellent food; but as they were a very fierce animal, we were obliged either to fhoot them, or to hunt them with large dogs, which we found upon the place at our landing, and which belonged to the detachment which was then upon the Ifland amaffing provifions for the garrifon of *Guam*. As thefe dogs had been purpofely trained to the killing of the wild hogs, they followed us very readily, and hunted for us; but though they were a large bold breed, the

2

hogs

hogs fought with fo much fury, that they frequently deftroyed them, fo that we by degrees loft the greateft part of them.

But this place was not only extremely grateful to us from the plenty and excellency of its frefh provifions, but was as much perhaps to be admired for its fruits and vegetable productions, which were moft fortunately adapted to the cure of the fea fcurvy, which had fo terribly reduced us For in the woods there were inconceivable quantities of coco-nuts, with the cabbages growing on the fame tree : There were befides guavoes, limes, fweet and fower oranges, and a kind of fruit, peculiar to thefe Iflands, called by the *Indians Rima*, but by us the *Bread-Fruit*, for it was conftantly eaten by us during our ftay upon the Ifland inftead of bread, and fo univerfally preferred to it, that no fhip's bread was expended during that whole interval.   It grew upon a tree which is fomewhat lofty, and which, towards the top, divides into large and fpreading branches. The leaves of this tree are of a remarkable deep green, are notched about the edges, and are generally from a foot to eighteen inches in length. The fruit itfelf grows indifferently on all parts of the branches ; it is in fhape rather eliptical than round, is covered with a rough rind, and is ufually feven or eight inches long ; each of them grows fingly and not in clufters. This fruit is fitteft to be ufed, when it is full grown, but is ftill green ; in which ftate, its tafte has fome diftant refemblance to that of an artichoke bottom, and its texture is not very different, for it is foft and fpungy. As it ripens it grows fofter and of a yellow colour, and then contracts a lufcious tafte, and an agreeable fmell, not unlike a ripe peach ; but then it is efteemed unwholefome, and is faid to produce fluxes. In the annexed view of the watering place, there is drawn one of the trees bearing this fruit, it being that marked with the letter (*c*). Befides the fruits already enumerated, there were many other vegetables extremely conducive to the cure of the malady we had long laboured under, fuch as water-melons, dandelion, creeping purflain, mint, fcurvy-grafs, and forrel ; all which, together with the frefh meats of the place, we devoured with great eagernefs, prompted
ed

A View of the Wat

Plate XXXIV.

ng Place at TENIAN.

ted thereto by the ſtrong inclination, which nature never fails of exciting in ſcorbutic diſorders for theſe powerful ſpecifics.

It will eaſily be conceived from what hath been already ſaid, that our cheer upon this Iſland was in ſome degree luxurious, but I have not yet recited all the varieties of proviſion which we here indulged in. Indeed we thought it prudent totally to abſtain from fiſh, the few we caught at our firſt arrival having ſurfeited thoſe who eat of them; but conſidering how much we had been inured to that ſpecies of food, we did not regard this circumſtance as a diſadvantage, eſpecially as the defect was ſo amply ſupplied by the beef, pork and fowls already mentioned, and by great plenty of wild fowl; for I muſt obſerve, that near the center of the Iſland there were two conſiderable pieces of freſh water, which abounded with duck, teal and curlew: Not to mention the whiſtling plover, which we found there in prodigious plenty.

And now perhaps it may be wondered at, that an Iſland, ſo exquiſitely furniſhed with the conveniencies of life, and ſo well adapted, not only to the ſubſiſtence, but likewiſe to the enjoyment of mankind, ſhould be entirely deſtitute of inhabitants, eſpecially as it is in the neighbourhood of other Iſlands, which in ſome meaſure depend upon this for their ſupport. To obviate this difficulty, I muſt obſerve, that it is not fifty years ſince the Iſland was depopulated. The *Indians* we had in our cuſtody aſſured us, that formerly the three Iſlands of *Tinian*, *Rota* and *Guam*, were all full of inhabitants; and that *Tinian* alone contained thirty thouſand ſouls: But a ſickneſs raging amongſt theſe Iſlands, which deſtroyed multitudes of the people, the *Spaniards*, to recruit their numbers at *Guam*, which were greatly diminiſhed by this mortality, ordered all the inhabitants of *Tinian* thither; where, languiſhing for their former habitations, and their cuſtomary method of life, the greateſt part of them in a few years died of grief. Indeed, independent of that attachment which all mankind have ever ſhown to the places of their birth and bringing up, it ſhould ſeem, from what has been already

ready

ready faid, that there were few countries more worthy to be regretted than this of *Tinian*.

Thefe poor *Indians* might reafonably have expected, at the great diftance from *Spain*, where they were placed, to have efcaped the violence and cruelty of that haughty Nation, fo fatal to a large proportion of the whole human race : But it feems their remote fituation could not protect them from fharing in the common deftruction of the weftern world, all the advantage they received from their diftance being only to perifh an age or two later. It may perhaps be doubted, if the number of the inhabitants of *Tinian*, who were banifhed to *Guam*, and who died there pining for their native home, was fo great, as what we have related above; but, not to mention the concurrent affertion of our prifoners, and the commodioufnefs of the Ifland, and its great fertility, there are ftill remains to be met with on the place, which evince it to have been once extremely populous : For there are, in all parts of the Ifland, a great number of ruins of a very particular kind ; they ufually confift of two rows of fquare pyramidal pillars, each pillar being about fix feet from the next, and the diftance between the rows being about twelve feet; the pillars themfelves are about five feet fquare at the bafe, and about thirteen feet high ; and on the top of each of them there is a femi-globe, with the flat part upwards ; the whole of the pillars and femi-globe is folid, being compofed of fand and ftone cemented together, and plaiftered over. This odd fabrick will be better underftood, by infpecting the view of the watering place inferted above, where an affemblage of thefe pillars is drawn, and is denoted by the letter (*a*). If the account our prifoners gave us of thefe ftructures was true, the Ifland muft indeed have been extremely populous ; for they affured us, that they were the foundations of particular buildings fet apart for thofe *Indians* only, who had engaged in fome religious vow ; and monaftic inftitutions are often to be met with in many Pagan nations. However, if thefe ruins were originally the bafis of the common

mon

mon dwelling-houfes of the natives, their numbers muſt have been confiderable; for in many parts of the Iſland they are extremely thick planted, and ſufficiently evince the great plenty of former inhabitants. But to return to the preſent ſtate of the Iſland.

Having mentioned the conveniencies of this place, the excellency and quantity of its fruits and proviſions, the neatneſs of its lawns, the ſtatelineſs, freſhneſs and fragance of its woods, the happy inequality of its ſurface, and the variety and elegance of the views it afforded, I muſt now obſerve that all theſe advantages were greatly enhanced by the healthineſs of its climate, by the almoſt conſtant breezes which prevail there, and by the frequent ſhowers which fall, and which, though of a very ſhort and almoſt momentary duration, are extremely grateful and refreſhing, and are perhaps one cauſe of the ſalubrity of the air, and of the extraordinary influence it was obſerved to have upon us in increaſing and invigorating our appetites and digeſtion. This was ſo remarkable, that thoſe amongſt our officers, who were at all other times ſpare and temperate eaters, who, beſides a ſlight breakfaſt, made but one moderate repaſt a day, were here, in appearance, transformed into gluttons; for inſtead of one reaſonable fleſh-meal, they were now ſcarcely ſatisfied with three, and each of them ſo prodigious in quantity, as would at another time have produced a fever or a ſurfeit: And yet our digeſtion ſo well correſponded with the keeneſs of our appetites, that we were neither diſordered nor even loaded by this repletion; for after having, according to the cuſtom of the Iſland, made a large beef breakfaſt, it was not long before we began to conſider the approach of dinner as a very deſirable, though ſomewhat tardy incident.

And now having been thus large in my encomiums on this Iſland, in which however, I conceive, I have not done it juſtice, it is neceſſary I ſhould ſpeak of thoſe circumſtances in which it is defective, whether in point of beauty or utility.

And

And firft, with refpect to its water. I muft own, that before I had feen this fpot, I did not conceive that the abfence of running water, of which it is entirely deftitute, could have been fo well re-placed by any other means, as it is in this Ifland; for though there are no ftreams, yet the water of the wells and fprings, which are to be met with every where near the furface, is extremely good; and in the midft of the Ifland there are two or three confiderable pieces of excellent water, whofe edges are as neat and even, as if they had been bafons purpofely made for the decoration of the place. It muft however be confeffed, that with regard to the beauty of the profpects, the want of rills and ftreams is a very great defect, not to be compenfated either by large pieces of ftanding water, or by the neighbourhood of the fea, though that, by reafon of the fmall-nefs of the Ifland, generally makes a part of every extenfive view.

As to the refidence upon the Ifland, the principal inconvenience attending it is the vaft numbers of mufcatos, and various other fpecies of flies, together with an infect called a tick, which, though principally attached to the cattle, would yet frequently faften upon our limbs and bodies, and if not perceived and removed in time, would bury its head under the fkin, and raife a painful inflamma-tion. We found here too centipedes and fcorpions, which we fup-pofed were venemous, but none of us ever received any injury from them.

But the moft important and formidable exception to this place re-mains ftill to be told. This is the inconvenience of the road, and the little fecurity there is at fome feafons for a fhip at anchor. The only proper anchoring place for fhips of burthen is at the S. W. end of the Ifland. As a direction for readily finding it, there is an-nexed a very accurate view of the S. W. fide of the Ifland, where (*a*) is the peak of *Saypan*, feen over the northern part of *Tinian*, and bearing N. N. E. $\frac{1}{4}$ E. And (*b*) is the anchoring place, di-ftant eight miles from the obferver. And as an additional affift-

ance,

a

A View of the I.

Side of TENIAN.

G:Muller sculp.

A view of the anchoring place at T

Plate XXXVI.

...AN *where the* CENTURION *water'd.*

Js. Muller Sculp.

ance, there is alfo added a near view of the anchoring place itfelf, which reprefents it fo exactly, that none hereafter can poffible miftake it.   In this place the *Centurion* anchored in twenty and twenty-two fathom water, oppofite to a fandy bay, and about a mile and an half diftant from the fhore.   The bottom of this road is full of fharp-pointed coral rocks, which, during four months of the year, that is, from the middle of *June* to the middle of *October*, renders it a very unfafe place to lie at.   This is the feafon of the weftern monfoons, when near the full and change of the moon, but more particularly at the change, the wind is ufually variable all round the compafs, and feldom fails to blow with fuch fury, that the ftouteft cables are not to be confided in ; what adds to the danger at thefe times, is the exceffive rapidity of the tide of flood which fets to the S. E, between this Ifland and that of *Aguiguan*, a fmall Ifland near the fouthern extremity of *Tinian*, which is reprefented in the general chart, hereafter inferted, only by a dot.   This tide runs at firft with a vaft head and overfall of water, and occafions fuch a hollow and overgrown fea, as is fcarcely to be conceived ; fo that (as will be hereafter more particularly mentioned) we were under the dreadful apprehenfion of being pooped by it, though we were in a fixty-gun fhip.   In the remaining eight months of the year, that is, from the middle of *October* to the middle of *June*, there is a conftant feafon of fettled weather, when, if the cables are but well armed, there is fcarcely any danger of their being fo much as rubbed : So that during all that interval, it is as fecure a road as could be wifhed for.   I fhall only add, that the anchoring bank is very fhelving, and ftretches along the S. W. end of the Ifland, and that it is entirely free from fhoals, except a reef of rocks which is vifible, and lies about half a mile from the fhore, and affords a narrow paffage into a fmall fandy bay, which is the only place where boats can poffibly land.   After this account of the Ifland, and its produce, it is neceffary to return to our own hiftory.

Our firft undertaking, after our arrival, was the removal of our fick on fhore, as hath been mentioned. Whilft we were thus em-

ployed,

ployed, four of the *Indians* on shore, being part of the *Spanish*
Serjeant's detachment, came and surrendered themselves to us, so
that with those we took in the proa, we had now eight of them
in our custody. One of the four who submitted undertook to show
us the most convenient place for killing cattle, and two of our
men were ordered to attend him on that service; but one of them
unwarily trusting the *Indian* with his firelock and pistol, the
Indian escaped with them into the woods: His countrymen, who
remained behind, were apprehensive of suffering for this perfidy
of their comrade, and therefore begged leave to send one of their
own party into the country, who they engaged should both bring
back the arms, and persuade the whole detachment from *Guam* to
submit to us. The Commodore granted their request; and one of
them was dispatched on this errand, who returned next day, and
brought back the firelock and pistol, but assured us, he had met
with them in a path way in the wood, and protested that he had
not been able to meet with any one of his countrymen: This report had so little the air of truth, that we suspected there was some
treachery carrying on, and therefore to prevent any future communication amongst them, we immediately ordered all the *Indians* who
were in our power on board the ship, and did not permit them to
return any more on shore.

When our sick were well settled on the Island, we employed
all the hands that could be spared from attending them, in arming
the cables with a good rounding, several fathom from the anchor,
to secure them from being rubbed by the coral rocks, which here
abounded: And this being compleated, our next attention was our
leak, and in order to raise it out of water, we, on the first of
September, began to get the guns aft to bring the ship by the stern;
and now the Carpenters, being able to come at it on the outside,
ripped of the old sheathing that was left, and caulked all the
seams on both sides the cut-water, and leaded them over, and then
new sheathed the bows to the surface of the water: By this means
we conceived the defect was sufficiently secured; but upon our be-

2   ginning

ginning to bring the guns into their places, we had the mortification to perceive, that the water rushed into the ship in the old
place, with as much violence as ever : Hereupon we were necessitated to begin again ; and that our second attempt might be more effectual, we cleared the fore store-room, and sent a hundred and
thirty barrels of powder on board the small *Spanish* bark we had
seized here, by which means we raised the ship about three feet out
of the water forwards, and the Carpenters ripped of the sheathing
lower down, and new caulked all the seams, and afterwards laid on
new sheathing; and then, supposing the leak to be effectually
stopped, we began to move the guns forwards; but the upper deck
guns were scarcely in their places, when, to our amazement, it
burst out again; and now, as we durst not cut away the lining
within board, least a but-end or a plank might start, and we might
go down immediately, we had no other resource left than chincing
and caulking within board; and indeed by this means the leak was
stopped for some time ; but when our guns were all in their places,
and our stores were taken on board, the water again forced its way
through a hole in the stem, where one of the bolts was driven in ;
and on this we desisted from all farther efforts, being now well af-
ssured, that the defect was in the stem itself, and that it was not to be
remedied till we should have an an opportunity of heaving down.

Towards the middle of *September*, several of our sick were tolerably recovered by their residence on shore; and, on the 12th of
September, all those who were so far relieved, since their arrival, as
to be capable of doing duty, were sent on board the ship: And
then the Commodore, who was himself ill of the scurvy, had a
tent erected for him on shore, where he went with the view of
staying a few days for the recovery of his health, being convinced
by the general experience of his people, that no other method but
living on the land was to be trusted to for the removal of this
dreadful malady. The place, where his tent was pitched on this
occasion, was near the well, whence we got all our water, and was
indeed a most elegant spot. A view of it hath been already inserted

under

under the title of the watering place, where (b) is the Commodore's tent, and (d) the well where we watered.

As the crew on board were now reinforced by the recovered hands returned from the Island, we began to send our cask on shore to be fitted up, which till now could not be done, for the Coopers were not well enough to work. We likewise weighed our anchors, that we might examine our cables, which we suspected had by this time received considerable damage. And as the new moon was now approaching, when we apprehended violent gales, the Commodore, for our greater security, ordered that part of the cables next to the anchors to be armed with the chains of the fire-grapnels; and they were besides cackled twenty fathom from the anchors, and seven fathom from the service, with a good rounding of a 4 ½ inch hawser; and to all these precautions we added that of lowering the main and fore-yard close down, that in case of blowing weather the wind might have less power upon the ship, to make her ride a strain.

Thus effectually prepared, as we conceived, we expected the new moon, which was the 18th of *September*, and riding safe that and the three succeeding days, (though the weather proved very squally and uncertain) we flattered ourselves (for I was then on board) that the prudence of our measures had secured us from all accidents; but, on the 22d, the wind blew from the eastward with such fury, that we soon despaired of riding out the storm; and therefore we should have been extremely glad that the Commodore and the rest of our people on shore, which were the greatest part of our hands, had been on board with us, since our only hopes of safety seemed to depend on our putting immediately to sea; but all communication with the shore was now effectually cut off, for there was no possibility that a boat could live, so that we were necessitated to ride it out, till our cables parted. Indeed it was not long before this happened, for the small bower parted at five in the afternoon, and the ship swung off to the best bower; and as the night came on, the violence of the wind still encreased; but notwithstanding its inexpres-
sible

fible fury, the tide ran with fo much rapidity, as to prevail over it ; for the tide having fet to the northward in the beginning of the ftorm, turned fuddenly to the fouthward about fix in the evening, and forced the fhip before it in defpight of the ftorm, which blew upon the beam : And now the fea broke moft furprizingly all round us, and a large tumbling fwell threatened to poop us ; the long boat, which was at this time moored a-ftern, was on a fudden canted fo high, that it broke the tranfom of the Commodore's gallery, whofe cabin was on the quarter-deck, and would doubtlefs have rifen as high as the tafferel, had it not been for this ftroke which ftove the boat all to pieces ; but the poor boat-keeper, though extremely bruifed, was faved almoft by miracle. About eight, the tide flackened, but the wind did not abate ; fo that at eleven, the beft bower cable, by which alone we rode, parted. Our fheet anchor, which was the only one we had left, was inftantly cut from the bow ; but before it could reach the bottom, we were driven from twenty-two into thirty-five fathom ; and after we had veered away one whole cable, and two thirds of another, we could not find ground with fixty fathom of line : This was a plain indication, that the anchor lay near the edge of the bank, and could not hold us long. In this preffing danger, Mr. *Saumarez*, our firft Lieutenant, who now commanded on board, ordered feveral guns to be fired, and lights to be fhown, as a fignal to the Commodore of our diftrefs ; and in a fhort time after, it being then about one o'clock, and the night exceffively dark, a ftrong guft, attended with rain and lightning, drove us off the bank, and forced us out to fea, leaving behind us, on the Ifland, Mr. *Anfon*, with many more of our officers, and great part of our crew, amounting in the whole to an hundred and thirteen perfons. Thus were we all, both at fea and on fhore, reduced to the utmoft defpair by this cataftrophe, thofe on fhore conceiving they had no means left them ever to leave the Ifland, and we on board utterly unprepared to ftruggle with the fury of the feas and winds, we were now expofed to, and expecting each moment to be our laft.

CHAP.

CHAP. III.

Tranfactions at *Tinian* after the departure of the *Centurion*.

THE ftorm, which drove the *Centurion* to fea, blew with too much turbulence to permit either the Commodore or any of the people on fhore from hearing the guns, which fhe fired as fignals of diftrefs; and the frequent glare of the light-ning had prevented the explofions from being obferved : So that, when at day-break it was perceived from the fhore that the fhip was miffing, there was the utmoft confternation amongft them : For much the greateft part of them immediately concluded that fhe was loft, and intreated the Commodore that the boat might be fent round the Ifland to look for the wreck; and thofe who believed her fafe, had fcarcely any expectation that fhe would ever be able to make the Ifland again : For the wind continued to blow ftrong at Eaft, and they knew how poorly fhe was manned and provided for ftruggling with fo tempeftuous a gale. And if the *Centurion* was loft, or fhould be incapable of returning, there appeared in either cafe no poffibility of their ever getting off the Ifland : For they were at leaft fix hundred leagues from *Macao*, which was their neareft port; and they were mafters of no other veffel than the fmall *Spanifh* bark, of about fifteen tun, which they feized at their firft arrival, and which would not even hold a fourth part of their number : And the chance of their being taken off the Ifland by the cafual arrival of any other fhip was altogether defperate; as perhaps no *European* fhip had ever anchored here before, and it were madnefs to expect that like incidents fhould fend another here in an hundred ages to come : So that their defponding thoughts could only fuggeft to them the melancholy profpect of fpending the

remainder

remainder of their days on this Island, and bidding adieu for ever to their country, their friends, their families, and all their domeſtic endearments.

Nor was this the worſt they had to fear : For they had reaſon to expect, that the Governor of *Guam*, when he ſhould be informed of their ſituation, might ſend a force ſufficient to overpower them, and to remove them to that Iſland; and then, the moſt favourable treatment they could hope for would be to be detained priſoners for life; ſince, from the known policy and cruelty of the *Spaniards* in their diſtant ſettlements, it was rather to be expected, that the Governor, if he once had them in his power, would make their want of commiſſions (all of them being on board the *Centurion*) a pretext for treating them as pirates, and for depriving them of their lives with infamy.

In the midſt of theſe gloomy reflections, Mr. *Anſon* had doubtleſs his ſhare of diſquietude; but he always kept up his uſual compoſure and ſteadineſs: And having ſoon projected a ſcheme for extricating himſelf and his men from their preſent anxious ſituation, he firſt communicated it to ſome of the moſt intelligent perſons about him; and having ſatisfied himſelf that it was practicable, he then endeavoured to animate his people to a ſpeedy and vigorous proſecution of it. With this view he repreſented to them, how little foundation there was for their apprehenſions of the *Centurion's* being loſt: That he ſhould have hoped, they had been all of them better acquainted with ſea-affairs, than to give way to the impreſſion of ſo chimerical a fright; and that he doubted not, but if they would ſeriouſly conſider what ſuch a ſhip was capable of enduring, they would confeſs that there was not the leaſt probability of her having periſhed: That he was not without hopes that ſhe might return in a few days; but if ſhe did not, the worſt that could be ſuppoſed was, that ſhe was driven ſo far to the leeward of the Iſland that ſhe could not regain it, and that ſhe would conſequently be obliged to bear away for *Macao* on the coaſt of *China* : That as it was neceſſary to be prepared againſt all events, he had, in this

T t

caſe,

cafe, confidered of a method of carrying them off the Ifland, and joining their old fhip the *Centurion* again at *Macao :* That this method was to hale the *Spanifh* bark on fhore, to faw her afunder, and to lengthen her twelve feet, which would enlarge her to near forty tun burthen, and would enable her to carry them all to *China :* That he had confulted the Carpenters, and they had agreed that this propofal was very feazible, and that nothing was wanting to execute it, but the united refolution and induftry of the whole body : He added, that for his own part he would fhare the fatigue and labour with them, and would expect no more from any man than what he, the Commodore himfelf, was ready to fubmit to ; and concluded with reprefenting to them the importance of faving time ; and that, in order to be the better prepared for all events, it was neceffary to fet to work immediately, and to take it for granted, that the *Centurion* would not be able to put back (which was indeed the Commodore's fecret opinion) ; fince if fhe did return, they fhould only throw away a few days application ; but if fhe did not, their fituation, and the feafon of the year, required their utmoft difpatch.

Thefe remonftrances, though not without effect, did not immediately operate fo powerfully as Mr. *Anfon* could have wifhed : He indeed raifed their fpirits, by fhowing them the poffibility of their getting away, of which they had before defpaired ; but then, from their confidence of this refource, they grew lefs apprehenfive of their fituation, gave a greater fcope to their hopes, and flattered themfelves that the *Centurion* would return and prevent the execution of the Commodore's fcheme, which they could eafily forefee would be a work of confiderable labour : By this means, it was fome days before they were all of them heartily engaged in the project ; but at laft, being in general convinced of the impoffibility of the fhip's return, they fet themfelves zealoufly to the different tafks allotted them, and were as induftrious and as eager as their Commander could defire, punctually affembling at day-break at the rendezvous, whence they were diftributed to their different employments,

ployments, which they followed with unusual vigour till night came on.

And here I must interrupt the course of this transaction for a moment, to relate an incident which for some time gave Mr. *Anson* more concern than all the preceding disasters. A few days after the ship was driven off, some of the people on shore cried out, *a sail*. This spread a general joy, every one supposing that it was the ship returning; but presently, a second sail was descried, which quite destroyed their first conjecture, and made it difficult to guess what they were. The Commodore eagerly turned his glass towards them, and saw they were two boats; on which it immediately occurred to him, that the *Centurion* was gone to the bottom, and that these were her two boats coming back with the remains of her people; and this sudden and unexpected suggestion wrought on him so powerfully, that, to conceal his emotion, he was obliged (without speaking to any one) instantly to retire to his tent, where he past some bitter moments, in the firm belief that the ship was lost, and that now all his views of farther distressing the enemy, and of still signalizing his expedition by some important exploit, were at an end.

But he was soon relieved from these disturbing thoughts, by discovering that the two boats in the offing were *Indian* proas; and perceiving that they stood towards the shore, he directed every appearance that could give them any suspicion to be removed, and concealed his people, in the adjacent thickets, prepared to secure the *Indians* when they should land: But, after the proas had stood in within a quarter of a mile of the land, they suddenly stopt short, and remaining there motionless for near two hours, they then made sail again, and stood to the southward. But to return to the projected enlargement of the bark.

If we examine how they were prepared for going through with this undertaking, on which their safety depended, we shall find, that, independent of other matters which were of as much importance, the lengthning of the bark alone was attended with great difficul-

ty.

ty. Indeed, in a proper place, where all the neceſſary materials and tools were to be had, the embarraſment would have been much leſs; but ſome of theſe tools were to be made, and many of the materials were wanting; and it required no ſmall degree of invention to ſupply all theſe deficiencies. And when the hull of the bark ſhould be compleated, this was but one article; and there were many others of equal weight, which were to be well conſidered: Theſe were the rigging it, the victualling it, and laſtly, the navigating it, for the ſpace of ſix or ſeven hundred leagues, thro' unknown ſeas, where no one of the company had ever paſſed before. In ſome of theſe particulars ſuch obſtacles occurred, that, without the intervention of very extraordinary and unexpected accidents, the poſſibility of the whole enterprize would have fallen to the ground, and their utmoſt induſtry and efforts muſt have been fruitleſs. Of all theſe circumſtances I ſhall make a ſhort recital.

It fortunately happened that the Carpenters, both of the *Glouceſter* and of the *Tryal*, with their cheſts of tools, were on ſhore when the ſhip drove out to ſea; the Smith too was on ſhore, and had with him his forge and ſome tools, but unhappily his bellows had not been brought from on board; ſo that he was incapable of working, and without his aſſiſtance they could not hope to proceed with their deſign: Their firſt attention therefore was to make him a pair of bellows, but in this they were for ſome time puzzled, by their want of leather; however, as they had hides in ſufficient plenty, and they had found a hogſhead of lime, which the *Indians* or *Spaniards* had prepared for their own uſe, they tanned ſome hides with this lime; and though we may ſuppoſe the workmanſhip to be but indifferent, yet the leather they thus made ſerved tolerably well, and the bellows (to which a gun-barrel ſerved for a pipe) had no other inconvenience, than that of being ſomewhat ſtrong ſcented from the imperfection of the Tanner's work.

Whilſt

Whilft the Smith was preparing the neceffary iron-work, others were employed in cutting down trees, and fawing them into planks; and this being the moft laborious tafk, the Commodore wrought at it himfelf for the encouragement of his people. As there were neither blocks nor cordage fufficient for tackles to hale the bark on fhore, it was propofed to get her up on rollers; and for thefe, the body of the coco-nut tree was extremely ufeful; for its fmoothnefs and circular turn prevented much labour, and fitted it for the purpofe with very little workmanfhip: A number of thefe trees were therefore felled, and the ends of them properly opened for the reception of hand-fpikes; and in the mean time a dry dock was dug for the bark, and ways laid from thence quite into the fea, to facilitate the bringing her up. And befides thofe who were thus occupied in preparing meafures for the future enlargement of the bark, a party was conftantly ordered for the killing and preparing of provifions for the reft: And tho' in thefe various employments, fome of which demanded confiderable dexterity, it might have been expected there would have been great confufion and delay; yet good order being once eftablifhed, and all hands engaged, their preparations advanced apace. Indeed, the common men, I prefume, were not the lefs tractable for their want of fpirituous liquors : For, there being neither wine nor brandy on fhore, the juice of the coco-nut was their conftant drink, and this, though extremely pleafant, was not at all intoxicating, but kept them very cool and orderly.

And now the officers began to confider of all the articles neceffary for the fitting out the bark; when it was found, that the tents on fhore, and the fpare cordage accidentally left there by the *Centurion*, together with the fails and rigging already belonging to the bark, would ferve to rig her indifferently well, when fhe was lengthened : And as they had tallow in plenty, they propofed to pay her bottom with a mixture of tallow and lime, which it was known was well adapted to that purpofe: So that with refpect to her equipment, fhe would not have been very defective. There was, however, one

exception,

exception, which would have proved extremely inconvenient, and that was her fize: For as they could not make her quite forty tun burthen, fhe would have been incapable of containing half the crew below the deck, and fhe would have been fo top-heavy, that if they were all at the fame time ordered upon deck, there would be no fmall hazard of her over-fetting; but this was a difficulty not to be removed, as they could not augment her beyond the fize already propofed. After the manner of rigging and fitting up the bark was confidered and regulated, the next effential point to be thought on was, how to procure a fufficient ftock of provifions for their voyage; and here they were greatly at a lofs what courfe to take; for they had neither grain nor bread of any kind on fhore, their bread fruit, which would not keep at fea, having all along fupplied its place: And though they had live cattle enough, yet they had no falt to cure beef for a fea-ftore, nor would meat take falt in that climate. Indeed, they had preferved a fmall quantity of jerked beef, which they found upon the place at their landing; but this was greatly difproportioned to the run of near fix hundred leagues, which they were to engage in, and to the number of hands they fhould have on board. It was at laft, however, refolved to take on board as many coco-nuts as they poffibly could; to make the moft of their jerked beef, by a very fparing diftribution of it; and to endeavour to fupply their want of bread by rice; to furnifh themfelves with which, it was propofed, when the bark was fitted up, to make an expedition to the Ifland of *Rota*, where they were told, that the *Spaniards* had large plantations of rice under the care of the *Indian* inhabitants: But as this laft meafure was to be executed by force, it became neceffary to examine what ammunition had been left on fhore, and to preferve it carefully; and on this enquiry, they had the mortification to find, that the utmoft that could be collected, by the ftricteft fearch, did not amount to more than ninety charges of powder for their firelocks, which was confiderably fhort of one a-piece for each of the company, and was indeed a very flender ftock of ammunition, for fuch as were to

I

eat

eat no grain or bread for a month, but what they were to procure by force of arms.

But the moſt alarming circumſtance, and what, without the providential interpoſition of very improbable events, had rendered all their ſchemes abortive, remains yet to be related. The general idea of the fabric and equipment of the veſſel was ſettled in a few days; and when this was done, it was not difficult to make ſome eſtimation of the time neceſſary to compleat her. After this, it was natural to expect that the officers would conſider on the courſe they were to ſteer, and the land they were to make. Theſe reflections led them to the diſheartning diſcovery, that there was neither compaſs nor quadrant on the Iſland. Indeed the Commodore had brought a pocket-compaſs on ſhore for his own uſe; but Lieutenant *Brett* had borrowed it to determine the poſition of the neighbouring Iſlands, and he had been driven to ſea in the *Centurion*, without returning it: And as to a quadrant, that could not be expected to be found on ſhore, for as it was of no uſe at land, there could be no reaſon for bringing it from on board the ſhip. It was eight days, from the departure of the *Centurion*, before they were in any degree relieved from this terrible perplexity: At laſt, in rumaging a cheſt belonging to the *Spaniſh* bark, they found a ſmall compaſs, which, though little better than the toys uſually made for the amuſement of ſchool-boys, was to them an invaluable treaſure. And a few days after, by a ſimilar piece of good fortune, they found a quadrant on the ſea-ſhore, which had been thrown overboard amongſt other lumber belonging to the dead: The quadrant was eagerly ſeized, but on examination, it unluckily wanted vanes, and therefore in its preſent ſtate was altogether uſeleſs; however, fortune ſtill continuing in a favourable mood, it was not long before a perſon out of curioſity pulling out the drawer of an old table, which had been driven on ſhore, found therein ſome vanes, which fitted the quadrant very well; and it being thus compleated, it was examined by the known latitude of the place, and was found to anſwer to a ſufficient degree of exactneſs.

And

And now, all these obstacles being in some degree removed, (which were always as much as possible concealed from the vulgar, that they might not grow remiss with the apprehension of labouring to no purpose) the work proceeded very successfully and vigorously: The necessary iron-work was in great forwardness; and the timbers and planks (which, though not the most exquisite performances of the Sawyer's art, were yet sufficient for the purpose) were all prepared; so that, on the 6th of *October*, being the 14th day from the departure of the ship, they haled the bark on shore, and, on the two succeeding days she was sawn asunder, (though with great care not to cut her planks) and her two parts were separated the proper distance from each other, and, the materials being all ready before hand, they, the next day, being the 9th of *October*, went on with great dispatch in their proposed enlargement of her; and by this time they had all their future operations so fairly in view, and were so much masters of them, that they were able to determine when the whole would be finished, and had accordingly, fixed the 5th of *November* for the day of their putting to sea. But their projects and labours were now drawing to a speedier and happier conclusion; for on the 11th of *October*, in the afternoon, one of the *Gloucester's* men, being upon a hill in the middle of the Island, perceived the *Centurion* at a distance, and running down with his utmost speed towards the landing place, he, in the way, saw some of his comrades, to whom he hollowed out with great extafy, *The ship, the ship*. This being heard by Mr. *Gordon*, a Lieutenant of marines, who was convinced by the fellow's transport that his report was true, Mr. *Gordon* ran towards the place where the Commodore and his people were at work, and being fresh and in breath, easily outstripped the *Gloucester's* man, and got before him to the Commodore, who, on hearing this happy and unexpected news, threw down his axe with which he was then at work, and by his joy broke through, for the first time, the equable and unvaried character which he had hitherto preserved; the others, who were with him, instantly ran down to the sea-side in a kind of frenzy, eager to feast themselves

with

with a fight they had fo ardently wifhed for, and of which they had now for a confiderable time defpaired. By five in the evening, the *Centurion* was vifible in the offing to them all ; and, a boat being fent off with eighteen men to reinforce her, and with frefh meat and fruits for the refrefhment of her crew, fhe, the next afternoon, happily came to an anchor in the road, where the Commodore immediately came on board her, and was received by us with the fincereft and heartieft acclamations : For, from the following fhort recital of the fears, the dangers and fatigues we in the fhip underwent, during our nineteen days abfence from *Tinian*, it may be eafily conceived, that a harbour, refrefhments, repofe, and the joining of our Commander and Shipmates, were not lefs pleafing to us, than our return was to them.

U u

CHAP.

CHAP. IV.

Proceedings on board the *Centurion*, when driven out to fea.

THE *Centurion* being now once more fafely arrived at *Tinian*, to the mutual refpite of the labours of our divided crew, it is high time that the reader, after the relation already given of the projects and employment of thofe left on fhore, fhould be apprized of the fatigues and diftreffes, to which we, who were driven off to fea, were expofed during the long interval of nineteen days that we were abfent from the Ifland.

It has been already mentioned, that it was the 22d of *September*, about one o'clock, in an extreme dark night, when by the united violence of a prodigious ftorm, and an exceeding rapid tide, we were driven from our anchors, and forced to fea. Our condition then was truly deplorable; we were in a leaky fhip, with three cables in our hawfes, to one of which hung our only remaining anchor; we had not a gun on board lafhed, nor a port barred in; our fhrowds were loofe, and our top-mafts unrigged, and we had ftruck our fore and main-yards clofe down, before the ftorm came on, fo that there were no fails we could fet, except our mizen. In this dreadful extremity we could mufter no more ftrength on board, to navigate the fhip, than an hundred and eight hands, feveral Negroes and *Indians* included: This was fcarcely the fourth part of our complement; and of thefe the greater number were either boys, or fuch as, being lately recovered from the fcurvy, had not yet arrived at half their former vigour. No fooner were we at fea, but by the violence of the ftorm, and the working of the fhip, we made a great quantity of water through our hawfe-holes, ports and fcuppers, which, added to the conftant effect of our leak, rendered

our

our pumps alone a fufficient employment for us all : But though this leakage, by being a fhort time neglected, would inevitably end in our deftruction.; yet we had other dangers then impending, which occafioned this to be regarded as a fecondary confideration only. For we all imagined, that we were driving directly on the neighbouring Ifland of *Aguiguan*, which was about two leagues diftant ; and as we had lowered our main and fore-yards clofe down, we had no fails we could fet but the mizen, which was altogether infufficient to carry us clear of this inftant peril : We therefore immediately applied ourfelves to work, endeavouring, by the utmoft of our efforts, to heave up the main and fore-yards, in hopes that, if we could but be enabled to make ufe of our lower canvafs, we might poffibly weather the Ifland, and thereby fave ourfelves from this impending fhipwreck. But after full three hours ineffectual labour, the jeers broke, and the men being quite jaded, we were obliged, by mere debility, to defift, and quietly to expect our fate, which we then conceived to be unavoidable : For we imagined ourfelves, by this time, to be driven juft upon the fhore, and the night was fo extremely dark, that we expected to difcover the Ifland no otherwife than by ftriking upon it ; fo that the belief of our deftruction, and the uncertainty of the point of time when it would take place, occafioned us to pafs feveral hours, under the moft ferious apprehenfions, that each fucceeding moment would fend us to the bottom. Nor did thefe continued terrors, of inftantly ftriking and finking, end but with the day-break ; when we with great tranfport perceived, that the Ifland, we had thus dreaded, was at a confiderable diftance, and that a ftrong northern current had been the caufe of our prefervation.

The turbulent weather, which forced us from *Tinian*, did not begin to abate, till three days after ; and then we fwayed up the fore-yard, and began to heave up the main-yard, but the jeers broke and killed one of our men, and prevented us at that time from proceeding. The next day, being the 26th of *September*, was a day of moft fevere fatigue to us all ; for it muft be remembred,

U u 2

that

that in thefe exigencies no rank or office exempted any perfon from the manual application and bodily labour of a common failor. The bufinefs of this day was no lefs than an attempt to heave up the fheet-anchor, which we had hitherto dragged at our bows with two cables an end. This was a work of great importance to our future prefervation: For, not to mention the impediment to our navigation, and the hazard it would be to our fhip, if we attempted to make fail with the anchor in its prefent fituation, we had this moft interefting confideration to animate us, that it was the only anchor we had left; and, without fecuring it, we fhould be under the utmoft difficulties and hazards, when ever we made the land again; and therefore, being all of us fully apprized of the confequence of this enterprize, we laboured at it with the fevereft application for full twelve hours, when we had indeed made a confiderable progrefs, having brought the anchor in fight; but, it then growing dark, and we being exceffively fatigued, we were obliged to defift, and to leave our work unfinifhed, till the next morning, when, by the benefit of a night's reft, we compleated it, and hung the anchor at our bow.

It was the 27th of *September* in the morning, that is, five days after our departure, when we thus fecured our anchor; and the fame day, we got up our main-yard: And having now conquered in fome degree the diftrefs and diforder which we were neceffarily involved in at our firft driving out to fea, and being enabled to make ufe of our canvafs, we fet our courfes, and for the firft time ftood to the eaftward, in hopes of regaining the Ifland of *Tinian*, and joining our Commodore in a few days: For we were then, by our accounts, only forty-feven leagues to the South Weft of *Tinian*; fo that on the firft day of *October*, having then run the diftance neceffary for making the Ifland according to our reckoning, we were in full expectation of feeing it; but we were unhappily difappointed, and were thereby convinced, that a current had driven us to the weftward. And as we could not judge how much we might hereby have deviated, and confequently how long we might

ftill

ftill expect to be at fea, we had great apprehenfions that our ftock of water might prove deficient; for we were doubtful about the quantity we had on board, and found many of our cafks fo decayed, as to be half leaked out. However, we were delivered from our uncertainty the next day by having a fight of the Ifland of *Guam*, by which we difcovered that the currents had driven us forty-four leagues to the weftward of our accounts. This fight of land having fatisfied us of our fituation, we kept plying to the eaftward, though with exceffive labour, for, the wind continuing fixed in the eaftern board, we were obliged to tack often, and our crew was fo weak, that, without the affiftance of every man on board, it was not in our power to put the fhip about : This fevere employment lafted till the 11th of *October*, being the nineteenth day from our departure; when arriving in the offing of *Tinian*, we were reinforced from the fhore, as hath been already mentioned; and on the evening of the fame day, we, to our inexpreffible joy, came to an anchor in the road, thereby procuring to our fhipmates on fhore, as well as to ourfelves, a ceffation from the fatigues and apprehenfions, which this difaftrous incident had given rife to.

CHAP.

C H A P. V.

Employment at *Tinian*, till the final departure of the *Centurion* from thence; with a description of the *Ladrones*.

WHEN the Commodore came on board the *Centurion*, on her return to *Tinian*, as already mentioned, he resolved to stay no longer at the Island than was absolutely necessary to compleat our stock of water, a work which we immediately set ourselves about. But the loss of our long-boat, which was staved against our poop, when we were driven out to sea, put us to great inconveniencies in getting our water on board; for we were obliged to raft off all our cask, and the tide ran so strong, that, besides the frequent delays and difficulties it occasioned, we more than once lost the whole raft. Nor was this our only misfortune; for, on the 14th of *October*, being but the third day after our arrival, a sudden gust of wind brought home our anchor, forced us off the bank, and drove the ship out to sea a second time. The Commodore, it is true, and the principal officers were now on board; but we had near seventy men on shore, who had been employed in filling our water, and procuring provisions: These had with them our two Cutters; but as they were too many for the Cutters to bring off at once, we sent the eighteen oared barge to assist them; and at the same time made a signal for all that could to embark. The two Cutters soon came off to us full of men; but forty of the company, who were employed in killing cattle in the wood, and in bringing them down to the landing-place, were left behind; and though the eighteen oared barge was left for their conveyance, yet, as the ship soon drove to a considerable distance, it was not in their power to join us. However, as the weather was

favourable,

favourable, and our crew was now ftronger than when we were firft driven out, we, in about five days time, returned again to an anchor at *Tinian*, and relieved thofe we had left behind us from their fecond fears of being deferted by their fhip.

On our arrival, we found that the *Spanifh* bark, the old objeƈ of their hopes, had undergone a new metaorphofis: For thofe we had left on fhore began to defpair of our return, and conceiving that the lengthening the bark, as formerly propofed, was both a toilfome and unneceffary meafure, confidering the fmall number they confifted of, they had refolved to join her again, and to reftore her to her firft ftate; and in this fcheme they had made fome progrefs; for they had brought the two parts together, and would have foon compleated her, had not our coming back put a period to their labours and difquietudes.

Thefe people we had left behind informed us, that, juft before we were feen in the offing, two proas had ftood in very near the fhore, and had continued there for fome time; but, on the appearance of our fhip, they crowded away, and were prefently out of fight. And, on this occafion, I muft mention an incident, which, though it happened during the firft abfence of the fhip, was then omitted, to avoid interrupting the courfe of the narration.

It hath been already obferved, that a part of the detachment, fent to this Ifland under the command of the *Spanifh* Serjeant, lay concealed in the woods; and we were the lefs folicitous to find them out, as our prifoners all affured us, that it was impoffible for them to get off, and confequently that it was impoffible for them to fend any intelligence about us to *Guam*. But when the *Centurion* drove out to fea, and left the Commodore on fhore, he one day, attended by fome of his officers, endeavoured to make the tour of the Ifland: In this expedition, being on a rifing ground, they perceived in the valley beneath them the appearance of a fmall thicket, which, by obferving more nicely, they found had a progreffive motion: This at firft furprized them; but they foon difcovered, that it was no more than feveral large coco bufhes, which were dragged along the

4   ground,

ground, by perfons concealed beneath them. They immediately concluded that thefe were fome of the Serjeant's party (which was indeed true); and therefore the Commodore and his people made after them, in hopes of finding out their retreat. The *Indians* foon perceived they were dicovered, and hurried away with precipitation; but Mr. *Anfon* was fo near them, that he did not lofe fight of them till they arrived at their cell, which he and his officers entering found to be abandoned, there being a paffage from it down a precipice contrived for the conveniency of flight. They found here an old firelock or two, but no other arms. However, there was a great quantity of provifions, particularly falted fparibs of pork, which were excellent; and from what our people faw here, they concluded, that the extraordinary appetite, which they had found at this Ifland, was not confined to themfelves alone; for, it being about noon, the *Indians* had laid out a very plentiful repaft confidering their numbers, and had their bread-fruit and coco-nuts prepared ready for eating, and in a manner which plainly evinced, that, with them too, a good meal was neither an uncommon nor an unheeded article. The Commodore having in vain endeavoured to difcover the path by which the *Indians* had efcaped, he and his officers contented themfelves with fitting down to the dinner, which was thus luckily fitted to their prefent appetites; after which, they returned back to their old habitation, difpleafed at miffing the *Indians*, as they hoped to have engaged them in our fervice, if they could have had any conference with them. But notwithftanding what our prifoners had afferted, we were afterwards affured, that thefe *Indians* were carried off to *Guam* long before we left the place. But to return to our hiftory.

On our coming to an anchor again, after our fecond driving off to fea, we laboured indefatigably in getting in our water; and having, by the 20th of *October*, compleated it to fifty tun, which we fuppofed would be fufficient for our paffage to *Macao*, we, on the next day, fent one of each mefs on fhore, to gather as large a quantity of oranges, lemons, coco-nuts, and other fruits of the Ifland, as

they

A view of the N W side of SAYPAN *o*

Plate XXXVII.

the LADRONES *or Marian* ISLANDS.

they poſſibly could, for the uſe of themſelves and their meſs-mates, when at ſea. And, theſe purveyors returning on board us on the evening of the ſame day, we then ſet fire to the bark and proa, hoiſted in our boats, and got under ſail, ſteering away for the South end of the Iſland of *Formoſa*, and taking our leaves, for the third and laſt time, of the Iſland of *Tinian* : An Iſland, which, whether we conſider the excellence of its productions, the beauty of its appearance, the elegance of its woods and lawns, the healthineſs of its air, or the adventures it gave riſe to, may in all theſe views be truly ſtiled romantic.

And now, poſtponing for a ſhort time our run to *Formoſa*, and thence to *Canton*, I ſhall interrupt the narration with a deſcription of that range of Iſlands, uſually called the *Ladrones*, or *Marian* Iſlands, of which this of *Tinian* is one.

Theſe Iſlands were diſcovered by *Magellan* in the year 1521 ; and by the account given of the two he firſt fell in with, it ſhould ſeem that they were the Iſlands of *Saypan* and *Tinian* ; for they are deſcribed in his expedition as very beautiful Iſlands, and as lying between 15 and 16 degrees of North latitude. Theſe characteriſtics are particularly applicable to the two above mentioned places; for the pleaſing appearance of *Tinian* hath occaſioned the *Spaniards* to give it the additional name of *Buenaviſta* ; and *Saypan*, which is in the latitude of 15° : 22′ North, affords no contemptible proſpect when ſeen from the ſea, as may be ſufficiently evinced from the annexed view of its North Weſt ſide, taken at three leagues diſtance.

There are uſually reckoned twelve of theſe Iſlands; but it will appear, from the chart of the North part of the *Pacific* Ocean hereafter inſerted, that if the ſmall iſlets and rocks are counted in, then their whole number will amount to above twenty. They were formerly moſt of them well inhabited; and, even not ſixty years ago, the three principal Iſlands, *Guam*, *Rota*, and *Tinian* together, are ſaid to have contained above fifty thouſand people : But ſince that time *Tinian* hath been entirely depopulated; and only two or

X x

three

three hundred *Indians* have been left at *Rota*, to cultivate rice for
the Ifland of *Guam*; fo that now no more than *Guam* can properly
be faid to be inhabited. This Ifland of *Guam* is the only fettlement of the *Spaniards*; here they keep a governor and garrifon,
and here the *Manila* fhip generally touches for refrefhment, in her
paffage from *Acapulco* to the *Philippines*. It is efteemed to be
about thirty leagues in circumference, and contains, by the *Spanifh*
accounts, near four thoufand inhabitants, of which a thoufand are
faid to live in the city of *San Ignatio de Agand*, where the Governor generally refides, and where the houfes are reprefented as confiderable, being built with ftone and timber, and covered with tiles,
a very uncommon fabric for thefe warm climates and favage countries: Befides this city, there are upon the Ifland thirteen or fourteen villages. As this is a poft of fome confequence, on account
of the refrefhment it yields to the *Manila* fhip, there are two
caftles on the fea-fhore; one is the caftle of St. *Angelo*, which lies
near the road, where the *Manila* fhip ufually anchors, and is but
an infignificant fortrefs, mounting only five guns eight pounders;
the other is the caftle of St. *Lewis*, which is N. E. from St. *Angelo*, and four leagues diftant, and is intended to protect a road
where a fmall veffel anchors, which arrives here every other year
from *Manila*. This fort mounts the fame number of guns as the
former: And befides thefe forts, there is a battery of five pieces of
cannon on an eminence near the fea-fhore. The *Spanifh* troops
employed on this Ifland, confift of three companies of foot, from
forty to fifty men each; and this is the principal ftrength the Governor has to depend on; for he cannot rely on any affiftance from the
Indian inhabitants, being generally upon ill terms with them, and
fo apprehenfive of them, that he has debarred them the ufe of fire-
arms or lances.

The reft of thefe Iflands, though not inhabited, do yet abound
with many kinds of refrefhment and provifion; but there is no
good harbour or road to be met with amongft them all: Of that
of *Tinian* we have treated largely already; nor is the road of *Guam*

2   much

much better; for it is not unufual for the *Manila* fhip, though fhe propofes to ftay there but twenty-four hours, to be forced to fea, and to leave her boat behind her. This is an inconvenience fo fenfibly felt by the commerce at *Manila*, that it is always recommended to the Governor of *Guam*, to ufe his beft endeavours for the difcovery of fome fafe port in this part of the world. How induftrious he may be to comply with his inftructions, I know not; but this is certain, that, notwithftanding the many Iflands already found out between the coaft of *Mexico* and the *Philippines*, there is not yet known any one fafe port in that whole tract; though in other parts of the world it is not uncommon for very fmall Iflands to furnifh moft excellent harbours.

From what has been faid it appears, that the *Spaniards*, on the Ifland of *Guam*, are extremely few, compared to the *Indian* inhabitants; and formerly the difproportion was ftill greater, as may be eafily conceived from what hath been faid, in another chapter, of the numbers heretofore on *Tinian* alone. Thefe *Indians* are a bold well-limbed people; and it fhould feem from fome of their practices, that they are no ways defective in underftanding; for their flying proas in particular, which have been for ages the only veffels ufed by them, are fo fingular and extraordinary an invention, that it would do honour to any nation, however dexterous and acute. For if we confider the aptitude of this proa to the particular navigation of thefe Iflands, which lying all of them nearly under the fame meridian, and within the limits of the trade-wind, require the veffels made ufe of in paffing from one to the other, to be particularly fitted for failing with the wind upon the beam; or, if we examine the uncommon fimplicity and ingenuity of its fabric and contrivance, or the extraordinary velocity with which it moves, we fhall, in each of thefe articles, find it worthy of our admiration, and meriting a place amongft the mechanical productions of the moft civilized nations, where arts and fciences have moft eminently flourifhed. As former Navigators, though they have mentioned thefe

veffels

veffels, have yet treated of them imperfectly, and, as I conceive,
that, befides their curiofity, they may furnifh both the fhipwright
and feaman with no contemptible obfervations, I fhall here infert a
very exact defcription of the built, rigging, and working of thefe
veffels, which I am well enabled to do, for one of them, as I
have mentioned, fell into our hands at our firft arrival at *Tinian*, and
Mr. *Brett* took it to pieces, on purpofe to delineate its fabric and dimenfions with greater accuracy : So that the following account may
be relied on.

The name of flying proa given to thefe veffels, is owing to the
fwiftnefs with which they fail. Of this the *Spaniards* affert fuch
ftories, as appear altogether incredible to thofe who have never feen
thefe veffels move ; nor are the *Spaniards* the only people who relate thefe extraordinary tales of their celerity. For thofe who fhall
have the curiofity to enquire at the dock at *Portfmouth*, about a
trial made there fome years fince, with a very imperfect one built
at that place, will meet with accounts not lefs wonderful than any
the *Spaniards* have given. However, from fome rude eftimations
made by our people, of the velocity with which they croffed the
horizon at a diftance, while we lay at *Tinian*, I cannot help believing, that with a brifk trade-wind they will run near twenty
miles an hour : Which though, greatly fhort of what the *Spaniards*
report of them, is yet a prodigious degree of fwiftnefs. But let us
give a diftinct idea of its figure.

The conftruction of this proa is a direct contradiction to the practice of all the reft of mankind. For as the reft of the world make
the head of their veffels different from the ftern, but the two fides
alike ; the proa, on the contrary, has her head and ftern exactly
alike, but her two fides very different ;' the fide, intended to be
always the lee-fide, being flat ; and the windward-fide made rounding, in the manner of other veffels : And, to prevent her overfetting,
which from her fmall breadth, and the ftraight run of her lee-
ward-fide, would, without this precaution, infallibly happen, there

2   is

Fig. 2.

Fig. 1.

A Scale of Feet.

Plate XXXVIII.

Flying Proa, taken at the Ladrone Islands.

Fig. 3.

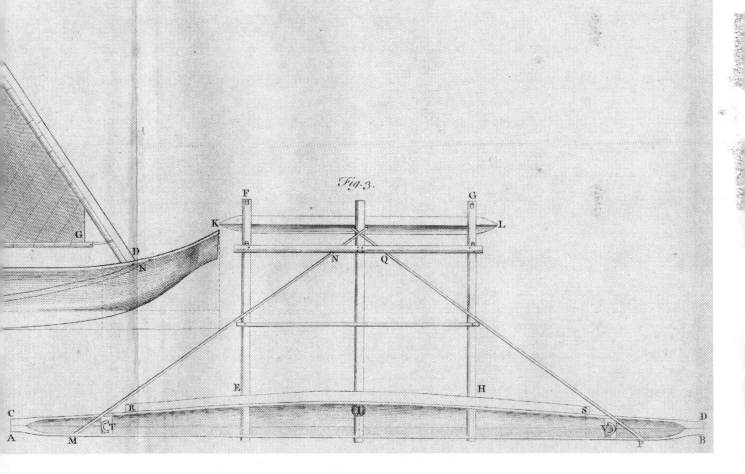

is a frame laid out from her to windward, to the end of which is faftened a log, fafhioned into the fhape of a fmall boat, and made hollow : The weight of the frame is intended to ballance the proa, and the fmall boat is by its buoyancy (as it is always in the water) to prevent her overfetting to windward; and this frame is ufually called an outrigger. The body of the proa (at leaft of that we took) is made of two pieces joined end-ways, and fowed together with bark, for there is no iron ufed about her: She is about two inches thick at the bottom, which at the gunwale is reduced to lefs than one : The dimenfions of each part will be better known from the uprights and views contained in the annexed plate, which were drawn from an exact menfuration; thefe I fhall endeavour to explain as minutely and diftinctly as I can.

Fig. 1. Reprefents the proa with her fail fet, as fhe appears when viewed from the leeward.

Fig. 2. Is a view of her from the head, with the outrigger to the windward.

Fig. 3. Is the plan of the whole; where (A B) is the lee-fide of the proa; (C D) the windward-fide; (E F G H) the outrigger or frame laid out to windward; (K L) the boat at the end of it; (M N P Q) two braces from the head and ftern to fteady the frame; (R S) a thin plank placed to windward, to prevent the proa from fhipping of water, and for a feat to the *Indian* who bales, and fometimes goods are carried upon it; (I) is the part of the middle outrigger, on which the maft is fixed : The maft itfelf is fupported (*Fig.* 2,) by the fhore (C D), and by the fhrowd (E F), and by two ftays, one of which may be feen, in *Fig.* 1, marked (C D), the other is hid by the fail: The fail (E F G), in *Fig.* 1, is made of matting, and the maft, yard, boom, and outriggers, are all made of bamboo: The heel of the yard is always lodged in one of the fockets (T) or (V), *Fig.* 3, according to the tack the proa goes on; and when fhe alters her tack, they bear away a little to bring her ftern up to the wind, then by eafing the halyard, and

raifing

raifing the yard, and carrying the heel of it along the lee-fide of
the proa, they fix it in the oppofite focket ; whilft the boom at the
fame time, by letting fly the fheet (M), and haling the fheet (N),
Fig. 1, fhifts into a contrary fituation to what it had before, and
that which was the ftern of the proa, now becomes the head, and
fhe is trimmed on the other tack. When it is neceffary to reef or
furl the fail, this is done by rolling it round the boom. The
proa generally carries fix or feven *Indians* ; two of which are placed
in the head and ftern, who fteer the veffel alternately with a pad-
dle according to the tack fhe goes on, he in the ftern being the
fteerfman ; the other *Indians* are employed either in baling out the
water which fhe accidentally fhips, or in fetting and trimming the
fail. From the defcription of thefe veffels it is fufficiently obvious,
how dexteroufly they are fitted for ranging this collection of Iflands
called the *Ladrones :* For as thefe Iflands lie nearly N. and S. of
each other, and are all within the limits of the trade-wind, the
proas, by failing moft excellently on a wind, and with either end
foremoft, can run from one of thefe Iflands to the other and back
again, only by fhifting the fail, without ever putting about ; and,
by the flatnefs of their lee-fide, and their fmall breadth, they are
capable of lying much nearer the wind than any other veffel hitherto known, and thereby have an advantage, which no veffels
that go large can ever pretend to : The advantage I mean is that of
running with a velocity nearly as great, and perhaps fometimes
greater than that with which the wind blows. This, however pa-
radoxical it may appear, is evident enough in fimilar inftances on
fhore : For it is well known, that the fails of a windmill often
move fafter than the wind ; and one great fuperiority of common
windmills over all others, that ever were, or ever will be contrived
to move with an horizontal motion, is analogous to the cafe we
have mentioned of a veffel upon a wind and before the wind : For
the fails of an horizontal windmill, the fafter they move, the more
they detract from the impulfe of the wind upon them ; whereas the

common

common windmills, by moving perpendicular to the torrent of air, are nearly as forcibly acted on by the wind, when they are in motion, as when they are at rest.

Thus much may suffice as to the description and nature of these singular embarkations. I must add, that vessels bearing some obscure resemblance to these, are to be met with in various parts of the *East-Indies*, but none of them, that I can learn, to be compared with those of the *Ladrones*, either in their construction or celerity; which should induce one to believe, that this was originally the invention of some genius of these Islands, and was afterwards imperfectly copied by the neighbouring nations: For though the *Ladrones* have no immediate intercourse with any other people, yet there lie to the S. and S. W. of them a great number of Islands, which are supposed to extend to the coast of *New Guinea*. These Islands are so near the *Ladrones*, that canoes from them have sometimes, by distress, been driven to *Guam*; and the *Spaniards* did once dispatch a bark for their discovery, which left two Jesuits amongst them, who were afterwards murthered: And the inhabitants of the *Ladrones*, with their proas, may, by like accident, have been driven amongst these Islands. Indeed I should conceive, that the same range of Islands extends to the S. E. as well as the S. W, and that to a prodigious distance: For *Schouten*, who traversed the South part of the *Pacific* Ocean in the year 1615, met with a large double canoe full of people, at above a thousand leagues distance from the *Ladrones* towards the S. E. If this double canoe was any distant imitation of the flying proa, which is no very improbable conjecture, this can only be accounted for, by supposing that there is a range of Islands, near enough to each other to be capable of an accidental communication, which is extended from the *Ladrones* thither. And indeed all those who have crossed from *America* to the *East-Indies* in a southern latitude, have never failed of meeting with several very small Islands scattered over that immense ocean.

And

And as there may be hence some reason to suppose, that the *Ladrones* are only a part of an extensive chain of Islands, spreading themselves to the southward, towards the unknown boundaries of the *Pacific* Ocean ; so it appears from the *Spanish* chart hereafter inserted, that the same chain is extended from the northward of the *Ladrones* to *Japan :* So that in this light the *Ladrones* will be only one small portion of a range of Islands reaching from *Japan*, perhaps to the unknown southern Continent. After this short account of these places, I shall now return to the prosecution of our voyage.

CHAP.

CHAP. VI.

From *Tinian* to *Macao.*

I HAVE already mentioned, that, on the 21ft of *October*, in the evening, we took our leave of the Ifland of *Tinian*, fteering the proper courfe for *Macao* in *China*. 'I'he eaftern monfoon was now, we reckoned, fairly fettled; and we had a conftant gale blowing right upon our ftern: So that we generally run from forty to fifty leagues a day. But we had a large hollow fea purfuing us, which occafioned the fhip to labour much; whence we received great damage in our rigging, which was grown very rotten, and our leak was augmented: But happily for us, our people were now in full health; fo that there were no complaints of fatigue, but all went through their attendance on the pumps, and every other duty of the fhip, with eafe and chearfulnefs.

Having now no other but our fheet-anchor left, except our prize anchors, which were ftowed in the hold and were too light to be depended on, we were under great concern how we fhould manage on the coaft of *China*, where we were all entire ftrangers, and where we fhould doubtlefs be frequently under the neceffity of coming to an anchor. Our fheet-anchor being obvioufly much too heavy for a coafting anchor, it was at length refolved, to fix two of our largeft prize anchors into one ftock, and to place between their fhanks two guns, four pounders, which was accordingly executed, and it was to ferve as a beft bower: And a third prize-anchor being in like manner joined with our ftream-anchor, with guns between them, we thereby made a fmall bower; fo that, befides our fheet-anchor, we had again two others at our bows, one of which weighed 3900, and the other 2900 pounds.

The

The 3d of *November*, about three in the afternoon, we faw an Ifland, which at firft we imagined to be the Ifland of *Botel Tobago Xima* : But on our nearer approach we found it to be much fmaller than that is ufually reprefented ; and about an hour after we faw another Ifland, five or fix miles farther to the weftward. As no chart, nor any journal we had feen, took notice of any other Ifland to the eaftward of *Formofa*, than *Botel Tobago Xima*, and as we had no obfervation of our latitude at noon, we were in fome perplexity, being apprehenfive that an extraordinary current had driven us into the neighbourhood of the *Bafhee Iflands* ; and therefore, when night came on, we brought to, and continued in this pofture till the next morning, which proving dark and cloudy, for fome time prolonged our uncertainty ; but it cleared up about nine o'clock, when we again difcerned the two Iflands above-mentioned ; we then preft forwards to the weftward, and by eleven got a fight of the fouthern part of the Ifland of *Formofa*. This fatisfied us that the fecond Ifland we faw was *Botel Tobago Xima*, and the firft a fmall ifland or rock, lying five or fix miles due Eaft from it, which, not being mentioned by any of our books or charts, was the occafion of our fears.

When we got fight of the Ifland of *Formofa*, we fteered W. by S, in order to double its extremity, and kept a good look-out for the rocks of *Vele Rete*, which we did not fee till two in the afternoon. They then bore from us W. N. W, three miles diftant, the South end of *Formofa* at the fame time bearing N. by W $\frac{1}{2}$ W, about five leagues diftant. To give thefe rocks a good birth, we immediately haled up S. by W, and fo left them between us and the land. Indeed we had reafon to be careful of them ; for though they appeared as high out of the water as a fhip's hull, yet they are environed with breakers on all fides, and there is a fhoal ftretching from them at leaft a mile and an half to the fouthward, whence they may be truly called dangerous. The courfe from *Botel Tobago Xima* to thefe rocks, is S. W. by W, and the diftance about twelve or thirteen leagues : And the South end of *Formofa*, off which
they

they lie, is in the latitude of 21° : 50' North, and in 23° : 50' West longitude from *Tinian*, according to our moſt approved reckonings, though by ſome of our accounts above a degree more.

While we were paſſing by theſe rocks of *Vele Rete*, there was an outcry of fire on the fore caſtle; this occaſioned a general alarm, and the whole crew inſtantly flocked together in the utmoſt confuſion, ſo that the officers found it difficult for ſome time to appeaſe the uproar : But having at laſt reduced the people to order, it was perceived that the fire proceeded from the furnace; and pulling down the brick-work, it was extinguiſhed with great facility, for it had taken its riſe from the bricks, which, being overheated, had begun to communicate the fire to the adjacent wood-work. In the evening we were ſurprized with a view of what we at firſt ſight conceived to have been breakers, but, on a ſtricter examination, we found them to be only a great number of fires on the Iſland of *Formoſa*. Theſe, we imagined, were intended by the inhabitants of that Iſland as ſignals for us to touch there, but that ſuited not our views, we being impatient to reach the port of *Macao* as ſoon as poſſible. From *Formoſa* we ſteered W. N. W, and ſometimes ſtill more northerly, propoſing to fall in with the coaſt of *China*, to the eaſtward of *Pedro Blanco*; for the rock ſo called is uſually eſteemed an excellent direction for ſhips bound to *Macao*. We continued this courſe till the following night, and then frequently brought to, to try if we were in ſoundings: But it was the 5th of *November*, at nine in the morning, before we ſtruck ground, and then we had forty-two fathom, and a bottom of grey ſand mixed with ſhells. When we had got about twenty miles farther W.N.W, we had thirty-five fathom and the ſame bottom, from whence our ſoundings gradually decreaſed from thirty-five to twenty-five fathom; but ſoon after, to our great ſurprize, they jumped back again to thirty fathom : This was an alteration we could not very well account for, ſince all the charts laid down regular ſoundings every where to the northward of *Pedro Blanco*; and for this reaſon we kept a very careful look-out, and altered our courſe to N. N. W, and having run

Y y 2

thirty-

thirty-five miles in this direction, our foundings again gradually diminished to twenty-two fathom, and we at laft, about midnight, got fight of the main land of *China*, bearing N. by W. four leagues diftant: We then brought the fhip to, with her head to the fea, propofing to wait for the morning; and before fun-rife we were furprized to find ourfelves in the midft of an incredible number of fifhing boats, which feemed to cover the furface of the fea as far as the eye could reach. I may well ftile their number incredible, fince I cannot believe, upon the loweft eftimate, that there were fo few as fix thoufand, moft of them manned with five hands, and none of thofe we faw with lefs than three. Nor was this fwarm of fifhing veffels peculiar to this fpot; for, as we ran on to the weftward, we found them as abundant on every part of the coaft. We at firft doubted not but we fhould procure a Pilot from them to carry us to *Macao*; but though many of them came clofe to the fhip, and we endeavoured to tempt them by fhowing them a number of dollars, a moft alluring bait for *Chinefe* of all ranks and profeffions, yet we could not entice them on board us, nor procure any directions from them; though, I prefume, the only difficulty was their not comprehending what we wanted them to do, for we could have no communication with them but by figns: Indeed we often pronounced the word *Macao*; but this we had reafon to fuppofe they underftood in a different fenfe; for in return they fometimes held up fifh to us, and we afterwards learnt, that the *Chinefe* name for fifh is of a fomewhat fimilar found. But what furprifed us moft was the inattention and want of curiofity, which we obferved in this herd of fifhermen: A fhip like ours had doubtlefs never been in thofe feas before; perhaps, there might not be one, amongft all the *Chinefe* employed in this fifhery, who had ever feen any *European* veffel; fo that we might reafonably have expected to have been confidered by them as a very uncommon and extraordinary object; but though many of their veffels came clofe to the fhip, yet they did not appear to be at all interefted about us, nor did they deviate in the leaft from their courfe to regard us; which

infenfibility,

infenfibility, efpecially in maritime perfons, about a matter in their own profeffion, is fcarcely to be credited, did not the general behaviour of the *Chinefe*, in other inftances, furnifh us with continual proofs of a fimilar turn of mind : It may perhaps be doubted, whether this caft of temper be the effect of nature or education ; but, in either cafe, it is an inconteftable fymptom of a mean and contemptible difpofition and is alone a fufficient confutation of the extravagant panegyrics, which many hypothetical writers have beftowed on the ingenuity and capacity of this Nation. But to return :

Not being able to procure any information from the *Chinefe* fifhermen about our proper courfe to *Macao*, it was neceffary for us to rely entirely on our own judgment ; and concluding from our latitude, which was 22° : 42′ North, and from our foundings, which were only feventeen or eighteen fathoms, that we were yet to the eaftward of *Pedro Blanco*, we ftood to the weftward : And for the affiftance of future Navigators, who may hereafter doubt about the parts of the coaft they are upon, I muft obferve, that befides the latitude of *Pedro Blanco*, which is 22° : 18′, and the depth of water, which to the weftward of that rock is almoft every where twenty fathoms, there is another circumftance which will give great affiftance in judging of the pofition of the fhip : This is the kind of ground ; for, till we came within thirty miles of *Pedro Blanco*, we had conftantly a fandy bottom ; but there the bottom changed to foft and muddy, and continued fo quite to the Ifland of *Macao* ; only while we were in fight of *Pedro Blanco* and very near it, we had for a fhort fpace a bottom of greenifh mud, intermixed with fand.

It was on the 5th of *November*, at midnight, when we firft made the coaft of *China* ; and the next day, about two o'clock, as we were ftanding to the weftward within two leagues of the coaft, and ftill furrounded by fifhing veffels in as great numbers as at firft, we perceived that a boat a-head of us waved a red flag, and blew a horn : This we confidered as a fignal made to us, either

to

to warn us of fome fhoal, or to inform us that they would fupply us with a Pilot, and in this belief we immediately fent our Cutter to the boat, to know their intentions; but we were foon made fenfible of our miftake, and found that this boat was the Commodore of the whole fifhery, and that the fignal fhe had made, was to order them all to leave off fifhing, and to return in fhore, which we faw them inftantly obey. On this difappointment we kept on our courfe, and foon after paffed by two very fmall rocks, which. lay four or five miles diftant from the fhore; but night came on before we got fight of *Pedro Blanco*, and we therefore brought to till the morning, when we had the fatisfaction to difcover it. It is a rock of a fmall circumference, but of a moderate height, and, both in fhape and colour, refembles a fugar loaf, and is about feven or eight miles from the fhore. We paffed within a mile and an half of it, and left it between us and the land, ftill keeping on to the weftward; and the next day, being the 7th, we were a-breaft of a chain of Iflands, which ftretched from Eaft to Weft. Thefe, as we afterwards found, were called the Iflands of *Lema*; they are rocky and barren, and are in all, fmall and great, fifteen or fixteen; and there are, befides, a great number of other Iflands between them and the main land of *China*. There is annexed a view of thefe Iflands, and likewife a view of the grand *Ladrone* hereafter mentioned, as it appears when (R), the weftermoft of the Iflands of *Lema*, bears W. N. W, at the diftance of a mile and half. Thefe Iflands we left on the ftarboard-fide, paffing within four miles of them, where we had twenty-four fathom water. We were ftill furrounded by fifhing boats; and we once more fent the Cutter on board one of them, to endeavour to procure a Pilot, but could not prevail; however, one of the *Chinefe* directed us by figns to fail round the weftermoft of the iflands or rocks of *Lema*, and then to hale up. We followed this direction, and in the evening came to an anchor in eighteen fathom; at which time, the rock (R) in the foregoing draught bore S. S. E. five miles diftant, and the grand *Ladrone* W. by S, about two leagues diftant. The rock (R) is a moft

excellent

R. *The westermost of the Rocks of* Lema.
A. *The grand* Ladrone.

The Isl

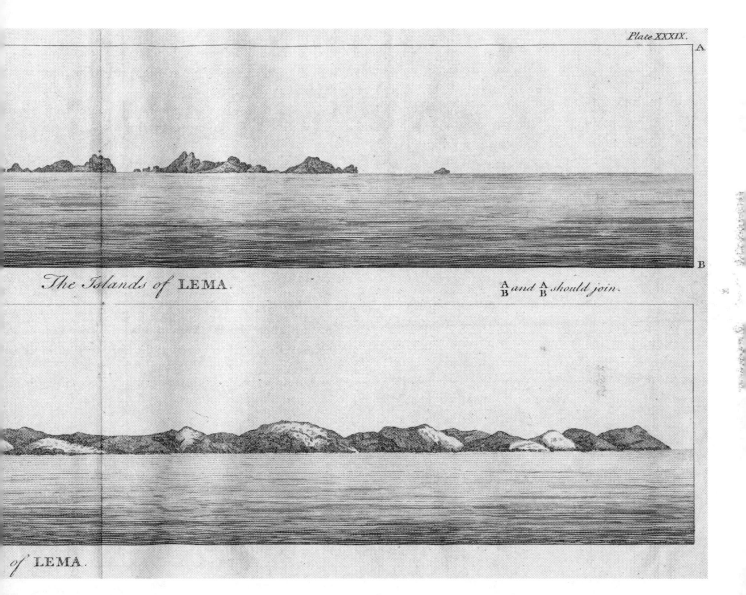

Plate XXXIX.

A

B

The Islands of LEMA.

$\frac{A}{B}$ and $\frac{A}{B}$ should join.

of LEMA.

excellent direction for fhips coming from the eaftward : Its latitude
is 21° : 52′ North, and it bears from *Pedro Blanco* S. 64° W, diftant twenty-one leagues.   You are to leave it on the ftarboard-fide,
and you may come within half a mile of it in eighteen fathom water : And then you muft fteer N. by W. ¼ W. for the channel, between the Iflands of *Cabouce* and *Bamboo,* which are to the northward of the grand *Ladrone.*

After having continued at anchor all night, we, on the 9th, at
four in the morning, fent our Cutter to found the channel, where we
propofed to pafs ; but before the return of the Cutter, a *Chinefe* Pilot put on board us, and told us, in broken *Portuguefe,* he would
carry us to *Macao* for thirty dollars : Thefe were immediately paid
him, and we then weighed and made fail ; and foon after, feveral
other Pilots came on board us, who, to recommend themfelves,
produced certificates from the Captains of feveral fhips they had pi-
lotted in, but we continued the fhip under the management of the
Chinefe who came firft on board.   By this time we learnt, that we
were not far diftant from *Macao,* and that there were in the river of
Canton, at the mouth of which *Macao* lies, eleven *European* fhips,
of which four were *Englifh.*   Our Pilot carried us between the
Iflands of *Bamboo* and *Cabouce,* but the winds hanging in the northern board, and the tides often fetting ftrongly againft us, we were
obliged to come frequently to an anchor, fo that we did not get
through between the two Iflands till the 12th of *November,* at two
in the morning.   In paffing through, our depth of water was from
twelve to fourteen fathom ; and as we ftill fteered on N. by W. ¼ W,
between a number of other Iflands, our foundings underwent little
or no variation till towards the evening, when they encreafed to
feventeen fathom ; in which depth (the wind dying away) we anchored not far from the Ifland of *Lantoon,* which is the largeft of all
this range of Iflands.   At feven in the morning we weighed again,
and fteering W. S. W. and S. W. by W, we at ten o'clock happily
anchored in *Macao* road, in five fathom water, the city of *Macao*
bearing W. by N, three leagues diftant ; the peak of *Lantoon*

2   E. by N,

E. by N, and the grand *Ladrone* S. by E, each of them about five leagues diftant. Thus, after a fatiguing cruife of above two years continuance, we once more arrived in an amicable port, in a civilized country; where the conveniencies of life were in great plenty; where the naval ftores, which we now extremely wanted, could be in fome degree procured; where we expected the inexpreffible fatisfaction of receiving letters from our relations and friends; and where our countrymen, who were lately arrived from *England*, would be capable of anfwering the numerous enquiries we were prepared to make, both about public and private occurrences, and to relate to us many particulars, which, whether of importance or not, would be liftned to by us with the utmoft attention, after the long fufpenfion of our correfpondence with our country, to which the nature of our undertaking had hitherto fubjected us.

CHAP.

CHAP. VII.

Proceedings at *Macao*.

THE city of *Macao*, in the road of which we came to an anchor on the 12th of *November*, is a *Portuguese* settlement, situated in an Island at the mouth of the river of *Canton*. It was formerly a very rich and populous city, and capable of defending itself against the power of the adjacent *Chinese* Governors : But at present it is much fallen from its antient splendor ; for though it is inhabited by *Portuguese*, and hath a Governor nominated by the King of *Portugal*, yet it subsists merely by the sufferance of the *Chinese*, who can starve the place, and dispossess the *Portuguese* whenever they please : This obliges the Governor of *Macao* to behave with great circumspection, and carefully to avoid every circumstance that may give offence to the *Chinese*. The river of *Canton*, at the mouth of which this city lies, is the only *Chinese* port, frequented by *European* ships ; and this river is indeed a more commodious harbour, on many accounts, than *Macao* : But the peculiar customs of the *Chinese*, only adapted to the entertainment of trading ships, and the apprehensions of the Commodore, least he should embroil the *East-India* Company with the Regency of *Canton*, if he should insist on being treated upon a different footing than the Merchantmen, made him resolve to go first to *Macao*, before he ventured into the port of *Canton*. Indeed, had not this reason prevailed with him, he himself had nothing to fear : For it is certain that he might have entered the port of *Canton*, and might have continued there as long as he pleased, and afterwards have left it again, although the whole power of the *Chinese* Empire had been brought together to oppose him.

The

The Commodore, not to depart from his ufual prudence, no fooner came to an anchor in *Macao* road, than he difpatched an officer with his compliments to the *Portuguefe* Governor of *Macao*, requefting his Excellency, by the fame officer, to advife him in what manner it would be proper to act, to avoid offending the *Chinefe*, which, as there were then four of our fhips in their power at *Canton*, was a matter worthy of attention. The difficulty, which the Commodore principally apprehended, related to the duty ufually paid by all fhips in the river of *Canton*, according to their tunnage. For as men of war are exempted in every foreign harbour from all manner of port charges, the Commodore thought it would be derogatory to the honour of his country, to fubmit to this duty in *China*: And therefore he defired the advice of the Governor of *Macao*, who, being an *European*, could not be ignorant of the privileges claimed by a *Britifh* man of war, and confequently might be expected to give us the beft lights for avoiding this perplexity. Our boat returned in the evening with two officers fent by the Governor, who informed the Commodore, that it was the Governor's opinion, that if the *Centurion* ventured into the river of *Canton*, the duty would certainly be demanded ; and therefore, if the Commodore approved of it, he would fend him a Pilot, who fhould conduct us into another fafe harbour called the *Typa*, which was every way commodious for careening the fhip (an operation we were refolved to begin upon as foon as poffible) and where the abovementioned duty would in all probability, be never afked for.

This propofal the Commodore agreed to, and in the morning we weighed anchor, and, under the direction of the *Portuguefe* Pilot, fteered for the intended harbour. As we entered two Iflands, which form the eaftern paffage to it, we found our foundings decreafed to three fathom and a half: But the Pilot affuring us that this was the leaft depth we fhould meet with, we continued our courfe, till at length the fhip ftuck faft in the mud, with only eighteen foot water abaft ; and, the tide of ebb making, the

2 water

water fewed to fixteen feet, but the fhip remained perfectly up-right; we then founded all round us, and finding the water deep-ned to the northward, we carried out our fmall bower with two hawfers an end, and at the return of the tide of flood hove the fhip a float; and a fmall breeze fpringing up at the fame inftant, we fet the fore-top-fail, and flipping the hawfer ran into the harbour, where we moored in about five fathom water. This harbour of the *Typa* is formed by a number of Iflands, and is about fix miles diftant from *Macao*. Here we faluted the caftle of *Macao* with eleven guns, which were returned by an equal number.

The next day the Commodore paid a vifit in perfon to the Go-vernor, and was faluted at his landing by eleven guns; which were returned by the *Centurion*. Mr. *Anfon*'s bufinefs in this vifit, was to folicit the Governor to grant us a fupply of provifions, and to furnifh us with fuch ftores as were neceffary to refit the fhip. The Governor feemed really inclined to do us all the fervice he could; and affured the Commodore, in a friendly manner, that he would privately give us all the affiftance in his power; but he, at the fame time, frankly owned, that he dared not openly furnifh us with any thing we demanded, unlefs we firft procured an order for it from the Viceroy of *Canton*; for that he neither received provifions for his garrifon nor any other neceffaries, but by permiffion from the *Chinefe* Government; and as they took care only to furnifh him from day to day, he was indeed no other than their vaffal, whom they could at all times compel to fubmit to their own terms, only by laying an embargo on his provifions.

On this declaration of the Governor, Mr. *Anfon* refolved him-felf to go to *Canton*, to procure a licence from the Viceroy; and he accordingly hired a *Chinefe* boat for himfelf and his attendants; but juft as he was ready to embark, the *Hoppo* or *Chinefe* Cuftom-houfe officer at *Macao* refufed to grant a permit to the boat, and ordered the watermen not to proceed at their peril. The Com-modore at firft endeavoured to prevail with the *Hoppo* to withdraw his injunction, and to grant a permit; and the Governor of *Macao*

Z z 2

em-

employed his intereſt with the *Hoppo* to the ſame purpoſe
Mr. *Anſon*, finding the officer inflexible, told him, the next day,
that if he longer refuſed to grant the permit, he would man and
arm his own boats, to carry him thither; aſking the *Hoppo*, at the
ſame time, who he imagined would dare to oppoſe him. This
threat immediately brought about what his intreaties had laboured
for in vain: The permit was granted, and Mr. *Anſon* went to
Canton. On his arrival there, he conſulted with the Supercargoes
and Officers of the *Engliſh* ſhips, how to procure an order from
the Viceroy for the neceſſaries he wanted: But in this he had reaſon to ſuppoſe, that the advice they gave him, though doubtleſs
well intended, was yet not the moſt prudent: For as it is the cuſtom with theſe Gentlemen, never to apply to the ſupreme Magiſ-
trate himſelf, whatever difficulties they labour under, but to tranſact
all matters relating to the Government, by the mediation of the
principal *Chineſe* Merchants, Mr. *Anſon* was adviſed to follow the
ſame method upon this occaſion, the *Engliſh* promiſing (in which
they were doubtleſs ſincere) to exert all their intereſt to engage the
Merchants in his favour. And when the *Chineſe* Merchants were
applied to, they readily undertook the management of it, and promiſed to anſwer for its ſucceſs; but after near a month's delay, and
reiterated excuſes, during which interval they pretended to be often
upon the point of compleating the buſineſs, they at laſt (being
preſſed, and meaſures being taken for delivering a letter to the Viceroy) threw off the maſk, and declared they neither had applied to
the Viceroy, nor could they; for he was too great a man, they
ſaid, for them to approach on any occaſion: And not contented with
having themſelves thus groſly deceived the Commodore, they now
uſed all their perſuaſion with the *Engliſh* at *Canton*, to prevent them
from intermeddling with any thing that regarded him, repreſenting
to them, that it would in all probability embroil them with the
Government, and occaſion them a great deal of unneceſſary trouble;
which groundleſs inſinuations had indeed but too much weight with
thoſe they were applied to.

It

It may be difficult to affign a reafon for this perfidious conduct o the *Chinese* Merchants: Intereft indeed is known to exert a boundlefs influence over the inhabitants of that Empire; but how their intereft could be affected in the prefent cafe is not eafy to difcover; unlefs they apprehended that the prefence of a fhip of force might damp their *Manila* trade, and therefore acted in this manner with a view of forcing the Commodore to *Batavia :* But it might be as natural in this light to fuppofe, that they would have been eager to have got him difpatched. I therefore rather impute their behaviour to the unparalleled pufillanimity of the Nation, and to the awe they are under of the Government: For as fuch a fhip as the *Centurion*, fitted for war only, had never been feen in thofe parts before, fhe was the horror of thefe daftards, and the Merchants were in fome degree terrified even with the idea of her, and could not think of applying to the Viceroy (who is doubtlefs fond of all opportunities of fleecing them) without reprefenting to themfelves the pretences which a hungry and tyrannical Magiftrate might poffibly find, for cenfuring their intermeddling in fo unufual a tranfaction, in which he might pretend the intereft of the State was immediately concerned. However, be this as it may, the Commodore was fatisfied that nothing was to be done by the interpofition of the Merchants, as it was on his preffing them to deliver a letter to the Viceroy, that they had declared they durft not intermeddle, and had confeffed, that notwithftanding all their pretences of ferving him, they had not yet taken one ftep towards it, Mr. *Anfon* therefore told them, that he would proceed to *Batavia*, and refit his fhip there; but informed them, at the fame time, that this was impoffible to be done, unlefs he was fupplied with a ftock of provifions fufficient for his paffage. The Merchants, on this, undertook to procure him provifions, but affured him, that it was what they durft not engage in openly, but propofed to manage it in a clandeftine manner, by putting a quantity of bread, flower and other provifion on board the *Englifh* fhips, which were now ready to fail; and thefe were to ftop at the mouth of the *Typa*, where the

Centurion's

Centurion's boats were to receive it. This article, which the Merchants reprefented as a matter of great favour, being fettled, the Commodore, on the 16th of *December*, returned from *Canton* to the fhip, feemingly refolved to proceed to *Batavia* to refit, as foon as he fhould get his fupplies of provifion on board.

But Mr. *Anfon* (who never intended going to *Batavia*) found, on his return to the *Centurion*, that her main-maft was fprung in two places, and that the leak was confiderably encreafed ; fo that, upon the whole, he was fully fatisfied, that though he fhould lay in a fufficient ftock of provifions, yet it would be impoffible for him to put to fea without refitting: For, if he left the port with his fhip in her prefent condition, fhe would be in the utmoft danger of foundring ; and therefore, notwithftanding the difficulties he had met with, he refolved at all events to have her hove down, before he left *Macao*. He was fully convinced, by what he had obferved at *Canton*, that his great caution not to injure the *Eaft-India* Company's affairs, and the regard he had fhown to the advice of their officers, had occafioned all his embarrafments. For he now faw clearly, that if he had at firft carried his fhip into the river of *Canton*, and had immediately applied himfelf to the *Mandarines*; who are the chief officers of State, inftead of employing the Merchants to apply for him, he would, in all probability, have had all his requefts granted, and would have been foon difpatched. He had already loft a month, by the wrong meafures he had been put upon, but he refolved to lofe as little more time as poffible ; and therefore, the 17th of *December*, being the next day after his return from *Canton*, he wrote a letter to the Viceroy of that place, acquainting him, that he was Commander in chief of a fquadron of his *Britannick* Majefty's fhips of war, which had been cruifing for two years paft in the *South Seas* againft the *Spaniards*, who were at war with the King his Mafter ; that, in his way back to *England*, he had put into the port of *Macao*, having a confiderable leak in his fhip, and being in great want of provifions, fo that it was impoffible for him to proceed on his voyage, till his fhip was repaired, and he was fupplied

plied with the neceffaries he wanted; that he had been at *Canton*, in hopes of being admitted to a perfonal audience of his Excellency; but being a ftranger to the cuftoms of the country, he had not been able to inform himfelf what fteps were neceffary to be taken to procure fuch an audience, and therefore was obliged to apply to him in this manner, to defire his Excellency to give orders, for his being permitted to employ Carpenters and proper workmen to refit his fhip, and to furnifh himfelf with provifions and ftores, thereby to enable him to purfue his voyage to *Great-Britain* with this monfoon, hoping, at the fame time, that thefe orders would be iffued with as little delay as poffible, leaft it might occafion his lofs of the feafon, and he might be prevented from departing till the next winter.

This letter was tranflated into the *Chinefe* language, and the Commodore delivered it himfelf to the *Hoppo* or chief officer of the Emperor's cuftoms at *Macao* defiring him to forward it to the Viceroy of *Canton*, with as much expedition as he could. The officer at firft feemed unwilling to take charge of it, and raifed many difficulties about it, fo that Mr. *Anfon* fufpected him of being in league with the Merchants of *Canton*, who had always fhown a great apprehenfion of the Commodore's having any immediate intercourfe with the Viceroy or *Mandarines*; and therefore the Commodore, with fome refentment, took back his letter from the *Hoppo*, and told him, he would immediately fend an officer with it to *Canton* in his own boat, and would give him pofitive orders not to return without an anfwer from the Viceroy. The *Hoppo* perceiving the Commodore to be in earneft, and fearing to be called to an account for his refufal, begged to be entrufted with the letter, and promifed to deliver it, and to procure an anfwer as foon as poffible. And now it was foon feen how juftly Mr. *Anfon* had at laft judged of the proper manner of dealing with the *Chinefe*; for this letter was written but the 17th of *December*, as hath been already obferved; and, on the 19th in the morning, a *Mandarine* of the firft rank, who was Governor of the city of *Janfon*, together with two *Mandarines* of an inferior clafs, and a great retinue of officers and

fervants,

fervants, having with them eighteen half gallies, decorated with a great number of ftreamers, and furnifhed with mufic, and full of men, came to grapnel a-head of the *Centurion*; whence the *Mandarine* fent a meffage to the Commodore, telling him, that he (the *Mandarine*) was ordered, by the Viceroy of *Canton*, to examine the condition of the fhip, and defiring the fhip's boat might be fent to fetch him on board. The *Centurion*'s boat was immediately difpatched, and preparations were made for receiving him; for a hundred of the moft fightly of the crew were uniformly dreft in the regimentals of the marines, and were drawn up under arms on the main-deck, againft his arrival. When he entered the fhip he was faluted by the drums, and what other military mufic there was on board; and paffing by the new-formed guard, he was met by the Commodore on the quarter-deck, who conducted him to the great cabbin. Here the *Mandarine* explained his commiffion, declaring, that his bufinefs was to examine all the particulars mentioned in the Commodore's letter to the Viceroy, and to confront them with the reprefentation that had been given of them; that he was particularly inftructed to infpect the leak, and had for that purpofe brought with him two *Chinefe* Carpenters; and that for the greater regularity and difpatch of his bufinefs, he had every head of enquiry feparately wrote down on a fheet of paper, with a void fpace oppofite to it, where he was to infert fuch information and remarks thereon, as he could procure by his own obfervation.

This *Mandarine* appeared to be a perfon of very confiderable parts, and endowed with more franknefs and honefty, than is to be found in the generality of the *Chinefe*. After the proper enquiries had been made, particularly about the leak, which the *Chinefe* Carpenters reported to be as dangerous as it had been reprefented, and confequently that it was impoffible for the *Centurion* to proceed to fea without being refitted, the *Mandarine* expreffed himfelf fatisfied with the account given in the Commodore's letter. And this Magiftrate, as he was more intelligent than any other perfon of his nation that came to our knowledge, fo likewife was he more curious

and

and inquifitive, viewing each part of the fhip with particular attention, and appearing greatly furprized at the largenefs of the lower
deck guns, and at the weight and fize of the fhot. The Commodore, obferving his aftonifhment, thought this a proper opportunity to convince the *Chinefe* of the prudence of granting
him a fpeedy and ample fupply of all he wanted : With this view
he told the *Mandarine,* and thofe who were with him, that, befides
the demands he made for a general fupply, he had a particular complaint againft the proceedings of the Cuftom-houfe of *Macao* ;
that at his firft arrival the *Chinefe* boats had brought on board
plenty of greens, and variety of frefh provifions for daily ufe, for
which they had always been paid to their full fatisfaction, but that
the Cuftom houfe officers at *Macao* had foon forbid them, by which
means he was deprived of thofe refrefhments which were of the
utmoft confequence to the health of his men, after their long and
fickly voyage ; that as they, the *Mandarines,* had informed
themfelves of his wants, and were eye-witneffes of the force and
ftrength of his fhip, they might be fatisfied it was not for want of
power to fupply himfelf, that he defired the permiffion of the Government to purchafe what provifions he ftood in need of ; that
they muft be convinced that the *Centurion* alone was capable of deftroying the whole navigation of the port of *Canton,* or of any other
port in *China,* without running the leaft rifque from all the force
the *Chinefe* could collect ; that it was true, this was not the manner of proceeding between nations in friendfhip with each other, but
it was likewife true, that it was not cuftomary for any nation to permit the fhips of their friends to ftarve and fink in their ports, when
thofe friends had money to fupply their wants, and only defired liberty to lay it out ; that they muft confefs, he and his people had hitherto behaved with great modefty and referve, but that,
as his wants were each day encreafing, hunger would at laft prove
too ftrong for any reftraint, and neceffity was acknowledged in all
countries to be fuperior to every other law ; and therefore it could
not be expected that his crew would long continue to ftarve in the

A a a   midft

midft of that plenty to which their eyes were every day witnefses:
To this the Commodore added, (though perhaps with a lefs ferious
air) that if by the delay of fupplying him with provifion his men
fhould be reduced to the neceffity of turning cannibals, and preying
upon their own fpecies, it was eafy to be forefeen that, independent
of their friendfhip to their comrades, they would, in point of lux-
ury, prefer the plump well fed *Chinefe* to their own immaciated
fhipmates. The firft *Mandarine* acquiefced in the juftnefs of this
reafoning, and told the Commodore, that he fhould that night proceed for *Canton* ; that on his arrival, a Council of *Mandarines* would
be fummoned, of which he himfelf was a Member, and that by
being employed in the prefent Commiffion, he was of courfe the
Commodore's Advocate; that, as he was fully convinced of the
urgency of Mr. *Anfon*'s neceffity, he did not doubt but, on his reprefentation, the Council would be of the fame opinion ; and that
all that was demanded would be amply and fpeedily granted : And
with regard to the Commodore's complaint of the Cuftom-houfe of
Macao, he undertook to rectify that immediately by his own authority; for defiring a lift to be given him of the quantity of provifion neceffary for the expence of the fhip for a day, he wrote a permit under it, and delivered it to one of his attendants, directing him
to fee that quantity fent on board early every morning ; and this order, from that time forwards, was punctually complied with.

When this weighty affair was thus in fome degree regulated, the
Commodore invited him and his two attendant *Mandarines* to dinner, telling them at the fame time, that if his provifion, either in
kind or quantity, was not what they might expect, they muft thank
themfelves for having confined him to fo hard an allowance. One
of his difhes was beef, which the *Chinefe* all diflike, though Mr. *Anfon* was not apprized of it; this feems to be derived from the *Indian* fuperftition, which for fome ages paft has made a great progrefs in *China*. However, his guefts did not entirely faft ; for the
three *Mandarines* compleatly finifhed the white part of four large
fowls. But they were extremely embarraffed with their knives and

2.   forks,

forks, and were quite incapable of making ufe of them : So that, after fome fruitlefs attempts to help themfelves, which were fufficiently awkward, one of the attendants was obliged to cut their meat in fmall pieces for them. But whatever difficulty they might have in complying with the *European* manner of eating, they feemed not to be novices in drinking. The Commodore excufed himfelf in this part of the entertainment, under the pretence of illnefs ; but there being another Gentleman prefent, of a florid and jovial complexion, the chief *Mandarine* clapped him on the fhoulder, and told him by the interpreter, that certainly he could not plead ficknefs, and therefore infifted on his bearing him company ; and that Gentleman perceiving, that after they had difpatched four or five bottles of *Frontiniac*, the *Mandarine* ftill continued unruffled, he ordered a bottle of citron-water to be brought up, which the *Chinefe* feemed much to relifh, and this being near finifhed, they arofe from table, in appearance cool and uninfluenced by what they had drank, and the Commodore having, according to cuftom, made the *Mandarine* a prefent, they all departed in the fame veffels that brought them.

After their departure, the Commodore with great impatience expected the refolution of the Council, and the neceffary licences for his refitment. For it muft be obferved, as hath already appeared from the preceding narration, that he could neither purchafe ftores nor neceffaries with his money, nor did any kind of workmen dare to engage themfelves to work for him, without the permiffion of the Government firft obtained. And in the execution of thefe particular injunctions, the Magiftrates never fail of exercifing great feverity, they, notwithftanding the fuftian elogiums beftowed upon them by the Catholic Miffionaries and their *European* copiers, being compofed of the fame fragile materials with the reft of mankind, and often making ufe of the authority of the law, not to fupprefs crimes, but to enrich themfelves by the pillage of thofe who commit them ; for capital punifhments are rare in *China*, the effeminate genius of the nation, and their ftrong attachment to lucre,

A a a 2   difpofing

difpofing them rather to make ufe of fines ; and hence arifes no inconfiderable profit to thofe who compofe their tribunals : Confe-quently prohibitions of all kinds, particularly fuch, as the alluring profpect of great profit may often tempt the fubject to infringe, cannot but be favourite inftitutions in fuch a Government. But to return :

Some time before this, Captain *Saunders* took his paffage to *England* on board a *Swedifh* fhip, and was charged with difpatches from the Commodore ; and foon after, in the month of *December*, Cap-tain *Mitchel*, Colonel *Cracherode*, and Mr. *Taffel*, one of the Agent-Victuallers, with his nephew Mr. *Charles Herriot*, embarked on board fome of our Company's fhips ; and I, having obtained the Commodore's leave to return home, embarked with them. I muft obferve too, (having omitted it before) that whilft we lay here at *Macao*, we were informed by fome of the officers of our *India-men*, that the *Severn* and *Pearl*, the two fhips of our fquadron, which had feparated from us off Cape *Noir*, were fafely arrived at *Rio Janeiro* on the coaft of *Brazil*. I have formerly taken notice, that at the time of their feparation, we apprehended them to be loft. And there were many reafons which greatly favoured this fuf-picion : For we knew that the *Severn* in particular was extreamly fickly ; and this was the more obvious to the reft of the fhips, as, in the preceding part of the voyage, her Commander Capt. *Legg* had been remarkable for his exemplary punctuality in keeping his ftation, till, for the laft ten days before his feparation, his crew was fo di-minifhed and enfeebled, that with his utmoft efforts it was not poffible for him to maintain his proper pofition with his wonted exactnefs. The extraordinary ficknefs on board him was by many imputed to the fhip, which was new, and on that account was believed to be the more unhealthy ; but whatever was the caufe of it, the *Severn* was by much the moft fickly of the fquadron : For before her departure from St. *Catherine's* fhe buried more men than any of them, info-much that the Commodore was obliged to recruit her with a num-ber of frefh hands ; and, the mortality ftill continuing on board her,

2

her, she was supplied with men a second time at sea, after our setting sail from St. *Julians*; and notwithstanding these different reinforcements, she was at last reduced to the distressed condition I have already mentioned : So that the Commodore himself was firmly persuaded she was lost; and therefore it was with great joy we received the news of her and the *Pearl*'s safety, after the strong persuasion, which had so long prevailed amongst us, of their having both perished. But to proceed with the transactions between Mr. *Anson* and the *Chinese*.

Notwithstanding the favourable disposition of the *Mandarine* Governor of *Janson*, at his leaving Mr. *Anson*, several days were elapsed before he had any advice from him; and Mr. *Anson* was privately informed there were great debates in Council upon his affair; partly perhaps owing to its being so unusual a case, and in part to the influence, as I suppose, of the intrigues of the *French* at *Canton*: For they had a countryman and fast friend residing on the spot, who spoke the language very well, and was not unacquainted with the venality of the Government, nor with the persons of several of the Magistrates, and consequently could not be at a loss for means of traversing the assistance desired by Mr. *Anson*. And this opposition of the *French* was not merely the effect of national prejudice or contrariety of political interests, but was in good measure owing to their vanity, a motive of much more weight with the generality of mankind, than any attachment to the public service of their community: For, the *French* pretending their *India-men* to be Men of War, their officers were apprehensive, that any distinction granted to Mr. *Anson*, on account of his bearing the King's Commission, would render them less considerable in the eyes of the *Chinese*, and would establish a prepossession at *Canton* in favour of ships of war, by which they, as trading vessels, would suffer in their importance: And I wish the affectation of endeavouring to pass for men of war, and the fear of sinking in the estimation of the *Chinese*, if the *Centurion* was treated in a different manner from themselves, had been confined to the officers of the *French* ships only,

only. However, notwithſtanding all theſe obſtacles, it ſhould ſeem, that the repreſentation of the Commodore to the *Mandarines* of the facility with which he could right himſelf, if juſtice were denied him, had at laſt its effect: For, on the 6th of *January*, in the morning, the Governor of *Janſon*, the Commodore's Advocate, ſent down the Viceroy of *Canton*'s warrant for the refitment of the *Centurion*, and for ſupplying her people with all they wanted; and, the next day, a number of *Chineſe* Smiths and Carpenters went on board, to agree for all the work by the great. They demanded at firſt, to the amount of a thouſand pounds ſterling for the neceſſary repairs of the ſhip, the boats, and the maſts: This the Commodore ſeemed to think an unreaſonable ſum, and endeavoured to perſuade them to work by the day; but that propoſal they would not hearken to; ſo it was at laſt agreed, that the Carpenters ſhould have to the amount of about ſix hundred pounds for their work; and that the Smiths ſhould be paid for their iron-work by weight, allowing them at the rate of three pounds a hundred nearly for the ſmall work, and forty-ſix ſhillings for the large.

This being regulated, the Commodore exerted himſelf to get this moſt important buſineſs compleated; I mean, the heaving down the *Centurion*, and examining the ſtate of her bottom: For this purpoſe the firſt Lieutenant was diſpatched to *Canton* to hire two country veſſels, called in their language junks, one of them being intended to heave down by, and the other to ſerve as a magazine for the powder and ammunition: At the ſame time the ground was ſmoothed on one of the neighbouring Iſlands, and a large tent was pitched for lodging the lumber and proviſions, and near a hundred *Chineſe* Caulkers were ſoon ſet to work on the decks and ſides of the ſhip. But all theſe preparations, and the getting ready the careening gear, took up a great deal of time; for the *Chineſe* Caulkers, though they worked very well, were far from being expeditious; and it was the 26th of *January* before the junks arrived; and the neceſſary materials, which were to be purchaſed at *Canton*, came down very ſlowly; partly from the diſtance of the place, and

and partly from the delays and backwardnefs of the *Chinefe* Merchants. And in this interval Mr. *Anfon* had the additional perplexity to difcover, that his fore-maft was broken afunder above the upper deck partners, and was only kept together by the fifhes which had been formerly clapt upon it.

However, the *Centurion*'s people made the moft of their time, and exerted themfelves the beft they could; and as, by clearing the fhip, the Carpenters were enabled to come at the leak, they took care to fecure that effectually, whilft the other preparations were going forwards. The leak was found to be below the fifteen foot mark, and was principally occafioned by one of the bolts being wore away and loofe in the joining of the ftem where it was fcarfed.

At laft, all things being prepared, they, on the 22d of *February*, in the morning, hove out the firft courfe of the *Centurion*'s ftarboard fide, and had the fatisfaction to find, that her bottom appeared found and good; and, the next day, (having by that time compleated the new fheathing of the firft courfe) they righted her again, to fet up anew the careening rigging which ftretched much. Thus they continued heaving down, and often righting the fhip from a fufpicion of their careening tackle, till the 3d of *March*; when, having compleated the paying and fheathing the bottom, which proved to be every where very found, they, for the laft time, righted the fhip to their great joy; for not only the fatigue of careening had been confiderable, but they had been apprehenfive of being attacked by the *Spaniards*, whilft the fhip was thus incapacitated for defence. Nor were their fears altogether groundlefs; for they learnt afterwards, by a *Portuguefe* veffel, that the *Spaniards* at *Manila* had been informed, that the *Centurion* was in the *Typa*, and intended to careen there; and that thereupon the Governor had fummoned his Council, and had propofed to them to endeavour to burn her, whilft fhe was careening, which was an enterprize, which, if properly conducted, might have put them in great danger: They were farther told, that this fcheme was not only propofed, but re-

solved

folved on ; and that a Captain of a veffel had actually undertaken to perform the bufinefs for forty thoufand dollars, which he was not to receive unlefs he fucceeded ; but the Governor pretending that there was no treafure in the royal cheft, and infifting that the Merchants fhould advance the money, and they refufing to comply with the demand, the affair was dropped : Perhaps the Merchants fufpected, that the whole was only a pretext to get forty thoufand dollars from them ; and indeed this was affirmed by fome who bore the Governor no good will, but with what truth it is difficult to afcertain.

As foon as the *Centurion* was righted, they took in her powder, and gunners ftores, and proceeded in getting in their guns as faft as poffible, and then ufed their utmoft expedition in repairing the fore-maft, and in compleating the other articles of her refitment. And being thus employed, they were alarmed, on the 10th of *March*, by a *Chinefe* Fifherman, who brought them intelligence that he had been on board a large *Spanifh* fhip off the grand *Ladrone*, and that there were two more in company with her : He added feveral particulars to his relation ; as that he had brought one of their officers to *Macao*, and that, on this, boats went off early in the morning from *Macao* to them : And the better to eftablifh the belief of his veracity, he faid he defired no money, if his information fhould not prove true. This was prefently believed to be the forementioned expedition from *Manila* ; and the Commodore immediately fitted his cannon and fmall arms in the beft manner he could for defence ; and having then his Pinnace and Cutter in the offing, who had been ordered to examine a *Portuguefe* veffel, which was getting under fail, he fent them the advice he had received, and directed them to look out ftrictly : But no fuch fhips ever appeared, and they were foon fatisfied, the whole of the ftory was a fiction ; though it was difficult to conceive what reafon could induce the fellow to be at fuch extraordinary pains to impofe on them.

It was the beginning of *April* before they had new-rigged the fhip, ftowed their provifions and water on board, and had fitted

her

her for the sea; and before this time the *Chinese* grew very uneasy, and extremely desirous that she should be gone; either not knowing, or pretending not to believe, that this was a point the Commodore was as eagerly set on as they could be. On the 3d of *April*, two *Mandarine* boats came on board from *Macao* to urge his departure; and this having been often done before, tho' there had been no pretence to suspect Mr. *Anson* of any affected delays, he at this last message answered them in a determined tone, desiring them to give him no further trouble, for he would go when he thought proper, and not before. On this rebuke the *Chinese* (though it was not in their power to compel him to be gone) immediately prohibited all provisions from being carried on board him, and took such care that their injunctions should be complied with, that from that time forwards nothing could be purchased at any rate whatever.

On the 6th of *April*, the *Centurion* weighed from the *Typa*, and warped to the southward; and, by the 15th, she was got into *Macao* road, compleating her water as she past along, so that there remained now very few articles more to attend to; and her whole business being finished by the 19th, she, at three in the afternoon of that day, weighed and made sail, and stood to sea.

Bbb   CHAP.

C H A P. VIII.

From *Macao* to Cape *Eſpiritu Santo*: The taking of the *Manila* galeon, and returning back again.

THE Commodore was now got to ſea, with his ſhip very well refitted, his ſtores repleniſhed, and an additional ſtock of proviſions on board: His crew too was ſomewhat reinforced; for he had entered twenty-three men during his ſtay at *Macao*, the greateſt part of which were Laſcars or *Indian* ſailors, and ſome few *Dutch*. He gave out at *Macao*, that he was bound to *Batavia*, and thence to *England*; and though the weſterly monſoon was now ſet in, when that paſſage is conſidered as impracticable, yet, by the confidence he had expreſſed in the ſtrength of his ſhip, and the dexterity of his people, he had perſuaded not only his own crew but the people at *Macao* likewiſe, that he propoſed to try this unuſual experiment; ſo that there were many letters put on board him by the inhabitants of *Canton* and *Macao* for their friends at *Batavia*.

But his real deſign was of a very different nature: For he knew, that inſtead of one annual ſhip from *Acapulco* to *Manila*, there would be this year, in all probability, two; ſince, by being before *Acapulco*, he had prevented one of them from putting to ſea the preceding feaſon. He therefore reſolved to cruiſe for theſe returning veſſels off Cape *Eſpiritu Santo*, on the Iſland of *Samal*, which is the firſt land they always make in the *Philippine* Iſlands. And as *June* is generally the month in which they arrive there, he doubted not but he ſhould get to his intended ſtation time enough to intercept them. It is true, they were ſaid to be ſtout veſſels, mounting forty-four guns apiece, and carrying above five hundred hands, and might be expected to return in company; and he himſelf had

but

but two hundred and twenty-feven hands on board, of which near thirty were boys : But this difproportion of ftrength did not deter him, as he knew his fhip to be much better fitted for a fea-engagement than theirs, and as he had reafon to expect that his men would exert themfelves in the moft extraordinary manner, when they had in view the immenfe wealth of thefe *Manila* galeons.

This project the Commodore had refolved on in his own thoughts, ever fince his leaving the coaft of *Mexico*. And the greateft mortification which he received, from the various delays he had met with in *China*, was his apprehenfion, left he might be thereby fo long retarded as to let the galeons efcape him. Indeed, at *Macao* it was incumbent on him to keep thefe views extremely fecret ; for there being a great intercourfe and a mutual connexion of interefts between that port and *Manila*, he had reafon to fear, that, if his defigns were difcovered, intelligence would be immediately fent to *Manila*, and meafures would be taken to prevent the galeons from falling into his hands : But being now at fea, and entirely clear of the coaft, he fummoned all his people on the quarter-deck, and informed them of his refolution to cruife for the two *Manila* fhips, of whofe wealth they were not ignorant. He told them he fhould chufe a ftation, where he could not fail of meeting with them ; and though they were ftout fhips, and full manned, yet, if his own people behaved with their accuftomed fpirit, he was certain he fhould prove too hard for them both, and that one of them at leaft could not fail of becoming his prize : He further added, that many ridiculous tales had been propagated about the ftrength of the fides of thefe fhips, and their being impenetrable to cannon-fhot ; that thefe fictions had been principally invented to palliate the cowardice of thofe who had formerly engaged them ; but he hoped there were none of thofe prefent weak enough to give credit to fo abfurd a ftory : For his own part, he did affure them upon his word, that, whenever he met with them, he would fight them fo near, that they fhould find, his bullets, inftead of being ftopped by one of their fides, fhould go through them both.

B b b 2   This

This speech of the Commodore's was received by his people with great joy: For no sooner had he ended, than they expressed their approbation, according to naval custom, by three strenuous cheers, and all declared their determination to succeed or perish, whenever the opportunity presented itself. And now their hopes, which since their departure from the coast of *Mexico*, had entirely subsided, were again revived; and they all persuaded themselves, that, notwithstanding the various casualties and disappointments they had hitherto met with, they should yet be repaid the price of their fatigues, and should at last return home enriched with the spoils of the enemy: For firmly relying on the assurances of the Commodore, that they should certainly meet with the vessels, they were all of them too sanguine to doubt a moment of mastering them; so that they considered themselves as having them already in their possession. And this confidence was so universally spread through the whole ship's company, that, the Commodore having taken some *Chinese* sheep to sea with him for his own provision, and one day enquiring of his Butcher, why, for some time past, he had seen no mutton at his table, asking him if all the sheep were killed, the Butcher very seriously replied, that there were indeed two sheep left, but that if his Honour would give him leave, he proposed to keep those for the entertainment of the General of the galeons.

When the *Centurion* left the port of *Macao*, she stood for some days to the westward; and, on the first of *May*, they saw part of the Island of *Formosa*; and, standing thence to the southward, they, on the 4th of *May*, were in the latitude of the *Bashee Islands*, as laid down by *Dampier*; but they suspected his account of inaccuracy, as they found that he had been considerably mistaken in the latitude of the South end of *Formosa*: For this reason they kept a good look-out, and about seven in the evening discovered from the mast-head five small Islands, which were judged to be the *Bashees*, and they had afterwards a sight of *Botel Tobago Xima*. By this means they had an opportunity of correcting the position of the *Bashee Islands*, which had been hitherto laid down twenty-five

leagues

A view of CAPE ESPIRITU SANTO, on SAMAL, one of the Phillipine I...
represented his Majestys Ship the CENTURION engag'd and took the Spanish Galeon cal...

Plate XL.

J. Mason Sculp.

's, in the latitude of 12:40ᵐ Nᵒ. Bearing W S W distant 6 leagues. In the position here.

NOSTRA SEIGNIORA DE CABADONGA, *from* ACAPULCO *bound to* MANILA

leagues too far to the weftward: For by their obfervations, they efteemed the middle of thefe Iflands to be in 21° : 4 North, and to bear from *Botel Tobago Xima* S. S. E. twenty leagues diftant, that Ifland itfelf being in 21° : 57′ North.

After getting a fight of the *Bafhee Iflands*, they ftood between the S. and S. W for Cape *Efpiritu Santo*; and, the 20th of *May* at noon, they firft difcovered that Cape, which about four o'clock they brought to bear S. S. W, about eleven leagues diftant. It appeared to be of a moderate height, with feveral round hummocks on it; and is exactly reprefented in the annexed plate. As it was known that there were centinels placed upon this Cape to make fignals to the *Acapulco* fhip, when fhe firft falls in with the land, the Commodore immediately tacked, and ordered the top-gallant fails to be taken in, to prevent being difcovered; and, this being the ftation in which it was refolved to cruife for the galeons, they kept the Cape between the South and the Weft, and endeavoured to confine themfelves between the latitude of 12° : 50, and 13° : 5, the Cape itfelf lying, by their obfervations, in 12° : 40 North, and in 4° of Eaft longitude from *Botel Tobago Xima*.

It was the laft of *May*, by the foreign ftile, when they arrived off this Cape; and, the month of *June*, by the fame ftile, being that in which the *Manila* fhips are ufually expected, the *Centurion*'s people were now waiting each hour with the utmoft impatience for the happy crifis which was to ballance the account of all their paft calamities. As from this time there was but fmall employment for the crew, the Commodore ordered them almoft every day to be exercifed in the management of the great guns, and in the ufe of their fmall arms. This had been his practice, more or lefs, at all convenient feafons during the whole courfe of his voyage; and the advantages which he received from it, in his engagement with the galeon, were an ample recompence for all his care and attention. Indeed, it fhould feem that there are few particulars of a Commander's duty of more importance than this, how much foever it may have been fometimes overlooked or mifunder-

2   ftood:

ftood: For it will, I fuppofe, be confeffed, that in two fhips of war, equal in the number of their men and guns, the difproportion of ftrength, arifing from a greater or lefs dexterity in the ufe of their great guns and fmall arms, is what can fcarcely be ballanced by any other circumftances whatever. For, as thefe are the weapons with which they are to engage, what greater inequality can there be betwixt two contending parties, than that one fide fhould perfectly underftand the ufe of their weapons, and fhould have the fkill to employ them in the moft effectual manner for the annoyance of their enemy, while the other fide fhould, by their awkward management of them, render them rather terrible to themfelves, than mifchievous to their antagonifts? This feems fo plain and natural a conclufion, that a perfon unacquainted with thefe affairs would fuppofe the firft care of a Commander to be the training his people to the ufe of their arms.

But human affairs are not always conducted by the plain dictates of common fenfe. There are many other principles which influence our tranfactions: And there is one in particular, which, though of a very erroneous complexion, is fcarcely ever excluded from our moft ferious deliberations; I mean cuftom, or the practice of thofe who have preceded us. This is ufually a power too mighty for reafon to grapple with; and is the moft terrible to thofe who oppofe it, as it has much of fuperftition in its nature, and purfues all thofe who queftion its authority with unrelenting vehemence. However, in thefe later ages of the world, fome lucky encroachments have been made upon its prerogative; and it may reafonably be hoped, that the Gentlemen of the Navy, whofe particular profeffion hath of late been confiderably improved by a number of new inventions, will of all others be the readieft to give up thofe practices, which have nothing to plead but prefcription, and will not fuppofe that every branch of their bufinefs hath already received all the perfection of which it is capable. Indeed, it muft be owned, that if a dexterity in the ufe of fmall arms, for inftance, hath been fometimes lefs attended to on board our fhips of

war,

war, than might have been wifhed for, it hath been rather owing to unfkilful methods of teaching it, than to negligence : For the common failors, how ftrongly foever attached to their own prejudices, are very quick fighted in finding out the defeds of others, and have ever fhewn a great contempt for the formalities practifed in the training of land troops to the ufe of their arms; but when thofe who have undertaken to inftruct the feamen have contented themfelves with inculcating only what was ufeful, and that in the fimpleft manner, they have conftantly found their people fufficiently docile, and the fuccefs hath even exceeded their expectation. Thus on board Mr. *Anfon*'s fhip, where they were only taught the fhorteft method of loading with cartridges, and were conftantly trained to fire at a mark, which was ufually hung at the yard-arm, and where fome little reward was given to the moft expert, the whole crew, by this management, were rendered extremely fkilful, quick in loading, all of them good markfmen, and fome of them moft extraordinary ones; fo that I doubt not but, in the ufe of fmall arms, they were more than a match for double their number, who had not been habituated to to the fame kind of exercife. But to return :

It was the laft of *May*, *N. S.* as hath been already faid, when the *Centurion* arrived off Cape *Efpiritu Santo* ; and confequently the next day began the month in which the galeons were to be expected. The Commodore therefore made all neceffary preparations for receiving them, having hoifted out his long boat, and lafhed her along fide, that the fhip might be ready for engaging, if they fell in with the galeons in the night. All this time too he was very folicitous to keep at fuch a diftance from the Cape, as not to be difcovered : But it hath been fince learnt, that, notwithftanding his care, he was feen from the land ; and advice of him was fent to *Manila*, where it was at firft difbelieved, but on reiterated intelligence (for it feems he was feen more than once) the Merchants were alarmed, and the Governor was applied to, who undertook (the commerce fupplying the neceffary fums) to fit out a force confift-

ing

ing of two ſhips of thirty-two guns, one of twenty guns, and two ſloops of ten guns each, to attack the *Centurion* on her ſtation: And ſome of theſe veſſels did actually weigh with this view; but the principal ſhip not being ready, and the monſoon being a-gainſt them, the Commerce and the Governor diſagreed, and the en-terprize was laid aſide. This frequent diſcovery of the *Centurion* from the ſhore was ſomewhat extraordinary; for the pitch of the Cape is not high, and ſhe uſually kept from ten to fifteen leagues diſtant; though once indeed, by an indraught of the tide, as was ſuppoſed, they found themſelves in the morning within ſeven leagues of the land.

As the month of *June* advanced, the expectancy and impatience of the Commodore's people each day encreaſed. And I think no better idea can be given of their great eagerneſs on this occaſion, than by copying a few paragraphs from the journal of an officer, who was then on board; as it will, I preſume, be a more natural picture of the full attachment of their thoughts to the buſineſs of their cruiſe, than can be given by any other means. The paragraphs I have ſelected, as they occur in order of time, are as follow:

" *May* 31, Exerciſing our men at their quarters, in great expecta-
" tion of meeting with the galeons very ſoon; this being the
" eleventh of *June* their ſtile."

" *June* 3, Keeping in our ſtations, and looking out for the
" galeons."

" *June* 5, Begin now to be in great expectation, this being the
" middle of *June* their ſtile."

" *June* 11, Begin to grow impatient at not ſeeing the galeons."

" *June* 13, The wind having blown freſh eaſterly for the forty-
" eight hours paſt, gives us great expectations of ſeeing the galeons
" ſoon."

" *June* 15, Cruiſing on and off, and looking out ſtrictly."

" *June* 19, This being the laſt day of *June, N. S.* the ga-
" leons, if they arrive at all, muſt appear ſoon."

From

From thefe famples it is fufficiently evident, how compleatly the treafure of the galeons had engroffed their imagination, and how anxioufly they paffed the latter part of their cruife, when the certainty of the arrival of thefe veffels was dwindled down to probability only, and that probability became each hour more and more doubtful. However, on the 20th of *June*, *O. S.* being juft a month from their arrival on their ftation, they were relieved from this ftate of uncertainty; when, at fun-rife, they difcovered a fail from the maft-head, in the S: E. quarter. On this, a general joy fpread through the whole fhip; for they had no doubt but this was one of the galeons, and they expected foon to fee the other. The Commodore inftantly ftood towards her, and at half an hour after feven they were near enough to fee her from the *Centurion*'s deck; at which time the galeon fired a gun, and took in her top-gallant fails, which was fuppofed to be a fignal to her confort, to haften her up; and therefore the *Centurion* fired a gun to leeward, to amufe her. The Commodore was furprized to find, that in all this time the galeon did not change her courfe, but continued to bear down upon him; for he hardly believed, what afterwards appeared to be the cafe, that fhe knew his fhip to be the *Centurion*, and refolved to fight him.

About noon the Commodore was little more than a league diftant from the galeon, and could fetch her wake, fo that fhe could not now efcape; and, no fecond fhip appearing, it was concluded that fhe had been feparated from her confort. Soon after, the galeon haled up her fore-fail, and brought too under top-fails, with her head to the northward, hoifting *Spanish* colours, and having the ftandard of *Spain* flying at the top-gallant maft-head. Mr. *Anfon*, in the mean time, had prepared all things for an engagement on board the *Centurion*, and had taken all poffible care, both for the moft effectual exertion of his fmall ftrength, and for the avoiding the confufion and tumult, too frequent in actions of this kind. He picked out about thirty of his choiceft hands and beft markf-men, whom he diftributed into his tops, and who fully anfwered

C c c   his

his expectation, by the fignal fervices they performed As he had
not hands enough remaining to quarter a fufficient number to each
great gun, in the cuftomary manner, he therefore, on his lower
tire, fixed only two men to each gun, who were to be folely employed in loading it, whilft the reft of his people were divided into
different gangs of ten or twelve men each, which were conftantly
moving about the decks, to run out and fire fuch guns as were
loaded. By this management he was enabled to make ufe of all
his guns; and inftead of firing broad-fides with intervals between
them, he kept up a conftant fire without intermiffion, whence he
doubted not to procure very fignal advantages; for it is common
with the *Spaniards* to fall down upon the decks when they fee a
broadfide preparing, and to continue in that pofture till it is given;
after which they rife again, and, prefuming the danger to be for
fome time over, work their guns and fire with great brifknefs,
till another broad-fide is ready: But the firing gun by gun, in
the manner directed by the Commodore, rendered this practice of
theirs impoffible.

The *Centurion* being thus prepared, and nearing the galeon a-
pace, there happened, a little after noon, feveral fqualls of wind
and rain, which often obfcured the galeon from their fight; but
whenever it cleared up, they obferved her refolutely lying to; and,
towards one o'clock, the *Centurion* hoifted her broad pendant and
colours, fhe being then within gun-fhot of the enemy. And the
Commodore obferving the *Spaniards* to have neglected clearing their
fhip till that time, as he then faw them throwing over-board cattle
and lumber, he gave orders to fire upon them with the chace guns,
to embarafs them in their work, and prevent them from compleating it, though his general directions had been not to engage till
they were within piftol fhot. The galeon returned the fire with
two of her ftern-chace; and, the *Centurion* getting her fprit-fail-
yard fore and aft, that if neceffary fhe might be ready for boarding, the *Spaniards* in a bravado rigged their fprit-fail-yard fore and
aft likewife. Soon after, the *Centurion* came abreaft of the enemy

4   within

within piftol-fhot, keeping to the leeward with a view of preventing them from putting before the wind, and gaining the port
of *Jalapay*, from which they were about feven leagues diftant.
And now the engagement began in earneft, and, for the firft half
hour, Mr. *Anfon* over-reached the galeon, and lay on her bow ;
where, by the great widenefs of his ports he could traverfe almoft
all his guns upon the enemy, whilft the galeon could only bring a
part of hers to bear.   Immediately, on the commencement of the
action, the mats, with which the galeon had ftuffed her netting,
took fire, and burnt violently, blazing up half as high as the mi-
zen-top.   This accident (fuppofed to be caufed by the *Centurion*'s
wads) threw the enemy into great confufion, and at the fame time
alarmed the Commodore, for he feared leaft the galeon fhould be
burnt, and leaft he himfelf too might fuffer by her driving on board
him : But the *Spaniards* at laft freed themfelves from the fire, by
cutting away the netting, and tumbling the whole mafs which
was in flames into the fea.   But ftill the *Centurion* kept her firft
advantageous pofition, firing her cannon with great regularity and
brifknefs, whilft at the fame time the galeon's decks lay open to her
topmen, who, having at their firft volley driven the *Spaniards* from
their tops, made prodigious havock with their fmall arms, killing or
wounding every officer but one that ever appeared on the quarter-
deck, and wounding in particular the General of the galeon himfelf.   And though the *Centurion*, after the firft half hour, loft her
original fituation, and was clofe along-fide the galeon, and the enemy continued to fire brifkly for near an hour longer, yet at laft the
Commodore's grape-fhot fwept their decks fo effectually, and the
number of their flain and wounded was fo confiderable, that they
began to fall into great diforder, efpecially as the General, who was
the life of the action, was no longer capable of exerting himfelf.
Their embarafment was vifible from on board the Commodore.
For the fhips were fo near, that fome of the *Spanifh* officers were
feen running about with great affiduity, to prevent the defertion of
their men from their quarters : But all their endeavours were in

vain :

vain ; for after having, as a laſt effort, fired five or ſix guns with more judgment than uſual, they gave up the conteſt ; and, the galeon's colours being ſinged off the enſign ſtaff in the beginning of the engagement, ſhe ſtruck the ſtandard at her main-top-gallant maſt-head, the perſon, who was employed to do it, having been in imminent peril of being killed, had not the Commodore, who perceived what he was about, given expreſs orders to his people to deſiſt from firing.

Thus was the *Centurion* poſſeſſed of this rich prize, amounting in value to near a million and half of dollars. She was called the *Noſtra Signora de Cabadonga*, and was commanded by the General *Don Jeronimo de Montero*, a *Portugueſe* by birth, and the moſt approved officer for ſkill and courage of any employed in that ſervice. The galeon was much larger than the *Centurion*, had five hundred and fifty men and thirty-ſix guns mounted for action, beſides twenty-eight pidreroes in her gunwale, quarters and tops, each of which carried a four pound ball. She was very well furniſhed with ſmall arms, and was particulary provided againſt boarding, both by her cloſe quarters, and by a ſtrong net-work of two inch rope, which was laced over her waiſt, and was defended by half pikes. She had ſixty-ſeven killed in the action, and eighty-four wounded, whilſt the *Centurion* had only two killed, and a Lieutenant and ſixteen wounded, all of whom but one recovered : Of ſo little conſequence are the moſt deſtructive arms in untutored and unpractiſed hands.

The treaſure thus taken by the *Centurion* having been for at leaſt eighteen months the great object of their hopes, it is impoſſible to deſcribe the tranſport on board, when, after all their reiterated diſappointments, they at laſt ſaw their wiſhes accompliſhed. But their joy was near being ſuddenly damped by a moſt tremendous incident : For no ſooner had the galeon ſtruck, than one of the Lieutenants coming to Mr. *Anſon* to congratulate him on his prize, whiſpered him at the ſame time, that the *Centurion* was dangerouſly on fire near the powder-room. The Commodore received this

I   dreadful

dreadful news without any apparent emotion, and, taking care not to alarm his people, gave the neceſſary orders for extinguiſhing it, which was happily done in a ſhort time, though its appearance at firſt was extremely terrible. It ſeems ſome cartridges had been blown up by accident between decks, whereby a quantity of oakum in the after-hatch way, near the after powder-room, was ſet on fire ; and the great ſmother and ſmoke of the oakum occaſioned the apprehenſion of a more extended and miſchievous fire. At the ſame inſtant too, the galeon fell on board the *Centurion* on the ſtarboard quarter, but ſhe was cleared without doing or receiving any conſiderable damage.

The Commodore made his firſt Lieutenant, Mr. *Saumarez*, Captain of this prize, appointing her a poſt-ſhip in his Majeſty's ſervice. Captain *Saumarez*, before night, ſent on board the *Centurion* all the *Spaniſh* priſoners, but ſuch as were thought the moſt proper to be retained to aſſiſt in navigating the galeon. And now the Commodore learnt, from ſome of theſe priſoners, that the other ſhip, which he had kept in the port of *Acapulco* the preceding year, inſtead of returning in company with the preſent prize as was expected, had ſet ſail from *Acapulco* alone much ſooner than uſual, and had, in all probability, got into the port of *Manila* long before the *Centurion* arrived off *Eſpiritu Santo* ; ſo that Mr. *Anſon*, notwithſtanding his preſent ſucceſs, had great reaſon to regret his loſs of time at *Macao*, which prevented him from taking two rich prizes inſtead of one.

The Commodore, when the action was ended, reſolved to make the beſt of his way with his prize for the river of *Canton*, being in the mean time fully employed in ſecuring his priſoners, and in removing the treaſure from on board the galeon into the *Centurion*. The laſt of theſe operations was too important to be poſtponed ; for as the navigation to *Canton* was through ſeas but little known, and where, from the ſeaſon of the year, much bad weather might be expected, it was of great conſequence that the treaſure ſhould be ſent on board the *Centurion*, which ſhip, by the preſence of the

Commander

Commander in Chief, the greater number of her hands, and her other advantages, was doubtlefs much fafer againft all the cafualties of winds and feas than the galeon: And the fecuring the prifoners was a matter of ftill more confequence, as not only the poffeffion of the treafure, but the lives of the captors depended thereon. This was indeed an article which gave the Commodore much trouble and difquietude ; for they were above double the number of his own people ; and fome of them, when they were brought on board the *Centurion*, and had obferved how flenderly fhe was manned, and the large proportion which the ftriplings bore to the reft, could not help expreffing themfelves with great indignation to be thus beaten by a handful of boys. The method, which was taken to hinder them from rifing, was by placing all but the officers and the wounded in the hold, where, to give them as much air as poffible, two hatchways were left open ; but then (to avoid all danger, whilft the *Centurion*'s people fhould be employed upon the deck) there was a fquare partition of thick planks, made in the fhape of a funnel, which enclofed each hatch-way on the lower deck, and reached to that directly over it on the upper deck ; thefe funnels ferved to communicate the air to the hold better than could have been done without them ; and, at the fame time, added greatly to the fecurity of the fhip ; for they being feven or eight feet high, it would have been extremely difficult for the *Spaniards* to have clambered up ; and ftill to augment that difficulty, four fwivel guns loaded with mufquet-bullets were planted at the mouth of each funnel, and a centinel with lighted match conftantly attended, prepared to fire into the hold amongft them, in cafe of any difturbance. Their officers, which amounted to feventeen or eighteen, were all lodged in the firft Lieutenant's cabbin, under a conftant guard of fix men ; and the General, as he was wounded, lay in the Commodore's cabbin with a centinel always with him ; and they were all informed, that any violence or difturbance would be punifhed with inftant death. And that the *Centurion*'s people might be at all times prepared, if, notwithftanding thefe regula-

tions,

tions, any tumult fhould arife, the fmall arms were conftantly kept loaded in a proper place, whilft all the men went armed with cut-laffes and piftols ; and no officer ever pulled off his cloaths, and when he flept had always his arms lying ready by him.

Thefe meafures were obvioufly neceffary, confidering the ha-zards to which the Commodore and his people would have been expofed, had they been lefs careful. Indeed, the fufferings of the poor prifoners, though impoffible to be alleviated, were much to be commiferated ; for the weather was extremely hot, the ftench of the hold loathfome beyond all conception, and their allowance of water but juft fufficient to keep them alive, it not being practicable to fpare them more than at the rate of a pint a day for each, the crew themfelves having only an allowance of a pint and an half. All this confidered, it was wonderful that not a man of them died during their long confinement, except three of the wounded, who died the fame night they were taken ; though it muft be con-feffed, that the greateft part of them were ftrangely metamorphifed by the heat of the hold ; for when they were firft taken, they were fightly robuft fellows ; but when, after above a month's imprifon-ment, they were difcharged in the river of *Canton*, they were re-duced to mere fkeletons ; and their air and looks correfponded much more to the conception formed of ghofts and fpectres, than to the figure and appearance of real men.

Thus employed in fecuring the treafure and the prifoners, the Commodore, as hath been faid, ftood for the river of *Canton* ; and, on the 30th of *June*, at fix in the evening, got fight of Cape *De-langano*, which then bore Weft ten leagues diftant ; and, the next day, he made the *Bafhee Iflands*, and the wind being fo far to the northward, that it was difficult to weather them, it was refolved to ftand through between *Grafton* and *Monmouth Iflands*, where the paffage feemed to be clear ; but in getting through, the fea had a very dangerous afpect, for it ripled and foamed, as if it had been full of breakers, which was ftill more terrible, as it was then night. But the fhips got through very fafe, (the prize always keep-

ing

ing a head) and it was found that the appearance which had alarmed them had been occasioned only by a strong tide. I must here observe, that though the *Bashee Islands* are usually reckoned to be no more than five, yet there are many more lying about them to the westward, which, as the channels amongst them are not at all known, makes it adviseable for ships, rather to pass to the northward or southward, than through them; and indeed the Commodore proposed to have gone to the northward, between them and *Formosa*, had it been possible for him to have weathered them. From hence the *Centurion* steering the proper course for the river of *Canton*, she, on the 8th of *July*, discovered the Island of *Supata*, the westermost of the *Lema* Islands, being the double peaked rock, particularly delineated in the view of the Islands of *Lema*, formerly inserted. This Island of *Supata* they made to be a hundred and thirty-nine leagues distant from *Grafton*'s Island, and to bear from it North 82° 37 West: And, on the 11th, having taken on board two *Chinese* Pilots, one for the *Centurion*, and the other for the prize, they came to an anchor off the city of *Macao*.

By this time the particulars of the cargoe of the galeon were well ascertained, and it was found that she had on board 1,313,843 pieces of eight, and 35,682 *oz.* of virgin silver, besides some cochineal, and a few other commodities, which, however, were but of small account, in comparison of the specie. And this being the Commodore's last prize, it hence appears, that all the treasure taken by the *Centurion* was not much short of 400,000 *l.* independent of the ships and merchandise, which she either burnt or destroyed, and which, by the most reasonable estimation, could not amount to so little as 600,000 *l.* more; so that the whole loss of the enemy, by our squadron, did doubtless exceed a million sterling. To which, if there be added the great expence of the Court of *Spain*, in fitting out *Pizarro*, and in paying the additional charges in *America*, incurred on our account, together with the loss of their men of war, the total of all these articles will be a most exorbitant sum, and is the strongest conviction of the utility of this expedition, which,

with

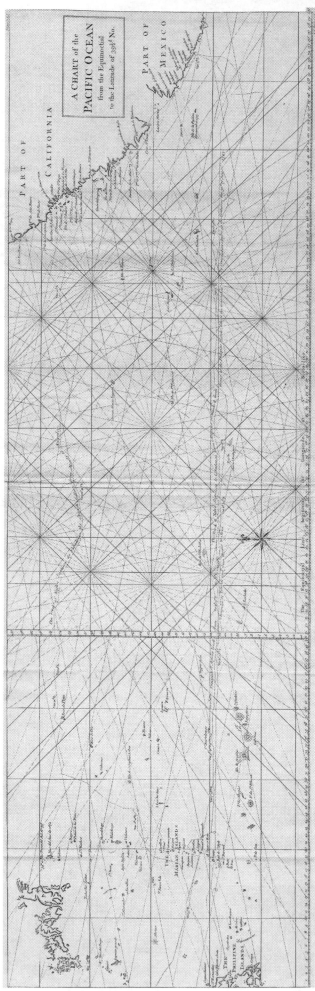

The material originally positioned here is too large for reproduction in this reissue. A PDF can be downloaded from the web address given on page iv of this book, by clicking on 'Resources Available'.

with all its numerous difadvantages, did yet prove fo extremely prejudicial to the enemy. I fhall only add, that there were taken on
board the galeon feveral draughts and journals, from fome of which
many of the particulars recited in the 10th chapter of the fecond
book are collected. Among the reft there was found a chart of all
the Ocean, between the *Philippines* and the coaft of *Mexico*, which
was what was made ufe of by the galeon in her own navigation.
A copy of this draught, corrected in fome places by our own obfervation, is here annexed, together with the route of the galeon
traced thereon from her own journals, and likewife the route of
the *Centurion*, from *Acapulco* through the fame Ocean. This is
the chart formerly referred to, in the account of the *Manila* trade :
And to render it ftill more compleat, the obferved variation of the
needle is annexed to feveral parts both of the *Spanifh* and *Englifh*
track ; which addition is of the greateft confequence, as no obfervations of this kind in the northern parts of the *Pacific* Ocean have
yet to my knowledge been publifhed, and as the quantity of the
variation fo nearly correfponds to what Dr. *Halley* predicted from
his Theory above fifty years ago. And with this digreffion I fhall
end this chapter, leaving the *Centurion* with her prize, at anchor
off *Macao*, preparing to enter the river of *Canton*.

CHAP.

CHAP. IX.

Tranſactions in the river of *Canton*.

THE Commodore, having taken Pilots on board, proceeded with his prize for the river of *Canton*; and, on the 14th of *July*, came to an anchor ſhort of the *Bocca Tigris*, which is a narrow paſſage forming the mouth of that river: This entrance he propoſed to ſtand through the next day, and to run up as far as *Tiger Iſland*, which is a very ſafe road, ſecured from all winds. But whilſt the *Centurion* and her prize were thus at anchor, a boat with an officer came off from the *Mandarine*, commanding the forts at *Bocca Tigris* to examine what the ſhips were, and whence they came. Mr. *Anſon* informed the officer, that his ſhip was a ſhip of war, belonging to the King of *Great Britain*; and that the other in company with him was a prize he had taken; that he was going into *Canton* river to ſhelter himſelf againſt the hurricanes which were then coming on; and that as ſoon as the monſoon ſhifted, he ſhould proceed for *England*. The officer then deſired an account of what men, guns, and ammunition were on board, a liſt of all which he ſaid was to be ſent to the Government of *Canton*. But when theſe articles were repeated to him, particularly when he was told that there were in the *Centurion* four hundred firelocks, and between three and four hundred barrels of powder, he ſhrugged up his ſhoulders, and ſeemed to be terrified with the bare recital, ſaying, that no ſhips ever came into *Canton* river armed in that manner; adding, that he durſt not ſet down the whole of this force, leaſt it ſhould too much alarm the Regency. After he had finiſhed his enquiries, and was preparing to depart, he deſired to leave two Cuſtom-houſe officers behind him; on which the Commodore told him, that though as a man of war

he

he was prohibited from trading, and had nothing to do with cuftoms or duties of any kind, yet, for the fatisfaction of the *Chinefe*, he would permit two of their people to be left on board, who might themfelves be witneffes how punctually he fhould comply with his inftructions. The officer feemed amazed when Mr. *Anfon* mentioned being exempted from all duties, and told him, that the Emperor's duty muft be paid by all fhips that came into his ports : And it is fuppofed, that on this occafion, private directions were given by him to the *Chinefe* Pilot, not to carry the Commodore through the *Bocca Tigris*; which makes it neceffary, more particularly, to defcribe that entrance.

The *Bocca Tigris* is a narrow paffage, little more than mufquet-fhot over, formed by two points of land, on each of which there is a fort, that on the ftarboard-fide being a battery on the water's edge, with eighteen embrafures, but where there were no more than twelve iron cannon mounted, feeming to be four or fix pounders; the fort on the larboard-fide is a large caftle, refembling thofe old buildings which here in *England* we often find diftinguifhed by that name; it is fituated on a high rock, and did not appear to be furnifhed with more than eight or ten cannon, none of which were fuppofed to exceed fix pounders. Thefe are the defences which fecure the river of *Canton*; and which the *Chinefe* (extremely defective in all military fkill) have imagined were fufficient to prevent any enemy from forcing his way through.

But it is obvious, from the defcription of thefe forts, that they could have given no obftruction to Mr. *Anfon*'s paffage, even if they had been well fupplied with gunners and ftores; and therefore, though the Pilot, after the *Chinefe* officer had been on board, refufed at firft to take charge of the fhip, till he had leave from the forts, yet as it was neceffary to get through without any delay, for fear of the bad weather which was hourly expected, the Commodore weighed on the 15th, and ordered the Pilot to carry him by the forts, threatening him that, if the fhip ran aground, he would inftantly hang him up at the yard-arm. The Pilot, awed

by

by thefe threats, carried the fhip through fafely, the forts not attempting to difpute the paffage. Indeed the poor Pilot did not efcape the refentment of his countrymen, for when he came on fhore, he was feized and fent to prifon, and was rigoroufly difciplined with the bamboo. However, he found means to get at Mr. *Anfon* afterwards, to defire of him fome recompence for the chaftifement he had undergone, and of which he then carried very fignificant marks about him ; and Mr. *Anfon*, in commiferation of his fufferings, gave him fuch a fum of money, as would at any time have enticed a *Chinefe* to have undergone a dozen baftinadings.

Nor was the Pilot the only perfon that fuffered on this occafion; for the Commodore foon after feeing fome royal junks pafs by him from *Bocca Tigris* towards *Canton*, he learnt, on enquiry, that the *Mandarine* commanding the forts was a prifoner on board them ; that he was already turned out, and was now carrying to *Canton*, where it was expected he would be feverely punifhed for having permitted the fhips to pafs ; and the Commodore urging the unreafonablenefs of this procedure, from the inability of the forts to have done otherwife, explaining to the *Chinefe* the great fuperiority his fhips would have had over the forts, by the number and fize of their guns, the *Chinefe* feemed to acquiefce in his reafoning, and allowed that their forts could not have ftopped him ; but they ftill afferted, that the *Mandarine* would infallibly fuffer, for not having done, what all his judges were convinced, was impoffible. To fuch indefenfible abfurdities are thofe obliged to fubmit, who think themfelves concerned to fupport their authority, when the neceffary force is wanting. But to return :

On the 16th of *July* the Commodore fent his fecond Lieutenant to *Canton*, with a letter to the Viceroy, informing him of the reafon of the *Centurion*'s putting into that port ; and that the Commodore himfelf foon propofed to repair to *Canton*, to pay a vifit to the Viceroy. The Lieutenant was very civilly received, and was promifed that an anfwer fhould be fent to the Commodore the next

2

day.

day. In the mean time Mr. *Anson* gave leave to feveral of the officers of the galeon to go to *Canton*, they engaging their parole to return in two days. When thefe prifoners got to *Canton*, the Regency fent for them, and examined them, enquiring particularly by what means they had fallen into Mr. *Anson*'s power. And on this occafion the prifoners were honeft enough to declare, that as the Kings of *Great-Britain* and of *Spain* were at war, they had propofed to themfelves the taking of the *Centurion*, and had bore down upon her with that view, but that the event had been contrary to their hopes: However, they acknowledged that they had been treated by the Commodore, much better than they believed they fhould have treated him, had he fallen into their hands. This confeffion from an enemy had great weight with the *Chinefe*, who, till then, though they had revered the Commodore's power, had yet fufpected his morals, and had confidered him rather as a lawlefs freebooter, than as one commiffioned by the State for the revenge of public injuries. But they now changed their opinion, and regarded him as a more important perfon; to which perhaps the vaft treafure of his prize might not a little contribute; the acquifition of wealth being a matter greatly adapted to the eftimation and reverence of the *Chinefe* Nation.

In this examination of the *Spanifh* prifoners, though the *Chinefe* had no reafon in the main to doubt of the account which was given them, yet there were two circumftances which appeared to them fo fingular, as to deferve a more ample explanation; one of them was the great difproportion of men between the *Centurion* and the galeon; the other was the humanity, with which the people of the galeon were treated after they were taken. The *Mandarines* therefore afked the *Spaniaras*, how they came to be overpowered by fo inferior a force; and how it happened, fince the two nations were at war, that they were not put to death when they came into the hands of the *Englifh*. To the firft of thefe enquiries the *Spaniards* replied, that though they had more hands than the *Centurion*, yet fhe being intended folely for war had a

great

great superiority in the size of her guns, and in many other articles, over the galeon, which was a vessel fitted out principally for traffic: And as to the second question, they told the *Chinese*, that amongst the nations of *Europe*, it was not customary to put to death those who submitted; though they readily owned, that the Commodore, from the natural bias of his temper, had treated both them and their countrymen, who had formerly been in his power, with very unusual courtesy, much beyond what they could have expected, or than was required by the customs established between nations at war with each other. These replies fully satisfied the *Chinese*, and at the same time wrought very powerfully in the Commodore's favour.

On the 20th of *July*, in the morning, three *Mandarines*, with a great number of boats, and a vast retinue, came on board the *Centurion*, and delivered to the Commodore the Viceroy of *Canton*'s order for a daily supply of provisions, and for Pilots to carry the ships up the river as far as the second bar; and at the same time they delivered him a message from the Viceroy, in answer to the letter sent to *Canton*. The substance of the message was, that the Viceroy desired to be excused from receiving the Commodore's visit, during the then excessive hot weather; because the assembling the *Mandarines* and soldiers, necessary to that ceremony, would prove extremely inconvenient and fatiguing; but that in *September*, when the weather would be more temperate, he should be glad to see both the Commodore himself, and the *English* Captain of the other ship, that was with him. As Mr. *Anson* knew that an express had been dispatched to the Court at *Pekin*, with an account of the *Centurion* and her prize being arrived in the river of *Canton*, he had no doubt, but the principal motive for putting off this visit was, that the Regency at *Canton* might gain time to receive the Emperor's instructions, about their behaviour on this unusual affair.

When the *Mandarines* had delivered their message, they began to talk to the Commodore about the duties to be paid by his ships;

but

but he immediately told them, that he would never submit to any demand of that kind ; that as he neither brought any merchandize thither, nor intended to carry any away, he could not be reasonably deemed to be within the meaning of the Emperor's orders, which were doubtless calculated for trading vessels only, adding, that no duties were ever demanded of men of war, by nations accustomed to their reception, and that his Master's orders expresly forbad him from paying any acknowledgement for his ships anchoring in any port whatever.

The *Mandarines* being thus cut short on the subject of the duty, they said they had another matter to mention, which was the only remaining one they had in charge ; this was a request to the Commodore, that he would release the prisoners he had taken on board the galeon ; for that the Viceroy of *Canton* apprehended the Emperor, his Master, might be displeased, if he should be informed, that persons, who were his allies, and carried on a great commerce with his subjects, were under confinement in his dominions. Mr. *Anson* was himself extremely desirous to get rid of the *Spaniards*, having, on his first arrival, sent about an hundred of them to *Macao*, and those who remained, which were near four hundred more, were, on many accounts, a great incumbrance to him. However, to inhance the favour, he at first raised some difficulties ; but permitting himself to be prevailed on, he at last told the *Mandarines*, that to show his readiness to oblige the Viceroy, he would release the prisoners, whenever they, the *Chinese*, would send boats to fetch them off. This matter being thus adjusted, the *Mandarines* departed ; and, on the 28th of *July*, two *Chinese* junks were sent from *Canton*, to take on board the prisoners, and to carry them to *Macao*. And the Commodore, agreeable to his promise, dismissed them all, and ordered his Purser to send with them eight days provision for their subsistence, during their sailing down the river ; this being dispatched, the *Centurion* and her prize came to her moorings, above the second bar, where they proposed to continue till the monsoon shifted. 2

Though

Though the ſhips, in conſequence of the Viceroy's permit, found no difficulty in purchaſing proviſions for their daily conſumption, yet it was impoſſible for the Commodore to proceed to *England*, without laying in a large quantity both of proviſions and ſtores for his uſe, during the voyage : The procuring this ſupply was attended with much embaraſment ; for there were people at *Canton* who had undertaken to furniſh him with biſcuit, and whatever elſe he wanted ; and his linguiſt, towards the middle of *September*, had aſſured him, from day to day, that all was ready, and would be ſent on board him immediately. But a fortnight being elapſed, and nothing being brought, the Commodore ſent to *Canton* to enquire more particularly into the reaſons of this diſappointment : And he had ſoon the vexation to be informed, that the whole was an illuſion ; that no order had been procured from the Viceroy, to furniſh him with his ſea-ſtores, as had been pretended ; that there was no biſcuit baked, nor any one of the articles in readineſs, which had been promiſed him ; nor did it appear, that the Contractors had taken the leaſt ſtep to comply with their agreement. This was moſt diſagreeable news, and made it ſuſpected, that the furniſhing the *Centurion* for her return to *Great-Britain* might prove a more troubleſome matter than had been hitherto imagined ; eſpecially too, as the month of *September* was nearly elapſed, without Mr. *Anſon*'s having received any meſſage from the Viceroy of *Canton*.

And here perhaps it might be expected that ſome ſatisfactory account ſhould be given of the motives of the *Chineſe* for this faithleſs procedure. But as I have already, in a former chapter, made ſome kind of conjectures about a ſimilar event, I ſhall not repeat them again in this place, but ſhall obſerve, that after all, it may perhaps be impoſſible for an *European*, ignorant of the cuſtoms and manners of that nation, to be fully apprized of the real incitements to this behaviour. Indeed, thus much may undoubtedly be aſſerted, that in artifice, falſhood, and an attachment to all kinds of lucre, many of the *Chineſe* are difficult to be paralleled by any other

people ;

people ; but then the combination of thefe talents, and the manner in which they are applied in particular emergencies, are often beyond the reach of a Foreigner's penetration : So that though it may be fafely concluded, that the *Chinefe* had fome intereft in thus amufing the Commodore, yet it may not be eafy to affign the individual views by which they were influenced. And that I may not be thought too fevere in afcribing to this Nation a fraudulent and felfifh turn of temper, fo contradictory to the character given of them in the legendary accounts of the *Roman* Miffionaries, I fhall here mention an extraordinary tranfaction or two, which I hope will be fome kind of confirmation of what I have advanced.

When the Commodore lay firft at *Macao*, one of his officers, who had been extremely ill, defired leave of him to go on fhore every day on a neighbouring Ifland, imagining that a walk upon the land would contribute greatly to the reftoring of his health : The Commodore would have diffuaded him, fufpecting the tricks of the *Chinefe*, but the officer continuing importunate, in the end the boat was ordered to carry him. The firft day he was put on fhore he took his exercife, and returned without receiving any moleftation, or even feeing any of the inhabitants ; but the fecond day, he was affaulted, foon after his arrival, by a great number of *Chinefe*, who had been hoeing rice in the neighbourhood, and who beat him fo violently with the handles of their hoes, that they foon laid him on the ground incapable of refiftance ; after which they robbed him, taking from him his fword, the hilt of which was filver, his money, his watch, gold-headed cane, fnuff-box, fleeve-buttons and hat, with feveral other trinkets : In the mean time the boat's crew, who were at fome little diftance, and had no arms of any kind with them, were incapable of giving him any affiftance ; till at laft one of them flew on the fellow who had the fword in his poffeffion, and wrefting it out of his hands drew it, and with it was preparing to fall on the *Chinefe*, fome of whom he could not have failed of killing ; but the officer, perceiving what he was about, immediately ordered him to defift, thinking it more prudent to fub-

E e e   mit

mit to the prefent violence, than to embroil his Commodore in an inextricable fquabble with the *Chinefe* Government, by the death of their fubjects; which calmnefs in this Gentleman was the more meritorious, as he was known to be a perfon of an uncommon fpirit, and of a fomewhat hafty temper : By this means the *Chinefe* recovered the poffeffion of the fword, which they foon perceived was prohibited to be made ufe of againft them, and carried off their whole booty unmolefted. No fooner were they gone, than a *Chinefe* on horfeback, very well dreffed, and who had the air and appearance of a Gentleman, came down to the fhore, and, as far as could be underflood by his figns, feemed to cenfure the conduct of his countrymen, and to commiferate the officer, being wonderfully officious to affift in getting him on board the boat: But notwithftanding this behaviour, it was fhrewdly fufpected that he was an accomplice in the theft, and time fully evinced the juftice of thofe fufpicions.

When the boat returned on board, and reported what had paffed to the Commodore, he immediately complained of it to the *Mandarine*, who attended to fee his fhip fupplied; but the *Mandarine* coolly replied, that the boat ought not to have gone on fhore, promifing, however, that if the thieves could be found out, they fhould be punifhed; though it appeared plain enough, by his manner of anfwering, that he would never give himfelf any trouble in fearching them out. However, a confiderable time afterwards, when fome *Chinefe* boats were felling provifions to the *Centurion*, the perfon who had wrefted the fword from the *Chinefs* came with great eagernefs to the Commodore, to affure him that one of the principal thieves was then in a provifion-boat along-fide the fhip; and the officer, who had been robbed, viewing the fellow on this report, and well remembring his face, orders were immediately given to feize him; and he was accordingly fecured on board the fhip, where ftrange difcoveries were now made.

This

This thief, on his being firſt apprehended, expreſſed ſo much fright in his countenance, that it was feared he would have died upon the ſpot; the *Mandarine* too, who attended the ſhip, had viſibly no ſmall ſhare of concern on the occaſion. Indeed he had reaſon enough to be alarmed, ſince it was ſoon evinced, that he had been privy to the whole robbery; for the Commodore declaring that he would not deliver up the thief, but would himſelf order him to be ſhot, the *Mandarine* immediately put off the magiſterial air, with which he had at firſt pretended to demand him, and begged his releaſe in the moſt abject manner: And the Commodore appearing inflexible, there came on board, in leſs than two hours time, five or ſix of the neighbouring *Mandarines*, who all joined in the ſame entreaty, and with a view of facilitating their ſuit, offered a large ſum of money for the fellow's liberty. Whilſt they were thus ſolliciting, it was diſcovered that the *Mandarine*, who was the moſt active amongſt them, and who ſeemed to be moſt intereſted in the event, was the very Gentleman, who came to the officer, juſt after the robbery, and who pretended to be ſo much diſpleaſed with the villany of his countrymen. And, on further inquiry it was found, that he was the *Mandarine* of the Iſland; and that he had, by the authority of his office, ordered the Peaſeants to commit that infamous action. And it ſeemed, as far as could be collected from the broken hints which were caſually thrown out, that he and his brethren, who were all privy to the tranſaction, were terrified with the fear of being called before the tribunal at *Canton*, where the firſt article of their puniſhment would be the ſtripping them of all they were worth; though their judges (however fond of inflicting a chaſtiſement ſo lucrative to themſelves) were perhaps of as tainted a complexion as the delinquents. Mr. *Anſon* was not diſpleaſed to have caught the *Chineſe* in this dilemma; and he entertained himſelf for ſome time with their perplexity, rejecting their money with ſcorn, appearing inexorable to their prayers, and giving out that the thief ſhould certainly be ſhot; but

as

as he then forefaw that he fhould be forced to take fhelter in their ports a fecond time, when the influence he might hereby acquire over the Magiftrates would be of great fervice to him, he at length permitted himfelf to be perfuaded, and as a favour releafed his prifoner, but not till the *Mandarine* had collected and returned all that had been ftolen from the officer, even to the minuteft trifle.

But notwithftanding this inftance of the good intelligence between the magiftrates and criminals, the ftrong addiction of the *Chinefe* to lucre often prompts them to break through this awful confederacy, and puts them on defrauding the authority that protects them of its proper quota of the pillage. For not long after the above-mentioned tranfaction, (the former *Mandarine*, attendant on the fhip, being, in the mean time, relieved by another) the Commodore loft a top-maft from his ftern, which, after the moft diligent enquiry, could not be traced : As it was not his own, but had been borrowed at *Macao* to heave down by, and was not to be replaced in that part of the world, he was extremely defirous to recover it, and publifhed a confiderable reward to any who would bring it him again. There were fufpicions from the firft of its being ftolen, which made him conclude a reward was the likelieft method of getting it back : Accordingly, foon after, the *Mandarine* told him, that fome of his, the *Mandarine*'s, people, had found the top-maft, defiring the Commodore to fend his boats to fetch it, which being done, the *Mandarine*'s people received the promifed reward ; but the Commodore told the *Mandarine*, that he would make him a prefent befides for the care he had taken in directing it to be fearched for ; and accordingly, Mr. *Anfon* gave a fum of money to his Linguift, to be delivered to the *Mandarine* ; but the Linguift knowing that the people had been paid, and ignorant that a further prefent had been promifed, kept the money himfelf : However, the *Mandarine* fully confiding in Mr. *Anfon*'s word, and fufpecting the Linguift, he took occafion, one morning, to admire the fize of the *Centurion*'s mafts, and thence, on a pretended fudden recollection, he made a

digreffion

digreffion to the top-maft which had been loft, and afked Mr. *An-son* if he had not got it again. Mr. *Anson* prefently perceived the bent of this converfation, and enquired of him if he had not received the money from the Linguift, and finding he had not, he offered to pay it him upon the fpot. But this the *Mandarine* refufed, having now fomewhat more in view than the fum which had been detained : For the next day the Linguift was feized, and was doubtlefs mulcted of all he had gotten in the Commodore's fervice, which was fuppofed to be little lefs than two thoufand dollars ; he was befides fo feverely baftinadoed with the bamboo, that it was with difficulty he efcaped with his life ; and when he was upbraided by the Commodore (to whom he afterwards came begging) with his folly, in rifquing all he had fuffered for fifty dollars, (the prefent intended for the *Mandarine*) he had no other excufe to make than the ftrong bias of his Nation to difhonefty, replying, in his broken jargon, *Chinefe man very great rogue truly, but have fafhion, no can help.*

It were endlefs to recount all the artifices, extortions and frauds which were practifed on the Commodore and his people, by this interefted race. The method of buying all things in *China* being by weight, the tricks made ufe of by the *Chinefe* to encreafe the weight of the provifion they fold to the *Centurion*, were almoft incredible. One time a large quantity of fowls and ducks being bought for the fhip's ufe, the greateft part of them prefently died : This alarmed the people on board with the apprehenfions that they had been killed by poifon ; but on examination it appeared, that it was only owing to their being crammed with ftones and gravel to encreafe their weight, the quantity thus forced into moft of the ducks being found to amount to ten ounces in each. The hogs too, which were bought ready killed of the *Chinefe* Butchers, had water injected into them for the fame purpofe ; fo that a carcafs, hung up all night for the water to drain from it, hath loft above a ftone of its weight ; and when, to avoid this cheat, the hogs were

2   bought

bought alive, it was found that the *Chinese* gave them falt to encreafe their thirft, and having by this means excited them to diink great quantities of water, they then took meafures to prevent them from difcharging it again by urine, and fold the tortured animal in this inflated ftate. When the Commodore firft put to fea from *Macao*, they practifed an artifice of another kind ; for as the *Chinefe* never object to the eating of any food that dies of itfelf, they took care, by fome fecret practices, that great part of his live feaftore fhould die in a fhort time after it was put on board, hoping to make a fecond profit of the dead carcaffes which they expected would be thrown overboard ; and two thirds of the hogs dying before the *Centurion* was out of fight of land, many of the *Chinefe* boats followed her, only to pick up the carrion. Thefe inftances may ferve as a fpecimen of the manners of this celebrated Nation, which is often recommended to the reft of the world as a pattern of all kinds of laudable qualities. But to return :

The Commodore, towards the end of *September*, having found out (as has been faid) that thofe, who had contracted to fupply him with fea-provifions and ftores, had deceived him, and that the Viceroy had not fent to him according to his promife, he faw it would be impoffible for him to furmount the embarafment he was under, without going himfelf to *Canton*, and vifiting the Viceroy ; and therefore, on the 27th of *September*, he fent a meffage to the *Mandarine*, who attended the *Centurion*, to inform him that he, the Commodore, intended, on the firft of *October*, to proceed in his boat to *Canton* ; adding, that the day after he got there, he fhould notify his arrival to the Viceroy, and fhould defire him to fix a time for his audience ; to which the *Mandarine* returned no other anfwer, than that he would acquaint the Viceroy with the Commodore's intentions. In the mean time all things were prepared for this expedition : And the boat's crew in particular, which Mr. *Anfon* propofed to take with him, were cloathed in an uniform drefs, refembling that of the Watermen on the *Thames* ; they were

2

in

in number eighteen and a Coxfwain; they had fcarlet jackets and blue filk waiftcoats, the whole trimmed with filver buttons, and with filver badges on their jackets and caps. As it was apprehended, and even afferted, that the payment of the cuftomary duties for the *Centurion* and her prize, would be demanded by the Regency of *Canton*, and would be infifted on previous to the granting a permiffion for victualling the fhip for her future voyage; the Commodore, who was refolved never to eftablifh fo difhonourable a precedent, took all poffible precaution to prevent the *Chinefe* from facilitating the fuccefs of their unreafonable pretenfions, by having him in their power at *Canton*: And therefore, for the fecurity of his fhip, and the great treafure on board her, he appointed his firft Lieutenant, Mr. *Brett*, to be Captain of the *Centurion* under him, giving him proper inftructions for his conduct; directing him, particularly, if he, the Commodore, fhould be detained at *Canton* on account of the duties in difpute, to take out the men from the *Centurion's* prize, and to deftroy her; and then to proceed down the river through the *Bocca Tigris*, with the *Centurion* alone, and to remain without that entrance, till he received further orders from Mr. *Anfon*.

Thefe neceffary fteps being taken, which were not unknown to the *Chinefe*, it fhould feem as if their deliberations were in fome fort embarafed thereby. It is reafonable to imagine, that they were in general very defirous of getting the duties to be paid them; not perhaps folely in confideration of the amount of thofe dues, but to keep up their reputation for addrefs and fubtlety, and to avoid the imputation of receding from claims, on which they had already fo frequently infifted. However, as they now forefaw that they had no other method of fucceeding than by violence, and that even againft this the Commodore was prepared, they were at laft difpofed, I conceive, to let the affair drop, rather than entangle themfelves in an hoftile meafure, which they found would only expofe them to the rifque of having the whole navigation of their port deftroyed, without any certain profpect of gaining their favourite point thereby.

However,

However, though there is reason to imagine that these were their thoughts at that time, yet they could not depart at once from the evasive conduct to which they had hitherto adhered. For when the Commodore, on the morning of the first of *October*, was preparing to set out for *Canton*, his Linguist came to him from the *Mandarine*, who attended his ship, to tell him that a letter had been received from the Viceroy of *Canton*, desiring the Commodore to put off his going thither for two or three days: But in the afternoon of the same day, another Linguist came on board, who, with much seeming fright, told Mr. *Anson*, that the Viceroy had expected him up that day, that the Council was assembled, and the troops had been under arms to receive him; and that the Viceroy was highly offended at the disappointment, and had sent the Commodore's Linguist to prison chained, supposing that the whole had been owing to the Linguist's negligence. This plausible tale gave the Commodore great concern, and made him apprehend that there was some treachery designed him, which he could not yet fathom; and though it afterwards appeared that the whole was a fiction, not one article of it having the least foundation, yet (for reasons best known to themselves) this falshood was so well supported by the artifices of the *Chinese* Merchants at *Canton*, that, three days afterwards, the Commodore received a letter signed by all the supercargoes of the *English* ships then at that place, expressing their great uneasiness at what had happened, and intimating their fears that some insult would be offered to his boat, if he came thither before the Viceroy was fully satisfied about the mistake. To this letter Mr. *Anson* replied, that he did not believe there had been any mistake; but was persuaded it was a forgery of the *Chinese* to prevent his visiting the Viceroy; that therefore he would certainly come up to *Canton* on the 13th of *October*, confident that the *Chinese* would not dare to offer him an insult, as well knowing it would be properly returned.

On

On the 13th of *October*, the Commodore continuing firm to his resolution, all the supercargoes of the *English*, *Danish*, and *Swedish* ships came on board the *Centurion*, to accompany him to *Canton*, for which place he set out in his barge the same day, attended by his own boats, and by those of the trading ships, which on this occasion came to form his retinue; and as he passed by *Wampo*, where the *European* vessels lay, he was saluted by all of them but the *French*, and in the evening he arrived safely at *Canton*. His reception at that city, and the most material transactions from henceforward, till his arrival in *Great-Britain*, shall be the subject of the ensuing chapter.

CHAP.

CHAP. X.

Proceedings at the city of *Canton*, and the return of the *Centurion* to *England*.

WHEN the Commodore arrived at *Canton*, he was vifited by the principal *Chinefe* Merchants, who affected to appear very much pleafed that he had met with no obftruction in getting thither, and who thence pretended to conclude, that the Viceroy was fatisfied about the former miftake, the reality of which they ftill infifted on; they added, that as foon as the Viceroy fhould be informed that Mr. *Anfon* was at *Canton*, (which they promifed fhould be done the next morning) they were perfuaded a day would be immediately appointed for the vifit, which was the principal bufinefs that had brought the Commodore thither.

The next day the Merchants returned to Mr. *Anfon*, and told him, that the Viceroy was then fo fully employed in preparing his difpatches for *Pekin*, that there was no getting admittance to him for fome days; but that they had engaged one of the officers of his Court to give them information, as foon as he fhould be at leifure, when they propofed to notify Mr. *Anfon*'s arrival, and to endeavour to fix the day of audience. The Commodore was by this time too well acquainted with their artifices, not to perceive that this was a falfehood; and had he confulted only his own judgment, he would have applied directly to the Viceroy by other hands: But the *Chinefe* Merchants had fo far prepoffeffed the fupercargoes of our fhips with chimerical fears, that they (the fupercargoes) were extremely apprehenfive of being embroiled with the Government, and of fuffering in their intereft, if thofe meafures were taken, which appeared to Mr. *Anfon* at that time to be the moft

2   prudential :

prudential: And therefore, leaſt the malice and double dealing of the *Chineſe* might have given riſe to ſome ſiniſter incident, which would be afterwards laid at his door, he reſolved to continue paſſive, as long as it ſhould appear that he loſt no time, by thus ſuſpending his own opinion. With this view, he promiſed not to take any immediate ſtep himſelf for getting admittance to the Viceroy, provided the *Chineſe*, with whom he contracted for proviſions, would let him ſee that his bread was baked, his meat ſalted, and his ſtores prepared with the utmoſt diſpatch : But ¨if by the time when all was in readineſs to be ſhipped off, (which it was ſuppoſed would be in about forty days) the Merchants ſhould not have procured the Viceroy's permiſſion, then the Commodore propoſed to apply for it himſelf. Theſe were the terms Mr. *Anſon* thought proper to offer, to quiet the uneaſineſs of the ſupercargoes ; and notwithſtanding the apparent equity of the conditions, many difficulties and objections· were urged ; nor would the *Chineſe* agree to them, till the Commodore had conſented to pay for every article he beſpoke before it was put in hand. However, at laſt the contract being paſt, it was ſome ſatisfaction to the Commodore to be certain that his preparations were now going on, and being himſelf on the ſpot, he took care to haſten them as much as poſſible.

During this interval, in which the ſtores and proviſions were getting ready, the Merchants continually entertained Mr. *Anſon* with accounts of their various endeavours to get a licence from the Viceroy, and their frequent diſappointments ; which to him was now a matter of amuſement, as he was fully ſatisfied there was not one word of truth in any thing they ſaid. But when all was compleated, and wanted only to be ſhipped, which was about the 24th of *November*, at which time too the N. E. monſoon was ſet in, he then reſolved to apply himſelf to the Viceroy to demand an audience, · as he was perſuaded that, without this ceremony, the procuring a permiſſion to ſend his ſtores on board would meet with great difficulty. On the 24th of *November*, therefore, Mr. *Anſon* ſent one of his officers to the *Mandarine*, who commanded the

guard

guard of the principal gate of the city of *Canton*, with a letter directed to the Viceroy. When this letter was delivered to the
Mandarine, he received the officer who brought it very civilly, and
took down the contents of it in *Chinese*, and promised that the Viceroy should be immediately acquainted with it; but told the officer,
it was not necessary for him to wait for an answer, because a meffage would be fent to the Commodore himfelf.

On this occasion Mr. *Anfon* had been under great difficulties about
a proper interpreter to fend with his officer, as he was well aware
that none of the *Chinese*, usually employed as Linguifts, could be
relied on: But he at laft prevailed with Mr. *Flint*, an *English* Gentleman belonging to the factory, who fpoke *Chinese* perfectly well,
to accompany his officer. This perfon, who upon this occasion and
many others was of fingular fervice to the Commodore, had been
left at *Canton* when a youth, by the late Captain *Rigby*. The
leaving him there to learn the *Chinese* language was a ftep taken
by that Captain, merely from his own perfuafion of the great advantages which the *Eaft-India* Company might one day receive from
an *English* interpreter; and though the utility of this meafure has
greatly exceeded all that was expected from it, yet I have not heard
that it has been to this day imitated: But we imprudently choofe
(except in this fingle inftance) to carry on the vaft tranfactions of
the port of *Canton*, either by the ridiculous jargon of broken *English*, which fome few of the *Chinese* have learnt, or by the fufpected
interpretation of the Linguifts of other Nations.

Two days after the fending the above-mentioned letter, a
fire broke out in the fuburbs of *Canton*. On the firft alarm, Mr.
Anfon went thither with his officers, and his boat's crew, to affift
the *Chinese*. When he came there, he found that it had begun in a
failor's fhed, and that by the flightnefs of the buildings, and the
awkwardnefs of the *Chinese*, it was getting head apace: But he
perceived, that by pulling down fome of the adjacent fheds it might
eafily be extinguifhed; and particularly obferving that it was running along a wooden cornifh, which would foon communicate it

to

to a great diftance, he ordered his people to begin with tearing a-
way that cornifh; this was prefently attempted, and would have
been foon executed; but, in the mean time, he was told, that, as
there was no *Mandarine* there to direct what was to be done, the
Chinefe would make him, the Commodore, anfwerable for whatever fhould be pulled down by his orders. On this his people defifted; and he fent them to the *Englifh* factory, to affift in fecuring the Company's treafure and effects, as it was eafy to forefee
that no diftance was a protection againft the rage of fuch a fire,
where fo little was done to put a ftop to it; for all this time the
Chinefe contented themfelves with viewing it, and now and then
holding one of their idols near it, which they feemed to expect
fhould check its progrefs: However, at laft, a *Mandarine* came out
of the city, attended by four or five hundred firemen: Thefe made
fome feeble efforts to pull down the neighbouring houfes; but by
this time the fire had greatly extended itfelf, and was got amongft
the Merchants warehoufes; and the *Chinefe* firemen, wanting both
fkill and fpirit, were incapable of checking its violence; fo that its
fury encreafed upon them, and it was feared the whole city would
be deftroyed. In this general confufion the Viceroy himfelf came
thither, and the Commodore was fent to, and was entreated to afford his affiftance, being told that he might take any meafures
he fhould think moft prudent in the prefent emergency. And now
he went thither a fecond time, carrying with him about forty of
his people; who, upon this occafion, exerted themfelves in fuch a
manner, as in that country was altogether without example: For
they were rather animated than deterred by the flames and falling
buildings, amongft which they wrought; fo that it was not uncommon to fee the moft forward of them tumble to the ground on the
roofs, and amidft the ruins of houfes, which their own efforts
brought down with them. By their boldnefs and activity the fire
was foon extinguifhed to the amazement of the *Chinefe*; and the
buildings being all on one floor, and the materials flight, the feamen,

men, notwithftanding their daring behaviour, happily efcaped with no other injuries, than fome confiderable bruifes.

The fire, though at laft thus luckily extinguifhed, did great mifchief during the time it continued; for it confumed an hundred fhops and eleven ftreets full of warehoufes, fo that the damage amounted to an immenfe fum ; and one of the *Chinefe* Merchants, well known to the *Englifh*, whofe name was *Succoy*, was fuppofed, for his own fhare, to have loft near two hundred thoufand pound fterling. It raged indeed with unufual violence, for in many of the warehoufes, there were large quantities of camphire, which greatly added to its fury, and produced a column of exceeding white flame, which fhot up into the air to fuch a prodigious height, that the flame itfelf was plainly feen on board the *Centurion*, though fhe was thirty miles diftant.

Whilft the Commodore and his people were labouring at the fire, and the terror of its becoming general ftill poffeffed the whole city, feveral of the moft confiderable *Chinefe* Merchants came to Mr. *Anfon*, to defire that he would let each of them have one of his foldiers (for fuch they ftiled his boat's crew from the uniformity of their drefs) to guard their warehoufes and dwelling houfes, which, from the known difhonefty of the populace, they feared would be pillaged in the tumult. Mr. *Anfon* granted them this requeft; and all the men that he thus furnifhed to the *Chinefe* behaved greatly to the fatisfaction of their employers, who afterwards highly applauded their great diligence and fidelity.

By this means, the refolution of the *Englifh* at the fire, and their truftinefs and punctuality elfewhere, was the general fubject of converfation amongft the *Chinefe*: And, the next morning, many of the principal inhabitants waited on the Commodore to thank him for his affiftance ; frankly owning to him, that they could never have extinguifhed the fire of themfelves, and that he had faved their city from being totally confumed. And foon after a meffage came to the Commodore from the Viceroy, appointing the 30th of *November*

vember for his audience; which sudden resolution of the Viceroy, in a matter that had been so long agitated in vain, was also owing to the signal services performed by Mr. *Anson* and his people at the fire, of which the Viceroy himself had been in some measure an eye-witness.

The fixing this business of the audience, was, on all accounts, a circumstance which Mr. *Anson* was much pleased with; as he was satisfied that the *Chinese* Government would not have determined this point, without having agreed among themselves to give up their pretensions to the duties they claimed, and to grant him all he could reasonably ask; for as they well knew the Commodore's sentiments, it would have been a piece of imprudence, not consistent with the refined cunning of the *Chinese*, to have admitted him to an audience, only to have contested with him. And therefore, being himself perfectly easy about the result of his visit, he made all necessary preparations against the day; and engaged Mr. *Flint*, whom I have mentioned before, to act as interpreter in the conference: Who, in this affair, as in all others, acquitted himself much to the Commodore's satisfaction; repeating with great boldness, and doubtless with exactness, all that was given in charge, a part which no *Chinese* Linguist would ever have performed with any tolerable fidelity.

At ten o'clock in the morning, on the day appointed, a *Mandarine* came to the Commodore, to let him know that the Viceroy was ready to receive him; on which the Commodore and his retinue immediately set out: And as soon as he entered the outer gate of the city, he found a guard of two hundred soldiers drawn up ready to attend him; these conducted him to the great parade before the Emperor's palace, where the Viceroy then resided. In this parade, a body of troops, to the number of ten thousand, were drawn up under arms, and made a very fine appearance, being all of them new cloathed for this ceremony: And Mr. *Anson* and his retinue having passed through the middle of them, he was then conducted

to

to the great hall of audience, where he found the Viceroy feated under a rich canopy in the Emperor's chair of State, with all his Council of *Mandarines* attending: Here there was a vacant feat prepared for the Commodore, in which he was placed on his arrival: He was ranked the third in order from the Viceroy, there being above him only the Head of the Law, and of the Treafury, who in the *Chinefe* Government take place of all military officers. When the Commodore was feated, he addreffed himfelf to the Viceroy by his interpreter, and began with reciting the various methods he had formerly taken to get an audience; adding, that he imputed the delays he had met with, to the infincerity of thofe he had employed, and that he had therefore no other means left, than to fend, as he had done, his own officer with a letter to the gate. On the mention of this the Viceroy ftopped the interpreter, and bid him affure Mr. *Anfon*, that the firft knowledge they had of his being at *Canton*, was from that letter. Mr. *Anfon* then proceeded, and told him, that the fubjects of the King of *Great-Britain* trading to *China* had complained to him, the Commodore, of the vexatious impofitions both of the Merchants and inferior Cuftom-houfe officers, to which they were frequently neceffitated to fubmit, by reafon of the difficulty of getting accefs to the *Mandarines*, who alone could grant them redrefs: That it was his, Mr. *Anfon*'s, duty, as an officer of the King of *Great-Britain*, to lay before the Viceroy thefe grievances of the *Britifh* fubjects, which he hoped the Viceroy would take into confideration, and would give orders, that for the future there fhould be no juft reafon for complaint. Here Mr. *Anfon* paufed, and waited fome time in expectation of an anfwer; but nothing being faid, he afked his interpreter if he was certain the Viceroy underftood what he had urged; the interpreter told him, he was certain it was underftood, but he believed no reply would be made to it. Mr. *Anfon* then reprefented to the Viceroy the cafe of the fhip *Haflingfield*, which, having been difmafted on the coaft of *China*,

2   had

had arrived in the river of *Canton* but a few days before. The people on board this veffel had been great fufferers by the fire ; the Captain in particular had all his goods burnt, and had loft befides, in the confufion, a cheft of treafure of four thoufand five hundred *Tahel*, which was fuppofed to be ftolen by the *Chinefe* boat-men. Mr. *Anfon* therefore defired that the Captain might have the affiftance of the Government, as it was apprehended the money could never be recovered without the interpofition of the *Mandarines*. And to this requeft the Viceroy made anfwer, that in fettling the Emperor's cuftoms for that fhip, fome abatement fhould be made in confideration of her loffes.

And now the Commodore having difpatched the bufinefs with which the officers of the *Eaft-India* Company had entrufted him, he entered on his own affairs ; acquainting the Viceroy, that the proper feafon was now fet in for returning to *Europe*, and that he waited only for a licence to fhip off his provifions and ftores, which were all ready ; and that as foon as this fhould be granted him, and he fhould have gotten his neceffaries on board, he intended to leave the river of *Canton*, and to make the beft of his way for *England*. The Viceroy replied to this, that the licence fhould be immediately iffued, and that every thing fhould be ordered on board the following day. And finding that Mr. *Anfon* had nothing farther to infift on, the Viceroy continued the converfation for fome time, acknowledging in very civil terms how much the *Chinefe* were obliged to him for his fignal fervices at the fire, and owning that he had faved the city from being deftroyed : And then obferving that the *Centurion* had been a good while on their coaft, he clofed his difcourfe, by wifhing the Commodore a good voyage to *Europe*. After which, the Commodore, thanking him for his civility and affiftance, took his leave.

As foon as the Commodore was out of the hall of audience, he was much preffed to go into a neighbouring apartment, where there was an entertainment provided ; but finding, on enquiry, that the

G g g   Viceroy

Viceroy himfelf was not to be prefent, he declined the invitation, and departed, attended in the fame manner as at his arrival ; only at his leaving the city he was faluted by three guns, which are as many as in that country are ever fired on any ceremony. Thus the Commodore, to his great joy, at laft finifhed this troublefome affair, which, for the preceding four months, had given him great difquietude. Indeed he was highly pleafed with procuring a licence for the fhipping of his ftores and provifions; for thereby he was enabled to return to *Great-Britain* with the firft of the monfoon, and to prevent all intelligence of his being expected : But this, though a very important point, was not the circumftance which gave him the greateft fatisfaction ; for he was more particularly attentive to the authentic precedent eftablifhed on this occafion, by which his Majefty's fhips of war are for the future exempted from all demands of duty in any of the ports of *China.*

In purfuance of the promifes of the Viceroy, the provifions were begun to be fent on board the day after the audience ; and, four days after, the Commodore embarked at *Canton* for the *Centurion*; and, on the 7th of *December*, the *Centurion* and her prize unmoored, and ftood down the river, paffing through the *Bocca Tigris* on the 10th. And on this occafion I muft obferve, that the *Chinefe* had taken care to man the two forts, on each fide of that paffage, with as many men as they could well contain, the greateft part of them armed with pikes and match-lock mufquets. Thefe garrifons affected to fhew themfelves as much as poffible to the fhips, and were doubtlefs intended to induce Mr. *Anfon* to think more reverently than he had hitherto done of the *Chinefe* military power : For this purpofe they were equipped with much parade, having a great number of colours expofed to view ; and on the caftle in particular there were laid confiderable heaps of large ftones ; and a foldier of unufual fize, dreffed in very fightly armour, ftalkt about on the parapet with a battle-ax in his hand, endeavouring to put on as important and martial an air as poffible,

though

though some of the observers on board the *Centurion* shrewdly suspected, from the appearance of his armour, that instead of steel, it was composed only of a particular kind of glittering paper.

The *Centurion* and her prize being now without the river of *Canton*, and consequently upon the point of leaving the *Chinese* jurisdiction, I beg leave, before I quit all mention of the *Chinese* affairs, to subjoin a few remarks on the disposition and genius of that extraordinary people. And though it may be supposed, that observations made at *Canton* only, a place situated in the corner of the Empire, are very imperfect materials on which to found any general conclusions, yet as those who have had opportunities of examining the inner parts of the country, have been evidently influenced by very ridiculous prepossessions, and as the transactions of Mr. *Anson* with the Regency of *Canton* were of an uncommon nature, in which many circumstances occurred, different perhaps from any which have happened before, I hope the following reflections, many of them drawn from these incidents, will not be altogether unacceptable to the reader.

That the *Chinese* are a very ingenious and industrious people, is sufficiently evinced, from the great number of curious manufactures which are established amongst them, and which are eagerly sought for by the most distant nations ; but though skill in the handicraft arts seems to be the most important qualification of this people, yet their talents therein are but of a second rate kind ; for they are much outdone by the *Japanese* in those manufactures, which are common to both countries ; and they are in numerous instances incapable of rivalling the mechanic dexterity of the *Europeans*. Indeed, their principal excellency seems to be imitation ; and they accordingly labour under that poverty of genius, which constantly attends all servile imitators. This is most conspicuous in works which require great truth and accuracy ; as in clocks, watches, fire-arms, &c. for in all these, though they can copy the different parts, and can form some resemblance of the whole, yet they never could arrive at such a justness in their fabric, as was

necessary

neceffary to produce the defired effect. And if we pafs from their manufacturers to artifts of a fuperior clafs, as painters, ftatuaries, &c. in thefe matters they feem to be ftill more defective, their painters, though very numerous and in great efteem, rarely fucceeding in the drawing or colouring of human figures, or in the grouping of large compofitions; and though in flowers and birds their performances are much more admired, yet even in thefe, fome part of the merit is rather to be imputed to the native brightnefs and excellency of the colours, than to the fkill of the painter; fince it is very unufual to fee the light and fhade juftly and naturally handled, or to find that eafe and grace in the drawing, which are to be met with in the works of *European* artifts. In fhort, there is a ftiffnefs and minutenefs in moft of the *Chinefe* productions, which are extremely difpleafing: And it may perhaps be afferted with great truth, that thefe defects in their arts are entirely owing to the peculiar turn of the people, amongft whom nothing great or fpirited is to be met with.

If we next examine the *Chinefe* literature, (taking our accounts from the writers, who have endeavoured to reprefent it in the moft favourable light) we fhall find, that on this head their obftinacy and abfurdity are moft wonderful: For though, for many ages, they have been furrounded by nations, to whom the ufe of letters was familiar, yet they, the *Chinefe* alone, have hitherto neglected to avail themfelves of that almoft divine invention, and have continued to adhere to the rude and inartificial method of reprefenting words by arbitrary marks; a method, which neceffarily renders the number of their characters too great for human memory to manage, makes writing to be an art that requires prodigious application, and in which no man can be otherwife than partially fkilled; whilft all reading, and underftanding of what is written, is attended with infinite obfcurity and confufion; for the connexion between thefe marks, and the words they reprefent, cannot be retained in books, but muft be delivered down from age to age by oral tradition: And how uncertain this muft prove in fuch a complicated

plicated

plicated fubject, is fufficiently obvious to thofe who have attended
to the variation which all verbal relations undergo, when they are
tranfmitted through three or four hands only. Hence it is eafy to
conclude, that the hiftory and inventions of paft ages, recorded by
thefe perplexed fymbols, muft frequently prove unintelligible ; and
confequently the learning and boafted antiquity of the Nation muft,
in numerous inftances, be extremely problematical.

But we are told by fome of the Miffionaries, that though the
fkill of the *Chinefe* in fcience is indeed much inferior to that of the
Europeans, yet the morality and juftice taught and practifed by
them are moft exemplary. And from the defcription given by
fome of thefe good fathers, one fhould be induced to believe, that
the whole Empire was a well-governed affectionate family, where
the only contefts were, who fhould exert the moft humanity
and beneficence: But our preceding relation of the behaviour of the Magiftrates, Merchants and Tradefmen at *Canton*, fufficiently refutes thefe jefuitical fictions. And as to their
theories of morality, if we may judge from the fpecimens exhibited in the works of the Miffionaries, we fhall find them folely
employed in recommending ridiculous attachments to certain im-
material points, inftead of difcuffing the proper criterion of human
actions, and regulating the general conduct of mankind to one another, on reafonable and equitable principles. Indeed, the only pre-
tenfion of the *Chinefe* to a more refined morality than their neighbours is founded, not on their integrity or beneficence, but folely
on the affected evennefs of their demeanor, and their conftant attention to fupprefs all fymptoms of paffion and violence. But it
muft be confidered, that hypocrify and fraud are often not lefs mifchievous to the general interefts of mankind, than impetuofity and
vehemence of temper: Since thefe, though ufually liable to the
imputation of imprudence, do not exclude fincerity, benevolence,
refolution, nor many other laudable qualities. And perhaps, if this
matter was examined to the bottom, it would appear, that the
calm and patient turn of the *Chinefe*, on which they fo much va-

2 luc

lue themfelves, and which diftinguifhes the Nation from all others, is in reality the fource of the moft exceptionable part of their character ; for it has been often obferved by thofe who have attended to the nature of mankind, that it is difficult to curb the more robuft and violent paffions, without augmenting, at the fame time, the force of the felfifh ones : So that the timidity, diffimulation, and difhonefty of the *Chinefe*, may, in fome fort, be owing to the compofure, and external decency, fo univerfally prevailing in that Empire.

Thus much for the general difpofition of the people : But I cannot difmifs this fubject, without adding a few words about the *Chinefe* Government, that too having been the fubject of boundlefs panegyric. And on this head I muft obferve, that the favourable accounts often given of their prudent regulations for the adminiftration of their domeftic affairs, are fufficiently confuted by their tranfactions with Mr. *Anfon* : For we have feen that their Magiftrates are corrupt, their people thievifh, and their tribunals crafty and venal. Nor is the conftitution of the Empire, or the general orders of the State lefs liable to exception : Since that form of Government, which does not in the firft place provide for the fecurity of the public againft the enterprizes of foreign powers, is certainly a moft defective inftitution : And yet this populous, this rich and extenfive country, fo pompoufly celebrated for its refined wifdom and policy, was conquered about an age fince by an handful of *Tartars* ; and even now, by the cowardice of the inhabitants, and the want of proper military regulations, it continues expofed not only to the attempts of any potent State, but to the ravages of every petty Invader. I have already obferved, on occafion of the Commodore's difputes with the *Chinefe*, that the *Centurion* alone was an overmatch for all the naval power of that Empire: This perhaps may appear an extraordinary pofition ; but to render it unqueftionable, there is exhibited in the annexed plate the draught of two of the veffels made ufe of by the *Chinefe*. The firft of thefe marked (A), is a junk of about a hundred and twenty tuns burthen,

A   B

Chine

Plate XLII.

C

Vessels.

then, and was what the *Centurion* hove down by; thefe are moft
ufed in the great rivers, though they fometimes ferve for fmall coafting voyages: The other junk marked (B) is about two hundred and
eighty tuns burthen, and is of the fame form with thofe in which
they trade to *Cochinchina, Manila, Batavia* and *Japan*, though
fome of their trading veffels are of a much larger fize; its head,
which is reprefented at (C) is perfectly flat; and when the veffel
is deep laden, the fecond or third plank of this flat furface is oft-
times under water. The mafts, fails, and rigging of thefe veffels
are ruder than their built; for their mafts are made of trees, no
otherwife fafhioned than by barking them, and lopping off their
branches. Each maft has only two fhrouds made of twifted rat-
tan, which are often both fhifted to the weather-fide; and the halyard, when the yard is up, ferves inftead of a third fhroud. The
fails are made of matt, ftrengthened every three feet by an horizontal rib of bamboo; they run upon the maft with hoops, as is
reprefented in the figure, and when they are lowered down, they
fold upon the deck. Thefe merchantmen carry no cannon; and it
appears, from this whole defcription, that they are utterly incapable of refifting any *European* armed veffel. Nor is the State provided with fhips of confiderable force, or of a better fabric, to protect them: For at *Canton*, where doubtlefs their principal naval
power is ftationed, we faw no more than four men of war junks,
of about three hundred tuns burthen, being of the make already
defcribed, and mounted only with eight or ten guns, the largeft of
which did not exceed a four pounder. This may fuffice to give
an idea of the defencelefs ftate of the *Chinefe* Empire. But it is
time to return to the Commodore, whom I left with his two fhips
without the *Bocca Tigris*, and who, on the 12th of *December*,
anchored before the town of *Macao*.

Whilft the fhips lay here, the Merchants of *Macao* finifhed
their agreement for the galeon, for which they had offered 6000
dollars; this was much fhort of her value but the impatience of
the Commodore to get to fea, to which the merchants were no

2 ftrangers,

ſtrangers, prompted them to inſiſt on ſo unequal a bargain. Mr. *Anſon* had learnt enough from the *Engliſh* at *Canton* to conjecture, that the war betwixt *Great-Britain* and *Spain* was ſtill continued; and that probably the *French* might engage in the aſſiſtance of *Spain*, before he could arrive in *Great-Britain*; and therefore, knowing that no intelligence could get to *Europe* of the prize he had taken, and the treaſure he had on board, till the return of the merchantmen from *Canton*, he was reſolved to make all poſſible expedition in getting back, that he might be himſelf the firſt meſſenger of his own good fortune, and might thereby prevent the enemy from forming any projects to intercept him : For theſe reaſons, he, to avoid all delay, accepted of the ſum offered for the galeon; and ſhe being delivered to the Merchants the 15th of *December* 1743, the *Centurion*, the ſame day, got under ſail, on her return to *England*. And, on the 3d of *January*, ſhe came to an anchor at *Prince's Iſland* in the Streights of *Sunaa*, and continued there wooding and watering till the 8th; when ſhe weighed and ſtood for *The Cape of Good Hope*, where, on the 11th of *March*, ſhe anchored in *Table-bay*.

The Cape of Good Hope is ſituated in a temperate climate, where the exceſſes of heat and cold are rarely known; and the *Dutch* inhabitants, who are numerous, and who here retain their native induſtry, have ſtock'd it with prodigious plenty of all ſort of fruits and proviſions; moſt of which, either from the equality of the ſeaſons, or the peculiarity of the ſoil, are more delicious in their kind than can be met with elſewhere : So that by theſe, and by the excellent water which abounds there, this ſettlement is the beſt provided of any in the known world, for the refreſhment of ſeamen after long voyages. Here the Commodore continued till the beginning of *April*, highly delighted with the place, which by its extraordinary accommodations, the healthineſs of its air, and the pictureſque appearance of the country, all enlivened by the addition of a civilized colony, was not diſgraced in an imaginary compariſon with the vallies of *Juan Fernandes*, and the lawns of *Tinian*. During

ring his stay he entered about forty new men; and having, by the 3d of *April* 1744, compleated his water and provision, he, on that day, weighed and put to sea; and, the 19th of the same month, they saw the Island of *Saint Helena*, which however they did not touch at, but stood on their way; and, on the 10th of *June*, being then in soundings, they spoke with an *English* ship from *Amsterdam* bound for *Philadelphia*, whence they received the first intelligence of a *French* war; the twelfth they got sight of the *Lizard*; and the fifteenth, in the evening, to their infinite joy, they came safe to an anchor at *Spithead*. But that the signal perils which had so often threatened them in the preceding part of the enterprize, might pursue them to the very last, Mr. *Anson*, learnt on his arrival, that there was a *French* fleet of considerable force cruising in the chops of the Channel, which, by the account of their position, he found the *Centurion* had run through, and had been all the time concealed by a fog. Thus was this expedition finished, when it had lasted three years and nine months, after having, by its event, strongly evinced this important truth, That though prudence, intrepidity, and perseverance united, are not exempted from the blows of adverse fortune; yet in a long series of transactions, they usually rise superior to its power, and in the end rarely fail of proving succesful.

F I N I S.

DIRECTIONS to the Bookbinder, for placing the Copper-Plates.

Directions to the Bookbinder, &c.

Printed in the United States
By Bookmasters